Advances in

CARBON NANOMATERIALS

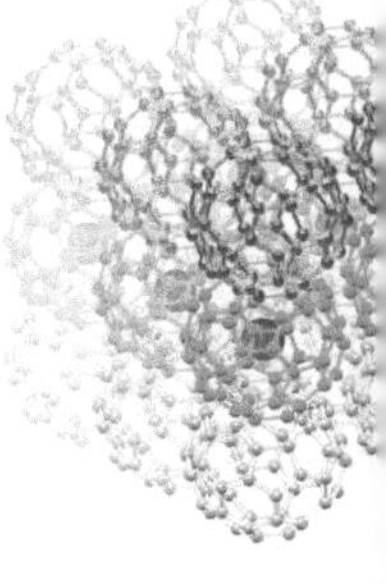

edited by
Nikos Tagmatarchis

Advances in CARBON NANOMATERIALS
Science and Applications

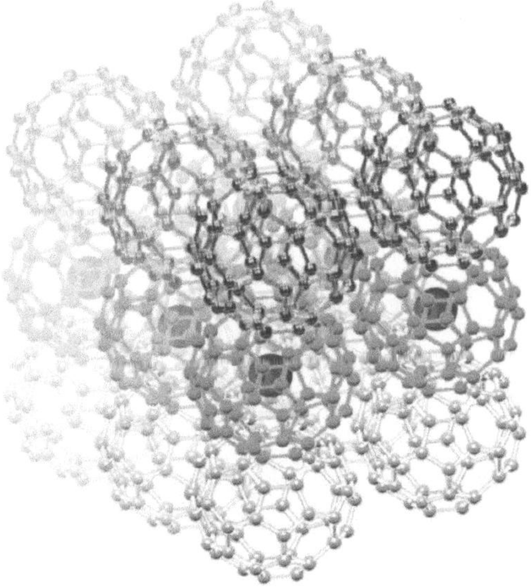

PAN STANFORD PUBLISHING

Published by

Pan Stanford Publishing Pte. Ltd.
Penthouse Level, Suntec Tower 3
8 Temasek Boulevard
Singapore 038988

Email: editorial@panstanford.com
Web: www.panstanford.com

British Library Cataloguing-in-Publication Data
A catalogue record for this book is available from the British Library.

Advances in Carbon Nanomaterials: Science and Applications

ISBN 978-981-426-78-78 (Hardcover)
ISBN 978-981-426-78-85 (eBook)

Printed in the USA

Contents

Preface

A promising class of nanostructured carbon-based materials, varied from spherical empty fullerenes and endohedral fullerenes encapsulating metal atoms to elongated carbon nanotubes and aggregated nanohorns, has led to an explosion of research associated with nanotechnology. *Advances in Carbon Nanomaterials* is a book that offers a wide range of diverse information. Rather than focusing on the latest developments in nanotechnology, the authors and editor of the book, through an appealing collection of nine chapters, offer a remarkably fresh and authoritative look at diverse areas and topics of nanocarbon materials to scientists, researchers and students.

In *Advances in Carbon Nanomaterials*, contributions by experts in diverse fields of chemistry, physics, materials science and medicine provide a comprehensive survey of the current state of knowledge of this constantly expanding subject. The book starts out with Chapter 1 in the form of an encyclopedia of carbon nanoforms, dealing with nomenclature and modelling of carbon nanomaterials, with special emphasis on the topology and morphology of those carbon nanostructures. Chapter 2 examines surfaces and thin films of fullerenes, while focusing on morphology, electronic structure, conduction and optical properties as well as phase transitions. Chapter 3 gives an insight into the structure of carbon nanotubes and the characterization of peapod materials with the aid of high-resolution transmission electron microscopy. Subsequently in Chapter 4, the novel electro-optical properties of carbon nanotubes are analysed through a wealth of spectroscopic evidence. Then, in Chapter 5, important advances in the field of fullerene-based electronics, together with an outline of the major electronic properties of fullerenes are presented. Moving into chemistry, Chapters 6 and 7 deal with the chemical functionalization of carbon nanohorns and endohedral metallofullerenes respectively Finally, applications in quantum computing and medicine conclude this fascinating overview of the field. Chapter 8 is dedicated to quantum computing with endohedral fullerenes, while Chapter 9 deals with the cell biology of carbon nanotubes

Finally, special acknowledgements go to all authors who contributed to this book.

Nikos Tagmatarchis
Theoretical and Physical Chemistry Institute
National Hellenic Research Foundation
Athens, Hellas

Chapter 1

Encyclopedia of Carbon Nanoforms

Irene Suarez-Martinez[1], Nicole Grobert[2], and Christopher P. Ewels[1]

[1] *Physics of Nanoscale Materials, Institut des Matériaux Jean Rouxel, CNRS UMR6502, BP32229, 44322 Nantes, France*

[2] *Department of Materials, University of Oxford, Parks Rd, Oxford, OX1 3PH, UK*

irene.suarez@cnrs-imn.fr; nicole.grobert@materials.ox.ac.uk; chris.ewels@cnrs-imn.fr

Since the discovery of C_{60} in 1985 and the paper on "Helical microtubules of graphitic carbon" in 1991, research into carbon nanotechnology has undergone a tremendous boom. As a result, a vast number of new carbon nanoforms have been identified, studied, and reported. Carbon nanostructures can range from structurally well-defined molecules to larger "macromolecules" of which the atomic arrangement cannot be described precisely. This chapter gives a comprehensive summary of different sp^2 and quasi-sp^2 carbon nanoforms, with special emphasis on their topology and morphology. We discuss briefly their various synthesis conditions and potential applications.

1.1 Introduction

In his book *The Periodic Table* Primo Levi says: "every element says something to someone (something different to each) [...] one must

Advances in Carbon Nanomaterials: Science and Applications
Edited by Nikos Tagmatarchis

ISBN 978-981-426-78-78 (Hardcover), 978-981-426-78-85 (eBook)
www.panstanford.com

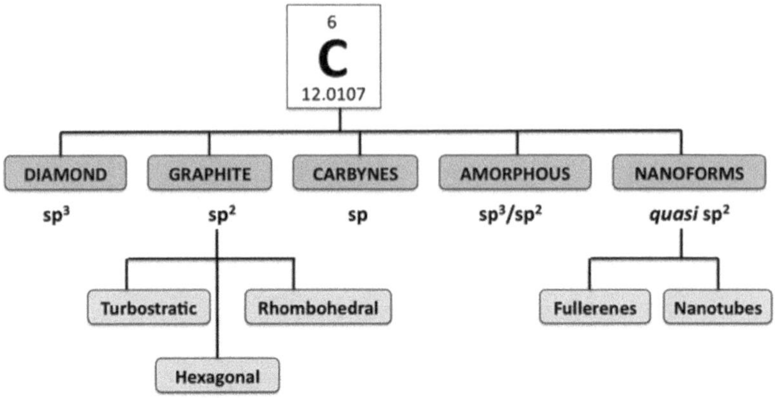

Figure 1.1. Summary of different carbon forms, taken from ref. 8.

perhaps make an exception for carbon, because it says everything to everyone."[1] Carbon is, indeed, an extraordinary element. The electronic configuration of $1s^2\ 2s^2\ 2p^2$ allows carbon atoms to form three different types of bonding, i.e., single, double, and triple bonds. This versatility of carbon to bond with other atoms is based on the fact that carbon can hybridize its 2s and 2p atomic orbitals in three different manners: sp^3 (for single bonding, tetrahedral), sp^2 (for double bonding, trigonal planar), and sp (for triple bonding, linear).

The carbon family tree traditionally covered graphite, diamond, and amorphous carbons, with the more recent addition of fullerenes and carbon nanotubes (see Fig. 1.1). However, in reality, due to the unique bonding versatility of carbon, the true range of carbon nanoforms is significantly richer than this.

Theoretical calculations and experimental studies predict outstanding physicochemical properties for many of these, which has led to an explosion of new carbon nanoforms being investigated. This exponential increase, in turn, has led to a bewildering growth in names (especially with view to sp^2-based carbon nanostructures), often with little or no attempt to standardize with other reports in the literature. The result is that it is increasingly difficult to identify the structure of a carbon nanomaterial based on its name. The same

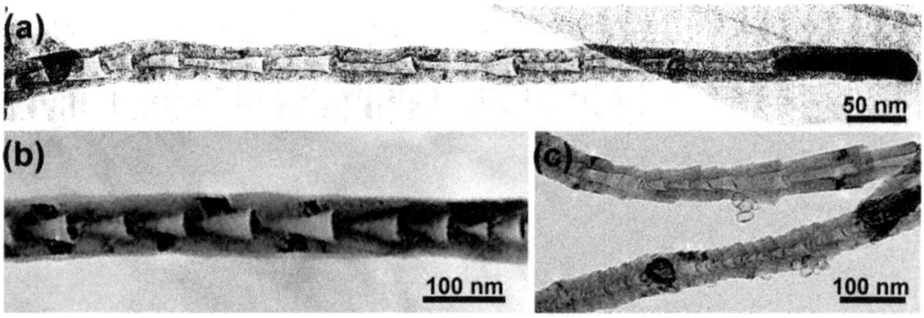

Figure 1.2. Transmission electron micrographs depicting a similar type of material named differently in different papers, (a) bamboo-shaped carbon tube,[4] (b) "stacked-cup"-like structure carbon nanotubes,[5] and (c) "stacked-cones."[6]

materials are sometimes referred to by different names depending on the authors (see Fig. 1.2), while in other cases the same name is used for different nanomaterials. In this chapter, we attempt to apply consistent naming to provide the grounds for objective comparison of the carbon nanomaterials.[2,3]

This chapter aims to provide a reference which will help researchers to quickly gain an overview of the different sp^2 and quasi-sp^2 carbon nanoforms reported in the literature. We describe the various carbon nanoforms identified and suggested to date including a brief summary of their morphology, topology, and properties. It is outside the scope of this chapter to provide an extensive description of the synthesis and applications of each sp^2 carbon nanoform; however, appropriate references are indicated for the reader who desires to learn more about a particular form. We place special emphasis on the nomenclature and the theoretical structure of each form, and try to establish a set of consistent nomenclature standards.[7] We explicitly exclude polymers, aromatic carbon molecules, and amorphous carbon-based films from this chapter, since they form specific families which are well documented elsewhere, as well as sp- and sp^3-dominated nanoobjects such as carbynes and nanodiamonds. For a good description of the wider world of carbon allotropes (including bulk forms such as graphite, diamond, and amorphous carbons) we recommend the article by E. H. L. Falcao *et al.*[8]

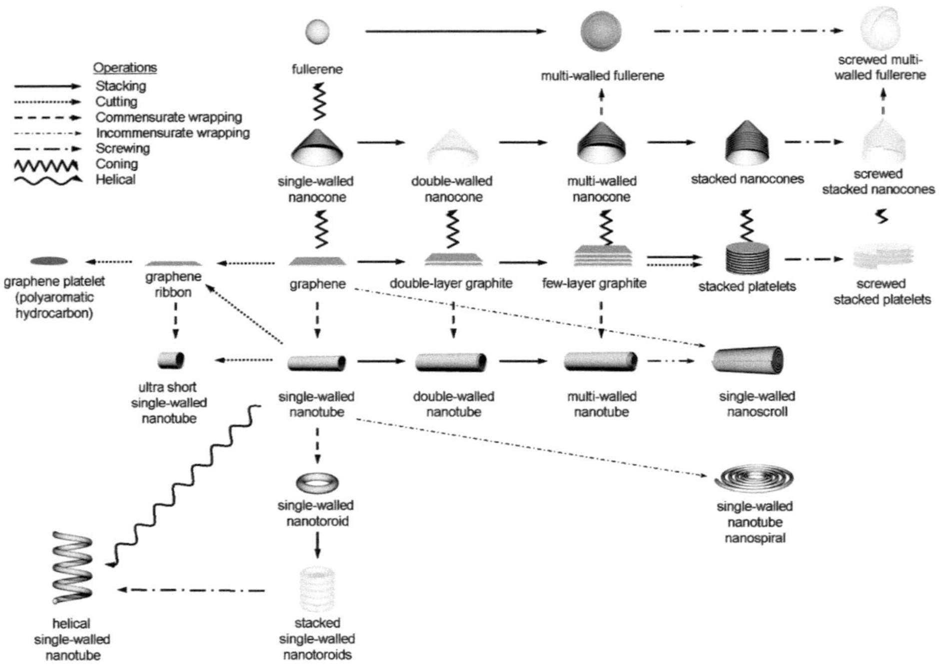

Figure 1.3. Schematic "family tree" depicting morphological relationships between different carbon nanoforms. (Faded carbon nanoforms have been predicted theoretically, but have not yet been observed experimentally.)

The nanoforms are ordered in this chapter based on atomic structure and overall morphology (following the "family tree" from Fig. 1.3). We begin with the structurally simplest sp^2 carbon nanoform: graphene, a quasi-two-dimensional (2D) single sheet of hexagonally arranged sp^2-bonded carbon atoms. We then examine quasi-one-dimensional (1D) forms (tubes, scrolls, etc.), and finally quasi-zero-dimensional (0D) forms such as cones, torii, and fullerenes. We finish with hybrid carbon nanoforms and those based on non-hexagonal carbon layers.

1.2 Graphene

[I1]

Graphene is a near planar sheet of sp^2-bonded carbon atoms distributed in a hexagonal network. It is a single carbon layer from graphite.

Graphene is both one of the "oldest" and also the "newest" of the carbon nanoforms. In principle, it is the simplest form of carbon – a single layer of carbon atoms. Its structural simplicity conceals some spectacular physics with the promise to revolutionize both fundamental and applied carbon science. Graphene has been the structural workhorse for computational calculations of carbon materials for many years and was thought for some time to be impossible to be isolated experimentally.[9,10] Recently, the group of Geim and Novoselov produced it through mechanical exfoliation in 2004[11] and 2005.[12] Since then the research effort and number of articles on graphene has increased exponentially, and the field is developing extremely rapidly, culminating in the award of the 2010 Nobel Prize for Physics. Good reviews of graphene science, e.g., by Geim *et al.*,[13] can be found at http://www.graphene.org/.

1.2.1 *The Structure of Graphene*

In graphite, all carbon atoms are sp^2-hybridized and have three equidistant neighbors forming a layer with a hexagonal honeycomb pattern. A single carbon layer of graphite is called graphene.[14] Three nearest neighbors form strong directional sigma bonds, while the fourth carbon electron forms an extended resonant π-bonded cloud above and beneath the sheet which serves to maintain the planarity of the layer. Topologically an arrangement of three equally spaced planar neighbors gives 120° bond angles and results in a planar array of tessellated hexagons.

The suffix "-ene" is related to fused polycyclic aromatic hydrocarbons, such as naphthalene, anthracene, and coronene. Graphene may be considered as the final member of this series, the largest member with quasi-infinite size. Graphene is the first truly 2D crystal ever known, and according to the Mermin–Wagner theorem it should not be completely planar at finite temperatures but intrinsically rippled.[15] Graphene presents a very unconventional electronic structure which is characterized by the linear dispersion of the π bands near the Fermi energy.[16] It is a zero-gap semiconductor.

1.2.2 *Synthesis Methods for Graphene*

Production techniques for graphene are undergoing rapid development at the time of writing. Current techniques can be divided roughly into three types. The first involves layer removal from graphite, via mechanical exfoliation (scotch tape method),[17] the use of surfactants to disperse layers of graphite,[18] or notably the formation of graphene oxide which can then be dispersed and reduced.[19] The second approach is based on exfoliation of graphene from SiC films via heating bulk SiC[20] whereby Si is removed from the areas closer to the surfaces and simultaneously graphene is formed at the resulting carbon-rich layer. The final approach makes use of epitaxial growth,[21] which appears the most promising for large-scale production. Notably large sheets of graphene can now be produced through chemical vapor deposition (CVD) of carbon species over monatomic nickel substrates whereby the Ni is dissolved in a second step to produce large freestanding sheets.[22]

1.2.3 *Terminology*

Terms such as "single graphene layer" or "single graphene sheet" are redundant. Graphene is always a single layer and therefore these terms should be avoided. Preferable terms are "graphene" or "graphene layer." Following the same argument, the term "few layers graphene" is not correct as graphene is never few layers but a single layer. The correct term is a "few layers of graphene" or "few layers graphite."

1.2.4 *Graphene-Related Forms: Graphene Nanowalls and Graphene Nanoribbons*

A graphene nanoribbon (GNR) is a strip of graphene of less than 100 nm width. GNRs are classified depending on the structure of the edge: armchair-like (aGNR), zigzag-like (zzGNR), and chiral (chGNR).

As for nanotubes, the number of graphene-based forms is in rapid expansion. Bi-layer, tri-layer, and few-layered graphite are the subject of many recent studies, with both commensurate and turbostratic ordering. Importantly, massless fermion behavior, as observed for graphene, is also observed for misoriented multi-layered systems.[23]

In the previous examples, the graphene layers are typically parallel to any substrate. "Vertically grown few-layered graphite" has also been produced, e.g., on a NiFe-coated sapphire substrate using microwave-enhanced plasma CVD[24] (referred to by the authors as "nanowalls," Fig. 1.4b). The growth process is similar to that of substrate-based multi-walled carbon nanotube (MWCNT) growth. In this case the "walls" are oriented almost perpendicularly to the substrate surface, are a few nanometers thick (typically less than 10 nm), and typically a micron long.[23] This material is expected to be of interest for, e.g., field emission.

When graphene is cut into a strip less than 100 nm wide, the term "graphene nanoribbon" applies. Depending on the direction of the cut, graphene nanoribbons (GNRs) are classified in armchair-like (aGNR), zigzag-like (zzGNR), and chiral (chGNR)[25] (see Fig. 1.4a). Zigzag-like edges have associated metallic states which give rise to a large peak at the Fermi level,[26,27] and the confinement induced by the edges can open the electronic gap. Nanoribbons can

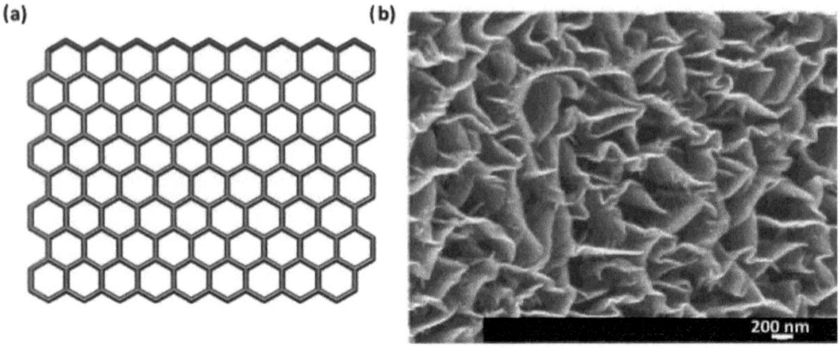

Figure 1.4. (a) Model of a graphene patch showing the two types of edges: armchair (highlighted in red) and zigzag (highlighted in blue). (b) Vertically grown few layers graphite[23] (referred to as "nanowalls" by the original authors).

be formed via lithography of larger graphene sheets,[28] or cutting open of carbon nanotubes.[29]

Various topological distortions of GNRs have been proposed in the literature. These include GNR rings and the same material with a single 180° twist in the graphene, resulting in a Möbius strip.[30] In the same way as ribbons could be produce by lithography of graphene, other shapes can be produced including triangles and other polygons, circles, etc.

We want to emphasize the role of graphene as the initial building block of a thought experiment to obtain other nanoforms. To move from the infinite 2D graphene, we typically need to introduce curvature and often dangling bonds at the edges of the nanostructures. There are three main ways to introduce curvature: first, by rolling or bending the graphene, second by introducing defects such as pentagons or heptagons within the sheet, and third by doping or functionalizing the layer.

1.2.5 *Applications of Graphene*

At the time of writing, applications of graphene are largely at the proposal stage. The linear dispersion at the Fermi level implies a zero electron effective mass, with associated remarkable carrier mobilities. Even given restrictions due to edge effects and defects mobilities of $\sim 10^4$ cm^2/Vs have been reported,[31] approaching that of isolated nanotubes.[32] Practical application interest currently focuses on transistor and device design,[33] and gas sensing applications,[34] although early work on graphene-based composites[35] shows promise for mechanical reinforcement and electronic percolation.

1.3 Carbon Nanotubes

A carbon nanotube is a tubular hollow-core sp^2-bonded carbon nanostructure with no axially oriented edges, where the tube walls are approximately parallel to the tube axis at all times. The number of walls define whether it is a single-, double-, triple-, few-, or multi-walled carbon nanotube. Nanotube diameters range from sub-nanometer for single-walled tubes to ~100 nm for large multi-walled tubes (elongated hollow/solid carbon nanostructures with diameters above 100 nm are referred to as carbon nanofibers and carbon nanorods).

Carbon nanotubes are hollow-core carbon tubes made from one or more carbon layers wrapped into a seamless tube about an axis. They have become practically synonymous with the term "nanotechnology" and are certainly the most famous of all nanomaterials. As early as the 1950s, hollow-core carbon fibers were reported by various groups.[36,37] The first clear nanotube observation was in the 1970s;[38] however, they were only brought to the attention of the wider scientific community with Sumio Iijima's seminal 1991 *Nature* article showing high-resolution transmission electron microscopy (HRTEM) images of multi-walled tubes[39] (for a more detailed description of the history of nanotubes, the interested reader is referred to ref. 40). Many books have been written on carbon nanotubes, and we particularly recommend the recently updated *Carbon Nanotube Science*.[41]

1.3.1 *The Structure of Carbon Nanotubes*

Structurally, carbon nanotubes can be visualized as seamless tubes made of graphene. The topological transformation required to obtain a single-walled carbon nanotube (SWCNT) from a graphene sheet of defined size is to roll it and to bond the two edges together. The length and direction of the rolling vector is known as the chiral vector (n,m)[42] (see Fig. 1.5 and Fig. 1.6). The chiral vector defines the nanotube diameter and the type of edge around the circumference. A classification based on the chiral vector gives armchair tubes when $n = m$, zigzag tubes when $m = 0$, and chiral tubes for all other n,m (see Fig. 1.7). The chiral vector also defines the electronic properties of SWCNTs.[43] Due to the folding of the conducting graphene for certain chiral vectors the resulting tube is metallic (for $n - m$ is a multiple of 3) while others are semiconductors. For semiconducting nanotubes, the band gap decreases with increasing diameter.[44] Multi-walled tubes are all metallic.

Nanotubes are classified based on the number of walls: SWCNTs, double-walled carbon nanotubes (DWCNTs), triple-walled carbon nanotubes, and MWCNTs. All consist of concentric cylinders with spacing between nanotube walls approximately the interlayer distance in turbostratic graphite, i.e., 0.34 nm. The number of walls can be determined – if the nanotube is isolated – by the number of

The cross-section of large diameter MWCNTs commonly becomes polygonized rather than spherical, where the localization of curvature is compensated for by improved commensurability in the layer stacking approaching that of AB-stacked graphite. Equally, large diameter SWCNTs and DWCNTs collapse to give "dog-bone" cross-sections, when the increased strain due to the curvature at the edge of the dog-bone structure is compensated for by the van der Waals interaction between the collapsed layers[47] (see Fig. 1.9b). lines in a transmission electron microscopy (TEM) image (see Fig. 1.8).

SWCNTs commonly form bundles (ropes) due to van der Waals interactions between neighboring tube walls. Although normally considered weak forces, van der Waals between two neighboring tubes can be high, $\sim$1.2 eV/nm along a nanotube interface,[48] as shown in Fig. 1.9a. Thus, the efficient separation and dispersion of

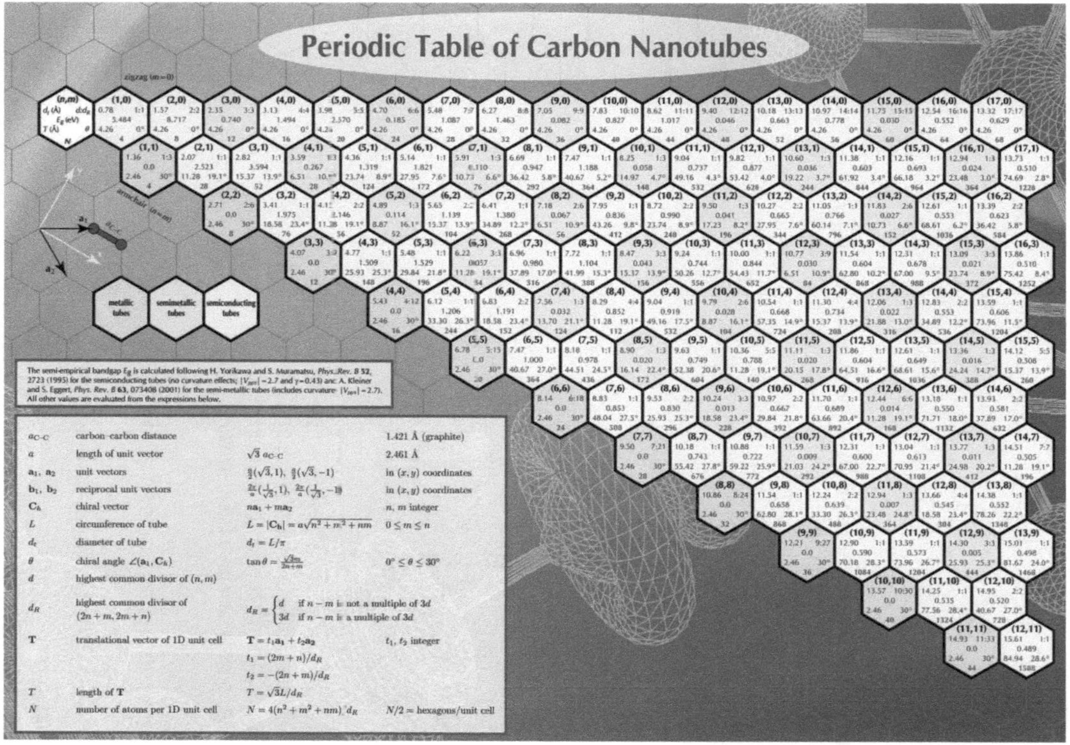

Figure 1.5. Periodic table of carbon nanotubes (reproduced with permission from quantumwise, www.quantumwise.com). A larger version of the table is freely downloadable from www.panstanford.com/books/9789814267878. See also Color Insert.

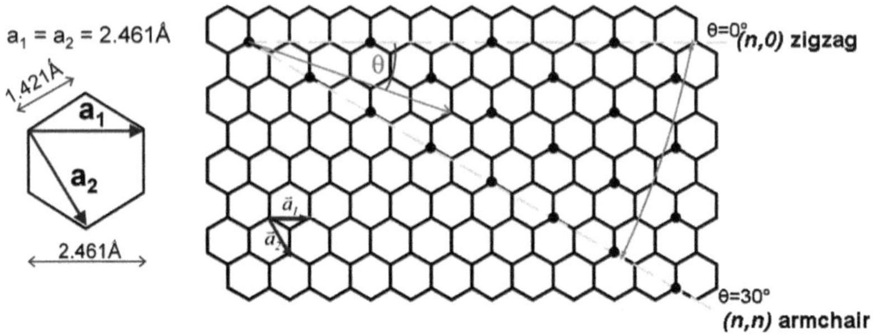

Figure 1.6. Schematic showing the graphene unit cell vectors, and definition of the chiral indices (Hamada indices) for a carbon nanotube (*n*,*m*), indicating the wrapping vector $na_1 + ma_2$ around the circumference of the nanotube. Thus, the vector marked with the arrow in the diagram would correspond to the circumference of a (4,2) nanotube. Black dots indicate metallic tubes.

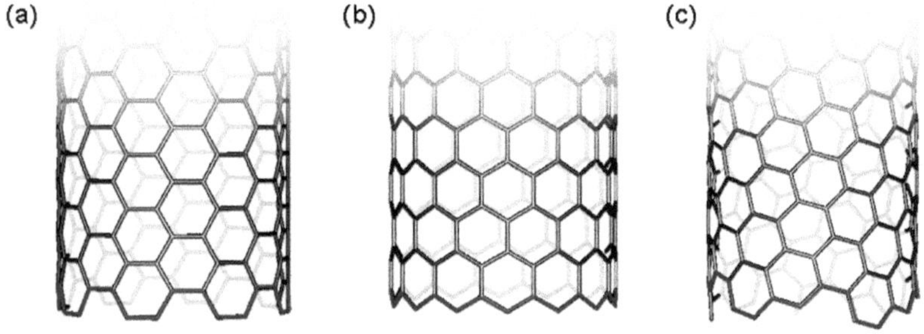

Figure 1.7. Different chirality single-walled carbon nanotubes, (a) armchair ($n = m$), (b) zigzag ($m = 0$), and (c) chiral (all other n, m) nanotubes. The names refer to the structure observed circumferentially around the tube (marked in red). See also Color Insert.

carbon nanotubes is an area of intense interest and one of the major obstacles to overcome (see, e.g., ref. 49), besides high production costs, before SWCNTs will become industrially viable in mainstream applications.

SWCNTs have been reported to reach lengths of up to 4 cm,[50] although they are more typically from microns to millimeters in length. In general SWCNT and DWCNT length is rather challenging to determine, unless they are grown perpendicular to substrates

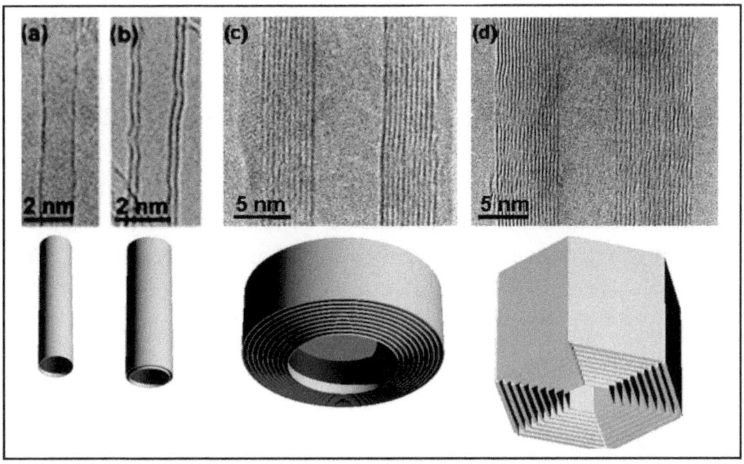

Figure 1.8. HRTEM images of (a) single-walled carbon nanotube (SWCNT), (b) double-walled carbon nanotube (DWCNT), (c) multi-walled carbon nanotube (MWCNT), and (d) polygonized MWCNT. Computer-generated images below show 3D representations of these forms. Polygonization in (d) can be observed through the difference in layer spacing on the left and the right, due to fortuitous alignment of the polygonized tube with respect to the electron beam. (c), (d) adapted from ref. 45.

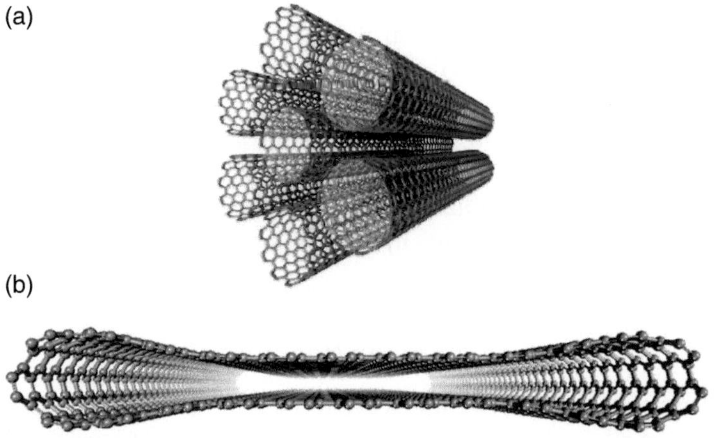

Figure 1.9. Bundles of (a) single-walled nanotubes (taken from ref. 46), (b) Dogbone image thanks V. V. Ivanovskaya (taken from ref. 47). See also Color Insert.

(so-called nanotube forests), due to their entangled nature. The length of MWCNTs grown on substrates via CVD, on the contrary, is easily measured using standard scanning electron microscopy (SEM). Such MWCNTs normally range between a few microns up to the centimeter range.[51]

At the other end of the scale, ultra-short nanotubes can be produced, normally through cutting of longer nanotubes, e.g., via fluorination followed by pyrolysis[52] or using oleum on SWCNT bundles.[53] Ultra-short can refer to anything between $\sim$7 nm[54] and $\sim$60 nm.[52] In this case, the nanotubes can be viewed as quasi-0D objects.

1.3.2 *Synthesis Methods for Carbon Nanotubes*

The properties of carbon nanotubes, such as diameter, number of walls, and length, are highly dependent on the production method used to make them.

Most carbon nanotube synthesis techniques involve the vaporization of carbon precursors in the form of either a graphite target (arc-discharge, laser ablation, and electrolysis) or hydrocarbons (CVD and plasma-enhanced CVD) in conjunction with metal catalysts. A detailed review of carbon nanotube synthesis can be found in ref. 3.

Carbon nanotubes can also be filtered and compacted into a film, often referred to confusingly as "buckypaper," which has been proposed for various applications included electromagnetic screening, cell-growth support, and filtration (we note that the films do not contain fullerenes, and a preferable name is "nanotube films"). It is also possible to create low-density nanotube "sponges," very light, highly porous hydrophobic materials which can be elastically deformed.[55]

1.3.3 *Applications of Carbon Nanotubes*

Carbon nanotube applications are too numerous and varied for the space available here, and the interested reader is referred to ref. 56 and recent books such as refs. 42 and 40, as well as several later chapters in this book. Current applications typically use MWCNTs

Table 1.1. Shows typical structural details of the most commonly used synthesis methods

Type		Arc-discharge	CVD	Laser ablation	Electrolysis
MWCNTs	General info	Only method which can produce carbon nanotubes without metal catalysts, highly graphitic MWCNTs	Suitable for floating catalyst growth, substrate growth, metal catalysts are essential, tubes are of relative high quality, but exhibit more defects than arc-discharge carbon nanotubes	Possible, but mainly used for SWCNTs production	Possible, but poor quality tubes are highly defective, high conc. byproducts, e.g., a-C, polyhedral particles
	Length	Several microns	Several microns to centimeter range	—	Tens of microns
	Diameter	Up to ca. 20 nm	Ca. 5–100 nm	—	Wide variety of types and sizes of nanoparticles and nanotubes
	No. of walls	2 to ca. 20	2 to ca. 50 or more	—	Highly defective walls, difficult to count
SWCNTs	General info	Mixed metal catalysts, e.g., mixtures of Ni and Y are necessary	Metal or mixed metal catalysts, e.g., mixtures of Co, Ni, Fe, and/or supported catalysts are necessary, diameters are usually larger for CVD SWCNTs than for those produced using other methods	Graphite targets containing mixed metal catalysts, e.g., mixtures of Co, Ni are necessary to form SWCNTs	Not yet observed
	Length	Due to entanglement difficult to measure, estimated micrometer range	Due to entanglement difficult to measure, estimated micrometer range, lengths of up 20 cm were reported, but not confirmed	Due to entanglement difficult to measure, estimated micrometer rang	
	Diameter	1–2 nm	1–5 nm	1–1.5 nm	
Metal-filled MWCNTs		Possible via anode doping with desired filling material, but limited filling yield	Possible using higher concentration of catalyst material, relatively high filling yield, dimensions are similar to non-filled tubes produced via CVD	Not yet observed	Low melting point metals, high filling yield, mainly amorphous carbon coating

and take advantage of their numerous interesting properties, be they mechanical (e.g., as reinforcement agents in composites, nanotube fibers), thermal (heat dissipation in microelectronics), electrical (percolative networks in non-conductive polymer matrices, field emitters, etc.), and physical (mechanical supports for catalysts, drug delivery, etc.). SWCNT applications are currently more limited but are of interest for optical (luminescence, absorption, strain characterization agents in matrices) and electrical (wires and devices in nanoelectronics, charge transport in solar capture devices, surface absorption for detection) applications, notably where their unique electronic structure coupled with 1D morphology is of particular benefit.

Products available on the consumer market at present are focused on high-end, high-value devices such as bike frames, golf clubs, and handheld X-ray devices, where the relatively high production cost of the nanotubes can be justified for the improved performance.[57] However, as nanotube prices drop this is beginning to change, with the recent arrival of a new wave of technologies such as touch-screen displays,[58] laptop heat dissipation,[59] and laptop batteries incorporating carbon nanotubes. To date, all carbon nanotube consumer applications make use of the bulk properties of the carbon nanotubes. Applications relying on the specific properties of individual carbon nanotubes are yet to be developed outsidelaboratories. The main obstacles that need to be overcome to viably generate single carbon nanotube products are the difficulty in synthesizing clean, uniform, and disperse carbon nanotubes, selectivity (e.g., isolated metallic or semiconducting SWCNTs), and the difficulty in manipulating individual carbon nanotubes at the industrial scale. There are also questions which remain regarding their potential toxicity.

1.4 Carbon Nanoscrolls

A carbon nanoscroll is a tubular hollow-core sp^2-bonded carbon nanostructure with two (or more) axially oriented edges, where the tube walls are approximately parallel to the tube axis at all times.

Soon after their discovery, two structural models for carbon nanotubes were proposed: the "Russian doll" model consisting of concentric cylinders (discussed in the previous section) and the "Swiss roll" model consisting of one or more sheets rolled up into a scroll.[38] Theoretical calculations predicted the Russian doll model to be more stable than the Swiss roll model due to the absence of dangling bonds. However, experiments have since shown the existence of this type of cylindrical structure, which may be more common than originally realized. Reference 60 provides a good review of carbon nanoscrolls (CNSs).

1.4.1 *The Structure of CNSs*

CNSs can be pictured as one or more sheets of graphene rolled into a scroll. CNSs cannot be determined uniquely by the chiral vector as SWCNTs are, they also require the "amount of overlap in the wrapping" to be specified. CNSs can also be interpreted as an edge dislocation in a MWCNT, where the dislocation line runs along the tube axis and the Burgers vector is perpendicular (i.e., along the radius of the nanotube).[61] However, strictly speaking, carbon nanotubes are not a crystalline solid as they are only periodic along the tube axis, and for this reason dislocation nomenclature should be used with care.

As for conventional carbon nanotubes, CNSs can be armchair, zigzag, or chiral depending on the orientation of the graphene sheet(s) with respect to the tube axis. Theoretical calculations predict armchair CNSs to be metallic or semi-metallic depending or their sizes, while zigzag CNSs are semiconductors but with energy gaps much smaller than the corresponding zigzag SWCNTs.[62]

Nanoscrolls are less stable than their equivalent length MWCNT due to the fixed energy cost associated with the two edges.[63] However, this energy cost is less significant once the edge-site dangling bonds are functionalized, and becomes negligible for nanoscrolls with many walls, as has been observed in the formation of carbon whiskers which are scrolls.[64] CNSs can also polygonize in the same way as large diameter MWCNTs, which can be characterized by a periodical arrangement of alternating bright spots along the nanotube length where the electron beam is parallel to the graphene layer.[65]

There are only a few papers in the literature that clearly establish the presence of nanoscrolls.[59] A different number of layers observed at each side of a tube in a TEM image is normally considered as a proof of the presence of a CNS.[66] However, the absence of this mismatch does not suffice to rule out the presence of CNS, as this mismatch will only be visible in TEM for certain scroll orientations. High correlation in the chirality of the interior tube walls is more solid proof. Assuming a random growth process, it is statistically extremely unlikely that the walls in a MWCNT will show the same chirality. However, in a CNS the walls must have uniform chirality since they are all derived from the wrapping of a single graphitic layer. Tube chirality is revealed by electron diffraction and in some cases by microscopy, and thus for CNS the distribution of (*hk0*) reflections gives a unique chiral angle,[59] as opposed to more annular powder-like pattern for a typical MWCNT (see Fig. 1.10).

Nanoscrolls can also be formed by rolling more than one graphene sheet (multiscrolls), or can wrap around conventional nanotubes.[69] There are even reports of a nanoscroll transforming into a MWCNT within the same tube.[70] In this case, the representation of a nanoscroll as a MWCNT helps for the visualization of the interface scroll-to-nanotube. As the dislocation line changes its direction, it can exit the tube walls perpendicular to the tube axis,[60] comparable to a screw dislocation in graphite. The conversion between the two forms can be achieved by the gliding of the screw dislocation (the so-called zipper mechanism).[60,62]

1.4.2 *Synthesis Method for CNSs*

The synthesis of carbon whiskers may be considered as the first production of scrolls. However, common arc-discharge carbon whiskers are not restricted to nanoscale diameters, typically ranging from a fraction of a micron to 5 μm.[63]

The production of nanoscrolls was first achieved through exfoliation and subsequent rolling of graphene sheets from graphite via K-intercalation,[71,72] and this has restored the interest of the carbon community in this nanoform. More recently, production of

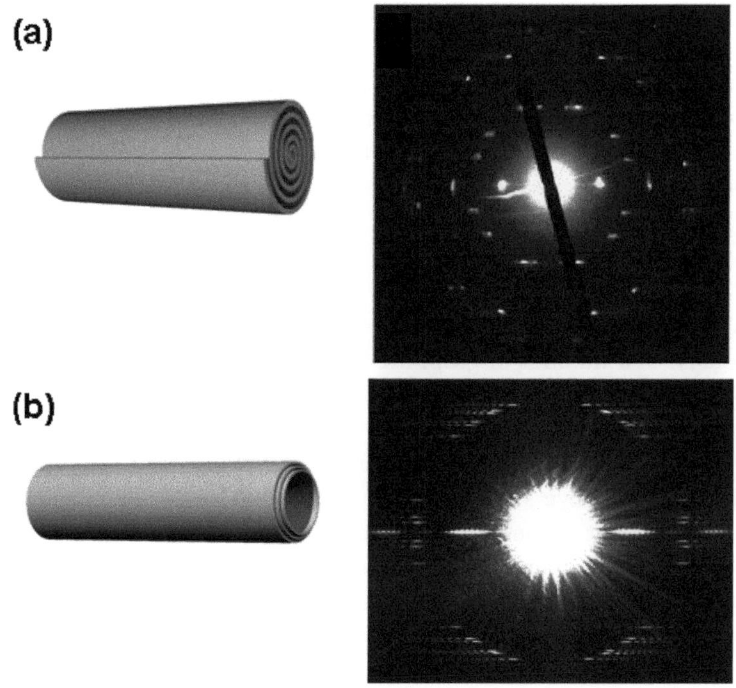

Figure 1.10. Electron diffraction patterns for (a) a nanoscroll with chiral vector 9.78°, (b) a triple-walled carbon nanotube. (b) consists of three sets of individual patterns due to the three nanotube shells, with chiral indices (35,14), (37,25), and (40,34), whereas (a) shows a single diffraction set showing all layers exhibit identical chirality. Diffraction images taken from refs. 67 and 68.

graphene through surfactant chemistry has also reported partial rolling of graphene sheets.[73] However, there appears to be far less interest within the literature in nanoscroll synthesis as compared to the nanoforms discussed above.

It is likely that many MWCNTs are in fact nanoscrolls, since without detailed diffraction studies it may be difficult to tell them apart. For example, CVD-grown nitrogen-doped multi-walled tubes were shown to have an extremely high degree of internal order, both in terms of the uniform chirality in the nanotube walls and of the crystallographic register between them,[74] and it is likely that these are actually scrolls. Equally many large MWCNTs may in fact be

MWCNTs surrounded by a nanoscroll, since at such large diameters nanoscrolls become energetically comparable with concentric tubes. For example, fluorination of large MWCNTs has been shown to double the interlayer spacing of the external tube walls,[75] an observation which is hard to explain without evoking the presence of a nanoscroll.

1.4.3 *Applications of CNSs*

A CNS resembles a MWCNT in that it is a cylinder whose walls consist of a number of graphitic layers. The mechanical properties of CNS and a MWCNT along the tube axis are relatively similar, e.g., similar Young modulus. However, unlike MWCNTs, a nanoscroll can vary its outer and inner diameter by rolling tighter or looser, and this may improve strain transfer to interior layers for mechanical reinforcement in composite applications.

In addition, nanoscrolls present a continuous, easily accessible, connected interlayer space, in contrast with the individual interwall spacing in MWCNTs. This unique characteristic of CNSs makes them a better potential candidate for hydrogen storage.[76]

The reactivity of nanoscrolls is increased compared to equivalent MWCNTs due to the presence of edges. In particular, these may be stabilized in the presence of nitrogen and may explain the observation of MWCNTs of uniform chirality in nitrogen-doped growth.[73]

1.5 Carbon Nanocones

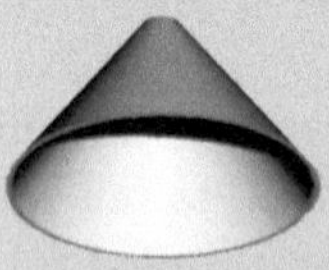

A carbon nanocone (CNC), also referred to as a nanohorn, is a conical object constructed from tri-coordinated carbon atoms. Nanocones can be classified by the number of layers as single (SWNC), double (DWNC), triple (TWNC), or few layers CNCs. SWNCs are often agglomerated (tips outwards) in what is referred to as a dahlia configuration.

1.5.1 *The Structure of Carbon Nanocones*

Carbon nanocones can be described as disclinations in graphene. The thought experiment consists of removing a wedge of material from graphene and reconnecting the dangling bonds, as shown in Fig. 1.11 (left), forming a cone. Due to the atomistic structure of graphene, this results in the formation of pentagons (marked in gray in Fig. 1.10). As the number of pentagons must be discrete (1 to 5), there is a discrete number of disclinations that can be produced, and therefore only five possible angles for the nanocone (see Fig. 1.11). The angle is easily related to the number of pentagons. The disclination angle is $n(\pi/3)$, with $0 \leq n \leq 5$, where n is the number of pentagons according to Euler's rule. The disclination angle is then related to the cone angle as $\theta = 2 \cdot \sin^{-1}(1 - n/6)$. Figure 1.11 (right) shows that while an integer number of up to 6 disclinations (and hence pentagons) can be removed from graphene, the precise position of the removed wedge (and hence pentagons) can be varied, resulting in an infinite number of nanocone structures. We note that neighboring pentagons are energetically unfavored due to chemical frustration (under-coordinated carbon atoms), referred to in fullerene chemistry as the isolated pentagon rule.[77]

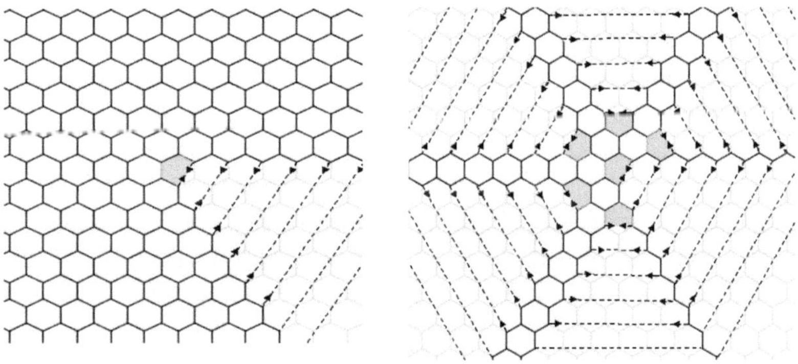

Figure 1.11. Representation of the construction of carbon nanocones by cutting a wedge (disclination) from graphene and reconnecting the resultant dangling bonds (dotted arrows). The pentagons thus created are colored in gray. (left) single pentagon cone, (right) up to 6 pentagons can be introduced; 6 pentagons results in a closed nanotube tip structure. Removing a wedge in this way necessarily results in a non-planar structure to maintain covalent C–C bond lengths.

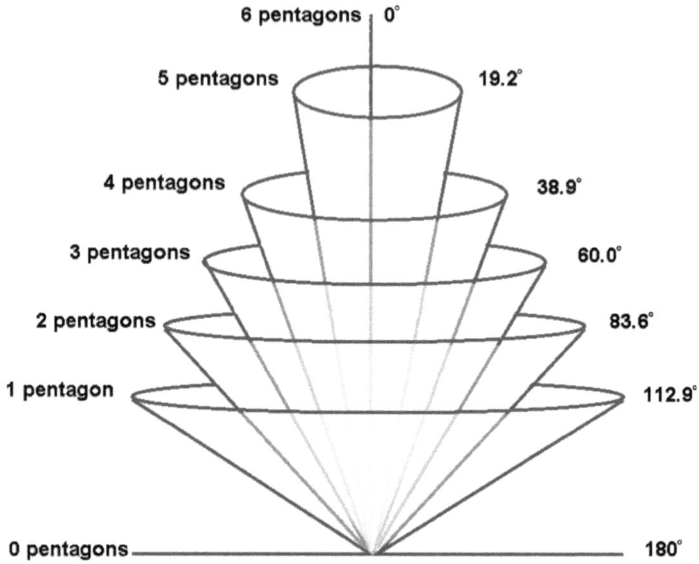

Figure 1.12. Representation of the possible nanohorns. On the right-hand side the angle of the cone is indicated, on the left-hand side the number of pentagons.

The structure with six pentagons is not a nanocone since the "tip angle" is zero. Instead the walls are parallel and the structure is thus a closed nanotube tip. The tip can be considered as half a fullerene.

Carbon nanocones are typically either multi-walled and individual,[78] or single-walled and agglomerated in larger clusters (tips outwards).[79] Depending on the protruding length, these agglomerates are classified into durian and dahlia configurations (see Fig. 1.13). When the cones protrude from the particle surfaces at heights of up to 20 nm, they are variously called durian, chestnut, or sea urchin structures (depending on the author's geographical origins!), since no tubular region are observed. However, in the dahlia structure, the nanocones have a more needle-like form.

1.5.2 *Terminology*

The nomenclature of this structure is not standardized in the literature. Theoretical modeling papers have often used the terms

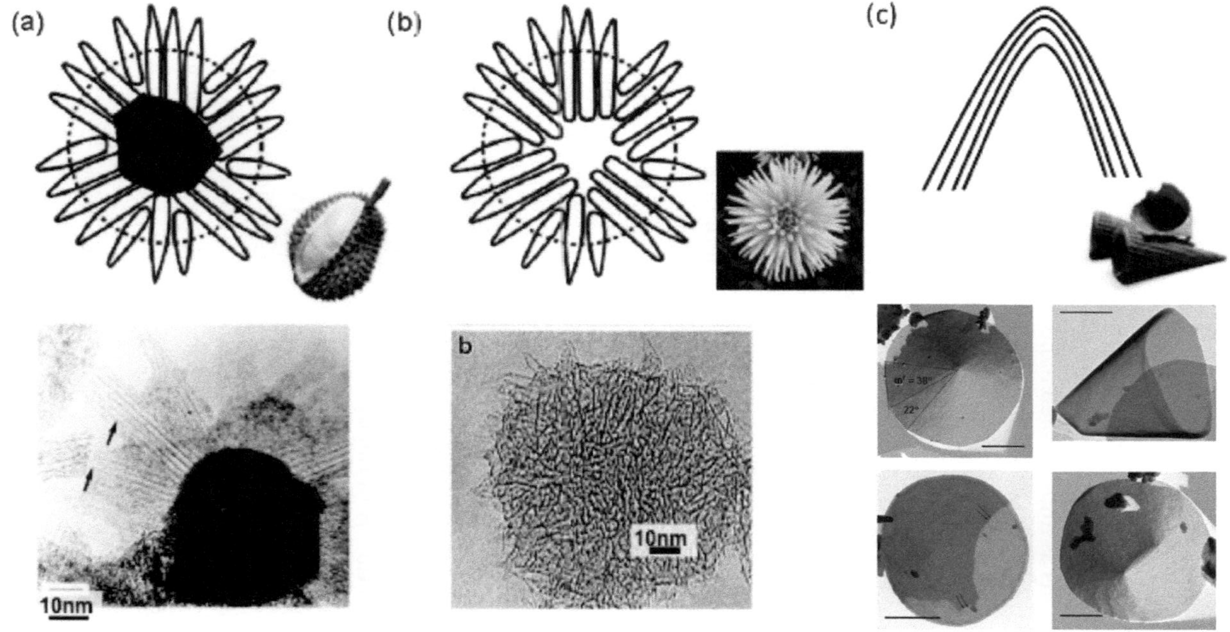

Figure 1.13. Representation (above) and microscopy image (below) of the (a) sea urchin/durian/chestnut nanocone aggregate, (b) dahlia aggregate nanocone structure, (c) isolated multi-walled nanocone. Images taken from refs. (a) 85, (b) Wikipedia: http://en.wikipedia.org/wiki/File:SWNH_Figs.jpg, (c) 78. See also Color Insert.

nanohorn and nanocone as synonymous. Experimental papers typically use "nanohorns" to refer to the clustered forms, and "nanocone" for the individual multi-walled structures. We propose for consistency to always use the term nanocone indicating the number of walls, with aggregate forms such as dahlias referred to as "aggregated" or "clustered" single-walled nanocones, and their isolated cousins as multi-walled nanocones.

1.5.3 *Synthesis of Carbon Nanocones*

First predicted in 1994,[78] isolated individual carbon nanocones have been reported only once, and consist of more than one layer.[79] These multi-walled carbon nanocones were synthesized in 1997 by pyrolysis of heavy oil in a carbon electric arc.

Clustered nanocones were first synthesized in 1999 by laser ablation of graphite.[80] There are many advantages to nanocone growth by this method as compared to nanotube growth. Resultant samples have 99.99% purity and no catalytic metal inclusions. The CO_2-laser has longer wavelength (10.6 μm) than typically used for nanotubes, and growth occurs at room temperature. Carbon nanocones can also be synthesized by other techniques such as electric arc-discharge in helium atmosphere at reduced pressure[81,82] or in liquid nitrogen,[83] torch arc,[84] or pulsed arc-discharge[85] in open air. Radial growth of these closed nanocone aggregates can occur either with[86] or without[79] a metallic catalytic particle at the cluster core. This typically determines whether the resultant structure is of the durian type (metal particle present) or dahlia type (no metal present), respectively (see Fig. 1.13).

The chemistry of nanocones is particularly interesting since the reactivity of the tip is very different to that of the side walls.[87] For example, it is known that the tip can be easily opened by mild acid etching. For more details of the chemistry of carbon nanocones, see Chapter 6 by Nikos Tagmatarchis.

1.6 Applications of Carbon Nanocones

The primary interest in nanocones to date has been as storage devices for hydrogen storage,[88] or as capsules for drug delivery.[89]

Nanocones have been suggested as a cheaper and more easily produced alternative to carbon nanotubes for those applications which require high surfaces areas, since while they have similar surface areas, nanocones require lower growth temperatures (typically room temperature) and no catalyst.

1.7 "Bamboo" Nanotubes

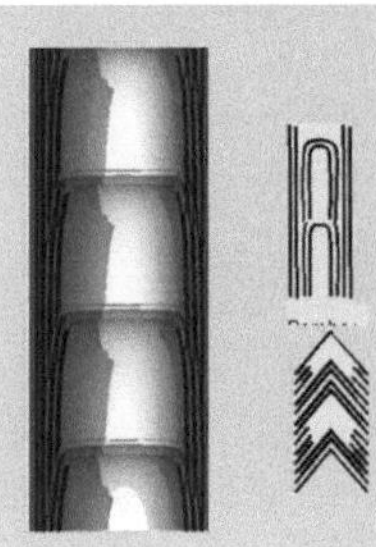

Bamboo nanotubes are tubulars tructures with a compartmented hollow core. Depending on the structure of the outer walls there are two possible bamboo structures: bamboo nanotubes where the external walls are almost parallel to the tube axis and herringbone-bamboo tubes where the layers are at an angle to the tube axis.

There are number of structural variants of MWCNTs. "Bamboo" nanotubes are tubes with approximately straight, parallel external walls, with the addition of regularly spaced internal compartments. When viewed with a TEM they resemble natural bamboo (see Fig. 1.14a). The partition walls are typically close to orthogonal to the nanotube axis. There exist also compartmented nanotubes consisting of stacked nanocones (see Fig. 1.14b), which are discussed further in Section 1.8 below.

1.7.1 *Synthesis of Bamboo Nanotubes*

The structure is normally associated with the introduction of heterogenous impurities,[92] notably when nitrogen or boron are present during synthesis in the CVD, by aerosol-based[93] or microwave plasma-assisted CVD,[94] or high-temperature routes such as arc-discharge[95] and laser ablation. A correlation is observed between the nitrogen content and the corrugation of the tubes. While there is much structural variation depending on precise growth conditions, bamboo tubes typically have nitrogen concentrations around ~15–20%[96] and can have local concentrations up to

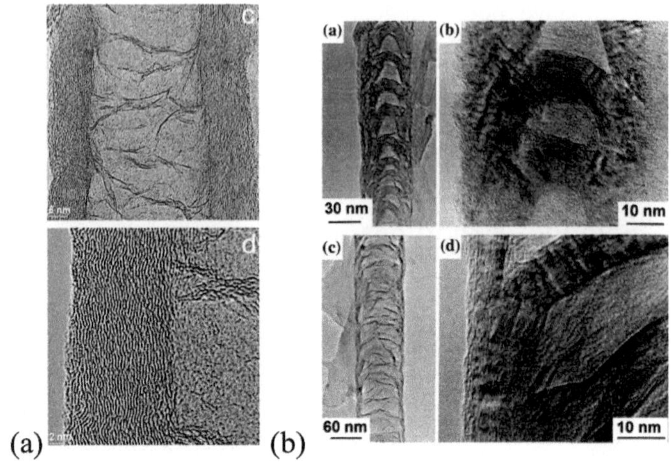

Figure 1.14. Different types of bamboo nanotubes, (a) conventional bamboo (taken from ref. 90) and (b) stacked-cone-type bamboo (taken from ref. 91).

25–30%,[97] particularly at interior surfaces. Increasing nitrogen concentration during synthesis produces shorter tubes with smaller diameter and an increase in the fraction of "bamboo"-shaped tubes.[98]

1.7.2 *Applications of Bamboo Nanotubes*

The introduction of nitrogen impurities means that bamboo N-doped are oxidized more easily than perfect tubes[97] since the surfaces are more reactive. However, this also renders them interesting for a number of applications, since they are more biocompatible,[99] and their increased chemical reactivity makes them interesting candidates for gas sensing[100] and Li storage.[101]

They disperse in solvents which are immiscible with undoped nanotubes,[102] and show improved functionalization behavior.[103] For all of the reasons above, "bamboo-type" (nitrogen doped) nanotubes are a subject of increasing interest within the nanotube field.

1.8 "Herringbone" Nanotubes

The term herringbone applies to cylindrical structures whose walls are parallel to each other, but not parallel to the tube axis. Depending on the structure of the core of the tube, herringbones can be either herringbone nanofibers (full core), herringbone nanotubes (continuous hollow core), or herringbone-bamboo nanotubes (compartmentalized hollow core).

Despite their relatively poor coverage within the nanocarbon scientific literature, herringbone-type nanotubes and nanofibers are one of the more commonly produced nanoforms. The name "herringbone" refers to their appearance when viewed in projection (e.g., in a TEM), as a series of stacked angled lines similar to the arrangement of bones down the back of a fish such as a herring.

1.8.1 *The Structure of Herringbone Nanotubes and Nanofibers*

Topologically there are two fundamental structural types of herringbone (see Fig. 1.15). In the first of these, herringbone nanotubes can be viewed as a stack of nanocones (for this reason they have also been named stacked-cups and stacked-cones). As for nanocones, the cone angle is restricted to specific angles (see Fig. 1.12) depending on the integer number of pentagons present at the tip.

The second type is similar but features a screw dislocation running along the core of the stack, i.e., each cone is "cut open" from its tip to its edge and connected to the layer above. The result is a single continuous layer which corkscrews around the stack axis. In this case, the discrete cone angle rule is relaxed. It is clear that these two structures have fundamentally different mechanical and electronic properties. Herringbones can either be filled or hollow

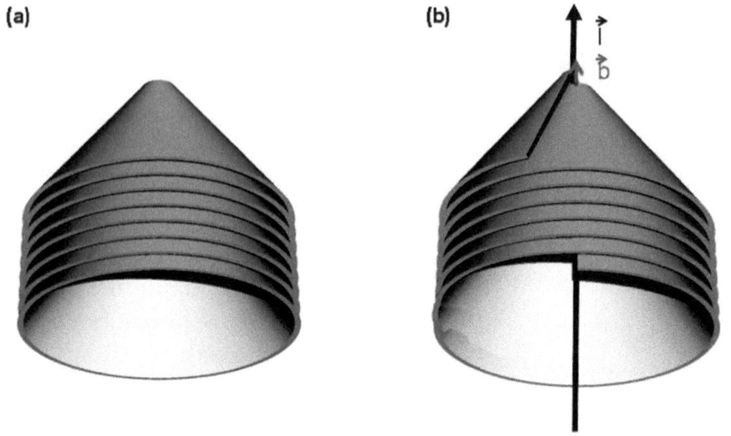

Figure 1.15. Two different herringbone fiber structures, (a) stacked nanocones and (b) stacked nanocones with a screw dislocation running along the stack core. While (a) consists of discrete nanoobjects, (b) is a single continuous surface.

(either through chemical etching, or during synthesis), resulting in *herringbone nanofibers* or *herringbone nanotubes* (Fig 1.16).

Finally, some herringbone nanofiber structures can exhibit partitions along their core similar to bamboo nanotubes, although in this case it is due to grouping of the component stacked-cone structures into small clusters (see Fig. 1.14b). Structures can also occur with repeating sections of filled and hollow-core cones, again resulting in a compartmentalized structure. These structures are therefore referred to as herringbone-bamboo (Fig. 1.16).[104]

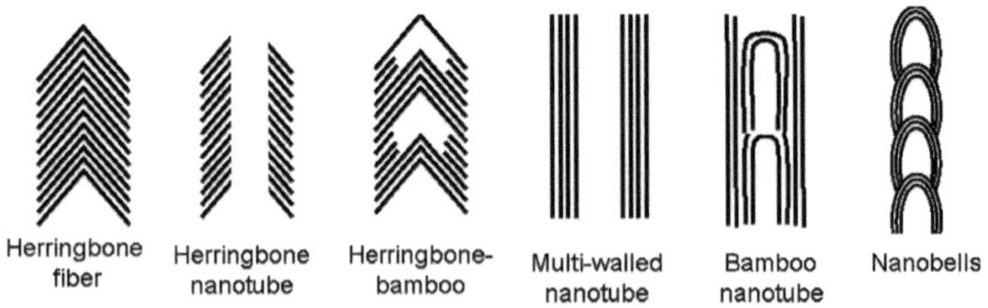

Figure 1.16. Schematic representation of the microscopy images produced by different linear carbon nanoforms.

1.8.2 *Herringbone Synthesis*

Herringbone structures are produced via CVD growth processes under very similar conditions to conventional MWCNT growth. They have typical diameters 50–150 nm (although can also be thinner) and lengths up to 200 μm,[105] or smaller (20–50 nm[106]). For smaller diameter herringbone tubes, it appears that the coning angle determines whether the tip is present (small cone angles) or missing (angles $> 30^\circ$).[105]

The structural variations for nanotubes and fibers during CVD are linked to a number of factors, notably the growth temperature and pressure, feed gas composition, presence of impurities, and choice and status of the catalyst particles. Bouchet-Fabre *et al.* have investigated the influence of different NH_3/Ar ratios in the gas flow on the growth of carpets of MWCNTs.[107] As the quantity of NH_3 increases, the morphology of the resultant samples changes from classical MWCNTs (small core, large number of walls, iron-based nanowires) at $[NH_3] < 10\%$, to bamboo nanotubes ($10\% < [NH_3] < 30\%$), and finally to highly compartmentalized nanobell-type structures ($30\% < [NH_3] < 40\%$).

1.8.3 *Herringbone Applications*

There are remarkably few studies of herringbone nanotube applications in the literature. The open wall stacked cone structure of herringbone tubes and fibers makes them interesting candidates for intercalation purposes such as hydrogen[108] and lithium storage. It was indeed found that their storage capacity of ~0.4 wt% at atmospheric pressure was higher than that of conventional MWCNTs.[109] Interestingly, herringbone nanotubes showed improved storage over herringbone nanofibers.

The stacked-cone herringbone structures are mechanically extremely weak, and mild mechanical treatment such as short time period ball milling completely destroys them. After ball milling (the resultant short segments curiously referred to as "nanobarrels"), the high surface area material has been successfully tested as a support in fuel cells[110] and in photoelectrochemical solar cells.[111]

1.9 Helical Nanotubes

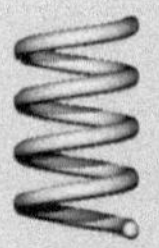

Coiled nanotubes are hollow-core sp^2 carbon-based concentric cylinders whose axis follows a helical pathway. As for other helical structures, they are defined by the ratio of the pitch, the diameter of the tube, and the diameter of the coil. Because they are in principle carbon nanotubes they can also be defined by the number of walls.

The ratio of the diameter of the coil to pitch is of importance for their applications. Therefore, a secondary classification applies: straight nanotubes (diameter/pitch = 0), coiled cord (diameter/pitch < 2), or coiled spring (diameter/pitch > 2).

Helical carbon structures can be classified into three groups: carbon microcoils, carbon nanocoils, and coiled carbon nanotubes (see Table 1.2). For recent reviews of the synthesis and mechanical applications of coiled carbon nanotubes, see refs. 112 and 113. Carbon microcoils were first seen in 1990.[114]

Table 1.2. Different coiled carbon structures and their corresponding dimensions (adapted from ref. 122)

	Carbon microcoils (μm)	Carbon nanocoils (nm)	Coiled carbon nanotube (nm)
Tube diameter	0.5–2	60–100	5–20
Coil pitch	1–5	120–150	20–100
Coil diameter	3–8	$\sim$ 100	50–80

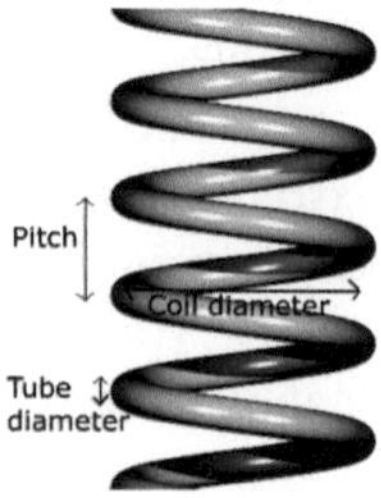

1.9.1 *Synthesis of Helical Nanotubes*

Helical nanotubes were proposed soon after the discovery of conventional nanotubes,[115] and were experimentally reported in 1994 by Zhang *et al.*[116] Helical nanotubes are typically grown in low-temperature catalytic CVD, where they can be obtained with high yields.[117,118]

Carbon microcoils and nanocoils are fibers (graphitic, solid-core structures), whose most important difference is the size. They are grown by, e.g., microwave plasma CVD of C_2H_2 over microsized Ni particles on SiC[119] or oxide catalysts[120] using H_2 and Ar carrier gas. By varying the temperature from 600 to 700°C it is possible to switch from majority nanocoil to microcoil growth.

Coiled nanotubes (also called helix-shaped or helical nanotubes) have crystalline graphitic structure and are hollow core (see Fig. 1.17). Coiled nanotubes are essentially standard MWCNTs whose axis follows a helical pathway, resembling a telephone cord. As such they can be "left-" or "right-"handed depending on the coil direction, and indeed can switch between these during growth.[121] An alternative topological description is that of a screw dislocation in a stack of multi-walled nanotori (see below).

Coiled nanotubes are normally only observed in catalytic CVD experiments at low temperatures (around 700°C) and often in the presence of nitrogen[122] or sulfur.[123] At higher temperatures (such as

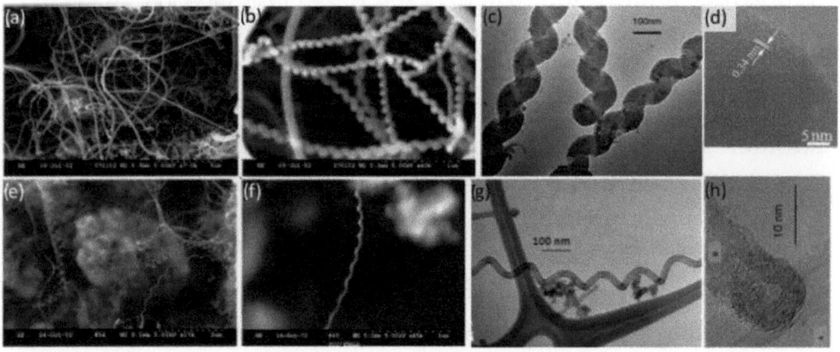

Figure 1.17. Typical SEM, TEM, and HRTEM images of (a–d) nanocoils and (e–h) coiled carbon nanotubes (a–c taken from ref. 119, d from ref. 121, and e-h from ref. 123).

during arc-discharge and laser ablation) nanotubes defects anneal and nanotubes tend to be more straight. Micro and nanocoils are also synthesized by CVD in the presence of a catalyst. Yield for one type or the other is obtained by controlling the temperature, flow rate, and catalytic particle size.[118]

1.9.2 *Topology of Helical Nanotubes*

Coiling in nanotubes is often linked with deliberate or accidental impurity doping.[91] Yudasaka and colleagues often observed coils at the tips of N-doped nanotubes,[124] and coiling has also been linked to choice of catalyst.[125] Their formation has been explained by non-uniform growth rates of the tube from the catalyst particle,[124] which is consistent with the presence of impurities within the catalyst.

There is a relationship between the coil pitch and the coil diameter. Coiled nanotubes are grouped in what is called "stability islands"[121] (with pitch of either $\sim$30 nm or 50–70 nm, diameters 30–50 nm). These stable groupings suggest that the helical shape has an intrinsic structural origin imposed by the atomic structure. The atomistic structure of coiled nanotubes has never been solved experimentally, but it is often explained by the presence of pentagons (in the outer part of the coil) and heptagons (in the inner part of the coil).[126] Such models have been extensively modeled, with the exact arrangement of pentagons and heptagons determining whether the tube is metallic, semi-metallic, or semiconducting.[127,128] Experimentally, electron diffraction shows successive offset 30° bends at regular intervals along the coil length[124,129] consistent with localized structural defects. However, another model based on pure hexagonal networks has been proposed.[130] In this case, the model is constructed by repeating the primitive unit cell of a SWCNT, each time shifting it slightly so as to keep the tube axis tangential to the axis of a helice. The resultant structure has slightly distorted C–C bonds, and is held together due to van der Waals interactions between the layers. This model seems more plausible, particularly since the experimentally observed helical tubes typically have large coil diameters; however, the model does not explain the offset bends described above.

Finally, there are also many theoretical studies examining the possibility that helical nanotubes are constructed, not from a periodic hexagonal array of carbon, but from layered carbon consisting of pentagons, heptagons, and optionally hexagons. These "Haeckelite" structures and their potential involvement in helical nanotube structure are discussed further below.

1.9.3 *Applications of Helical Nanotubes*

Coiled multi-walled tubes have been shown to have strength comparable to SWCNTs[131] and have been proposed as a suitable filler for composite reinforcement; in principle, they should be superior to straight nanotubes due to improved anchoring into the embedding matrix and better load transfer. They may also act as "molecular springs," providing greater energy absorption and shock resistance.

The theoretically described helical nanotubes containing pentagons and heptagons as well as hexagons can be metallic, semi-metallic, or semiconducting,[126] and have been proposed as having potential for nanoelectronic mechanical systems, or indeed as electrical inductors.[132] However, until small coil diameter single-walled helical nanotubes can be clearly synthesized, identified, and characterized experimentally, these possible applications remain speculative.

1.10 "Necklace" Tubes/Nanobells

Carbon nanobells are multi-walled tubular structure based on the repetition of semicircular units (bell-like), with orthogonal connection between planes of adjacent nanobells.

Carbon nanobells are constructed from a series of multi-walled open-ended carbon spheres connected along one direction where the connecting walls between units are almost orthogonal. Each unit resembles a "bell"-like structure. The outer surface of the tube appears undulating (see Fig. 1.16 and Fig. 1.18).

This structure is also referred to in the literature as carbon nanonecklaces, necklaces of pearls structure, necklace-like hollow carbon nanospheres, and surface-modulated spherical layered nanotubes. Although in some cases the structure has been denoted as a fiber, nanobells present a non-continuous hollow core and therefore it is bamboo-like. The tubes can be several micrometer long (up to 50 bell-like units) and diameter of 50–100 nm.[133]

They have been synthesized by thermal plasma process at >1700°C (and therefore liquid catalytic particles),[132] carbon evaporation at high gas pressure,[134] and by H_2 plasma followed by grinding of N-doped tubes.[135] In all cases, nitrogen impurities were present (either already in the tubes or as the gas carrier).

It is often observed that the metal catalyst particle is encapsulated in the final bell;[132] however, EDX data indicate that the metallic particles are not distributed along the rest of the tube. Raman spectroscopy shows bands related to graphitic (G, 2D, and 2D) and defected (D) structures in addition to a unassigned peak at

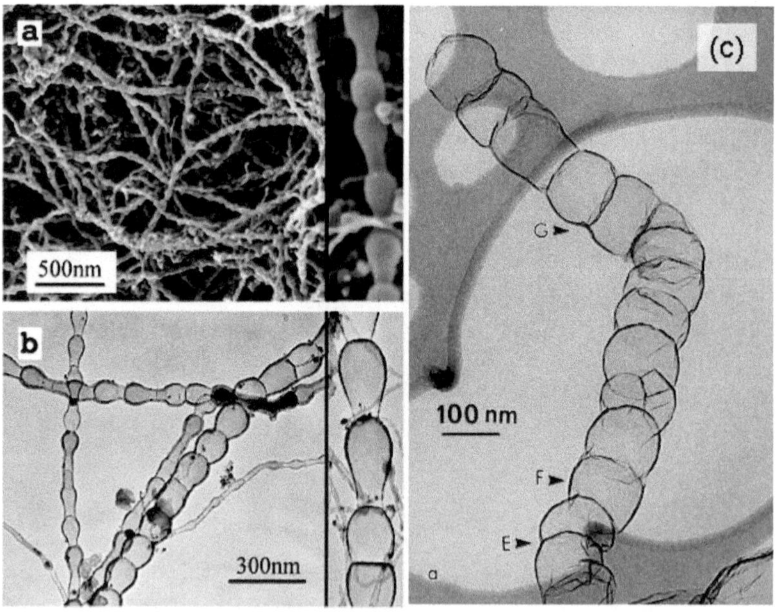

Figure 1.18. Examples of nanobell structures observed in the literature. (a,b) taken from ref. 132, and (c) from ref. 133.

179 cm^{-1}. It appears that the structure is the result of an unusual growth process, with graphitic walls forming over the surface of the metallic catalyst particle, which is ultimately ejected from its carbon "shell" before recommencing to grow new layers.

The atomistic structure of this form has been suggested to be a rolled up Haeckelite sheet in the (0,*n*) direction.[136] The bands of heptagons result in a negative curvature while the bands of pentagons result in positive curvature giving an overall aspect of periodic necks. Against this model is the fact that grinding results in the separation of individuals bells and showing a weak connection between the nanobells.[137] In addition, the orthogonal connection between planes of adjacent nanobells suggests the Haeckelite model is not correct.

Various authors have observed nanotubes of irregular diameter (called beaded carbon nanotubes), e.g., produced by vapor phase processes at $>1300^{\circ}$C.[138] These are not nanobells because the bell-like structure was not periodic. Both hollow and empty "beads" have been observed by different authors, and these have been cited as examples where carbon vapor-liquid-solid-type growth processes may be active, if such beads represent solidified remnants of a liquid carbon phase.

1.11 Fullerenes

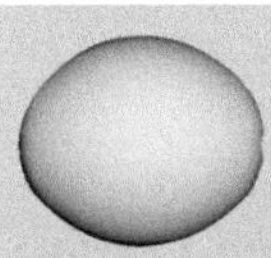

Fullerenes are closed single-walled cage molecules exclusively made of carbon, containing 12 pentagons and varying numbers of hexagons. To identify isomers the symmetry can be indicated, where not specified the highest symmetry is assumed. The I_h symmetry C_{60} is given the specific name Buckminsterfullerene. More recently, the term has been broadened to include any closed-cage structure constructed with purely threefold-coordinated carbon atoms.

Fullerenes were discovered in 1985, and were the nanoform which launched the revolution in carbon nanomaterials.[139] They are closed-cage molecules exclusively made of carbon. All fullerenes contain 12 pentagons following Euler's law, and any number of hexagons. The smallest of these to obey the isolated pentagon rule[140] (i.e., no carbon atoms occurring in more than one pentagon) is the famous C_{60}, Buckminsterfullerene, where the atoms and bonds delineate a truncated icosahedron. While various names were proposed early in its history (footballene,[141] soccerene, etc.), in 1995 IUPAC[1] confirmed fullerene as the standardized name for these molecules (with (C_{60}-I_h)[5,6]fullerene referring to Buckminsterfullerene), along with a standardized numbering convention for atomic sites using a 2D Schlegel projection of the fullerene cage (see Fig. 1.19). Along with carbon onions these are the only genuinely molecular forms of carbon since all other structures in this chapter are non-closed and hence have terminated dangling bonds.

The most complete reference for fullerene structure is the *Atlas of Fullerenes* by P. Fowler and D.E. Manolopoulus,[143] which provides a detailed catalogue of fullerene structures and tabulates their properties. Various free programs are available on the web for generating fullerene atomic coordinates.[144] Bond order in fullerenes is more polarized than that of most other nanoforms discussed here, with pentagonal bonds strongly single-bond in character (1.458 Å in C_{60}) and hexagon–hexagon bonds more double-bond in character (1.401 Å in C_{60}).[145]

The next fullerene able to fulfill the isolated pentagon rule after C_{60} is C_{70}. Increasing carbon content results in structures which vary from spherical (C_{60}), through pill-shaped (C_{70}, C_{84}, etc.) to rounded cages and eventually faceted[146] polygonal structures. There have been attempts to stabilize fullerenes with fused-pentagons (i.e., in breach of the isolated pentagon rule), using substitutional dopants[147] or metallic endoclusters within the fullerene cage.[148]

Fullerenes adopt an fcc molecular crystal structure (a = 14.117 Å) and in this form are referred to as fullerites (a HCP phase has also been identified with $a = 9.756$ Å, $c = 17.084$ Å[149]). There are

[1]International Union of Pure and Applied Chemistry: http://goldbook.iupac.org/F02547.html

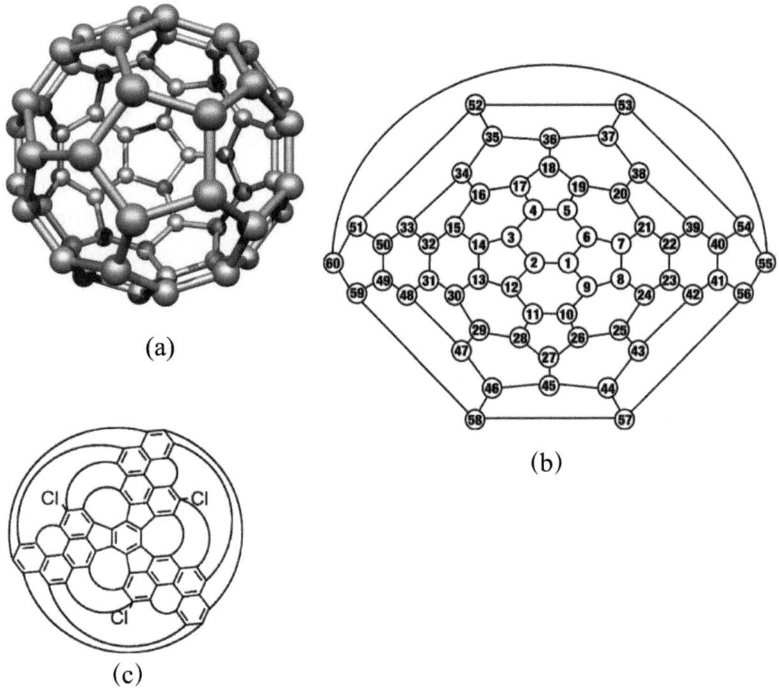

Figure 1.19. (a) Ball-and-stick image of C_{60} Buckminsterfullerene and (b) a Schlegel 2D projection of the same molecule (often used to show bond chemistry), with standard atom numbering as adopted by IUPAC. $C_{60}H_{27}Cl_3$, a synthetic precursor used for rational chemical synthesis of C_{60}.[149]

also many fullerene-based ionic crystals such as K_3C_{60}, often studied for superconducting behavior.

1.11.1 *Fullerene Synthesis*

The use of a focused pulsed laser of ~30 mJ onto a graphite target in a He atmosphere was the pioneer technique for the first synthesis of fullerenes. The mass spectra of the powder produced in the chamber showed carbon clusters up to 190 carbon atoms. The most predominant peak corresponded to a cluster of 60 carbon atoms. This cluster, C_{60}, was later denominated as Buckminsterfullerene.

The first group to produce solid C_{60} (a matter of days before the Kroto group in Sussex, to their eternal chagrin!) was Krätschmer *et al.*, using an electric arc between two graphite rods under vacuum to produce large quantities of fullerenes, which were then dissolved and crystallized in toluene.[150] Arc-discharge is now the standard production route for fullerenes. In 2002, a rational chemical synthesis route was developed for C_{60}.[151] A molecular polycyclic aromatic precursor bearing chlorine substituents at key positions forms C_{60} when subjected to flash vacuum pyrolysis at 1100°C. Rational routes for production of fullerene fragments such as "buckybowls" also exist; see section on small molecules below.

1.11.2 *Fullerene Chemistry*

Fullerene chemistry is the most developed of all the carbon nanoforms, and much of what has been learned with fullerenes has been later transferred to nanotubes, nanocones, etc. Fullerene functionalization and chemistry is now a discipline in its own right and is too vast a subject for coverage here. We refer instead the interested reader to later Chapter 2 by Petra Rudolf and Chapter 9 by Thomas Anthopoulos and ref. 152.

As well as surface functionalization, fullerene cages can be used to encapsulate other materials such as metals (the "Metallo-fullerenes,"[153] discussed further in Chapter 7 by Takeshi akasaka and Chapter 8 by Kyriakos Porfyrakis), hydrogen molecules,[154] and even molecular complexes such as Sc_3N.[155] These are collectively referred to as *endohedral fullerenes*, described using the standard notation of $X@C_n$, where X indicates the encapsulated species within fullerene C_n.

1.11.3 *Fullerene Applications*

Fullerene applications are various, including low-temperature superconductivity of fullerite-based phases[156] (despite the setback the field received with the fraudulent claims of gate-induced high-temperature superconductivity[157]). Photovoltaic films based around functionalized fullerenes in polymer matrices are the most efficient organic photovoltaics to date,[158] and fullerenes are also under consideration for use in fuel cell and battery electrodes.[159]

There is also interest in fullerene use in detectors, sensors, and even spintronics (notably using N@C_{60} as an individual qubit[160]). The optical response of fullerenes makes them interesting for optical limiting applications.[161]

Finally, fullerenes have a potentially bright future for medicinal applications.[162] The strong antioxidant nature of C_{60} makes it an effective radical scavenger,[163] yet under UV excitation it can lead to singlet oxygen production, of interest for biological damage applications such as controlled DNA cleavage. Its versatility under functionalization makes it an appropriate drug delivery agent, and it has even been shown to fit the hydrophobic cavity of HIV proteases, providing a new route to inhibit enzyme activity. Clearly, however, for such applications to be realized, our understanding of potential toxicological hazards associated with fullerenes and fullerene derivatives needs to be developed further.[164]

1.11.4 *Ultra-Hard Fullerites*

At high temperatures and pressure (up to 2100 K and 6–13 GPa[165,166]), fullerene crystals can fuse, resulting in a series of ultra-hard phases.[167] These ultra-hard fullerites have remarkable mechanical properties, e.g., they are the hardest materials known ($>$170 GPa), capable of scratching diamond and cubic boron nitride.[163]

Fullerenes polymerization can result in a range of different structures such as chains, 2D sheets, or three-dimensional (3D) solid forms. Interfullerene bonding occurs either via [2+2] cycloaddition between hexagon–hexagon bonds on neighboring cages, or single covalent bonds (Fig. 1.20).[168]

1.12 Onions

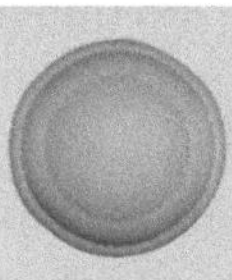

Onions are a family closed multi-shell cage molecules exclusively made of carbon. As they consist of concentric fullerene molecules, their nomenclature follows the rules of hybrids materials (see below): C_{60}@C_{240}@C_{540}...

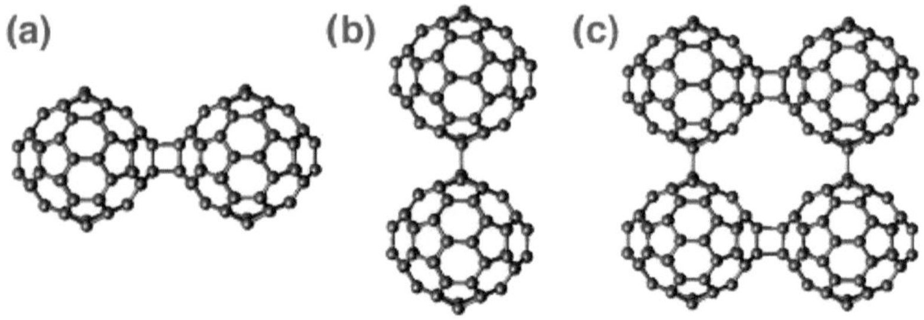

Figure 1.20. Interfullerene C-C bonding in polymeric fullerides. (a) [2+2] cycloaddition in AC_{60}, (b) single C-C covalent bonds in Na_2RbC_{60}, (c) mixed bonding in Li_4C_{60}, reproduced from reference [156].

Carbon onions were originally observed within sputtered amorphous carbon films by Sumio Iijima in 1980,[170] although at the time (pre-fullerenes) the curvature was assigned to tetrahedrally bonded carbon, and only later reassigned correctly.[171] Freestanding spherical carbon onions were first observed in 1992.[172] They consist of multiple fullerenes, one inside the other, with intersphere distance approximately that of the graphite interlayer spacing.[167] They are produced via electron irradiation of carbon soot or polygonized carbon particles, which leads to the formation of spherical multi-layered structures (such as the example in Fig. 1.21.b). Polygonized onions with facetted surfaces (see Fig. 1.21.b) are commonly observed as byproducts during MWCNT synthesis by the arc-electric route.[173] For a recent review of carbon onions see ref. 174. Spherical onion cores always follow a single configuration, $C_{60}@C_{240}@\ldots C_{60*n*n}\ldots$ (discussed further in spiroids section below).[175,176]

There can be significant variation in onion structure, including multi-core onions,[177] and quite commonly metal-catalyst-filled onions[178] especially during CVD growth.[179] Other more exotic production routes include ball milling,[180,181] and carbon ion implantation into high-temperature metal targets,[182] and underwater arc-discharge.[183] Intense irradiation of onions can lead to diamond formation in the onion core,[184] originally explained in terms of internal structural pressure but later revisited with a model based on the higher radiation stability of diamond as compared to

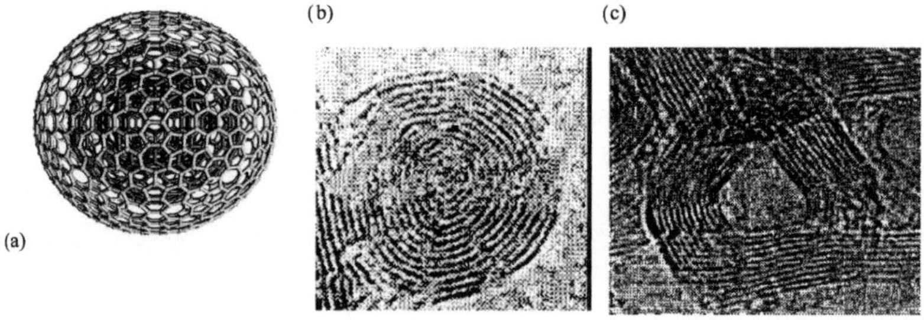

Figure 1.21. (a) Atomistic model of a spherical carbon onion, $C_{60}@C_{240}@C_{560}$. (b) Spherical carbon onion produced via electron irradiation.[169] (c) Polygonal carbon onion typical of byproducts during arc-electric nanotube growth.[171]

graphite.[185] Electron irradiation can equally be used to convert nanodiamonds into carbon onions.[186]

The UV absorption spectra of onions matches that seen for interstellar dust.[187] While metal-filled onions are of interest for their electromagnetic response,[188] the primary interest in carbon onions is for tribiological applications. The onion's spherical shape means they should serve as useful low-friction lubricants, while the multi-layer structure makes them mechanically more robust than simple single-layer fullerenes. When mixed with oils they have been shown to reduce friction and wear.[189–191]

The nautilus-shell was proposed as a possible growth mechanism for multi-shell fullerenes (carbon onions).[192] Starting with a hemisphere, the hemisphere is completed to form a sphere, but the radius is uniformly increased during this completion. The result is the two edges, which would normally fuse to form a closed cage, are instead separated by a radial distance of 3.4 Å. The outer "lip" can then continue in the same fashion, forming a second and subsequent layers to the structure (see Fig. 1.22).

We note that the comparison of the Nautilus-shell structure with a carbon onion is the 0D equivalent of the comparison between a MWCNT and a nanoscroll. Smaller "bowl-shaped" sp^2 Carbon molecules, precursors to the Nautilus but with their outer lip hydrogen terminated, have been successfully synthesized and isolated.[193]

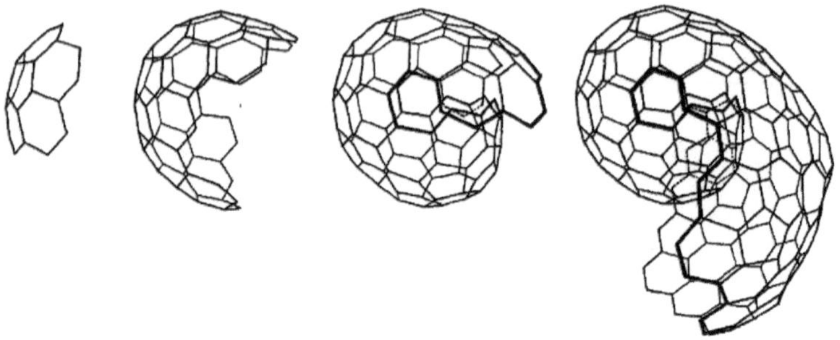

Figure 1.22. The hypothetical "Nautilus-shell" structure, showing snapshots during proposed structural growth (from ref. 190).

While the Nautilus structure has never been observed experimentally, molecular spiral carbon structures have been synthesized, and have been given the name *spiroids*[194] (following the geometric term *helicoid*, a "warped surface generated by a moving straight line which always passes through or touches a fixed helix"[195]). An example of a spiroid is shown in Fig. 1.23. Spiroids form under the same conditions as spherical carbon onions, i.e., under electron irradiation of carbon nanoparticles. Their continuous surface follows an Archimedean spiral with equal spacing between

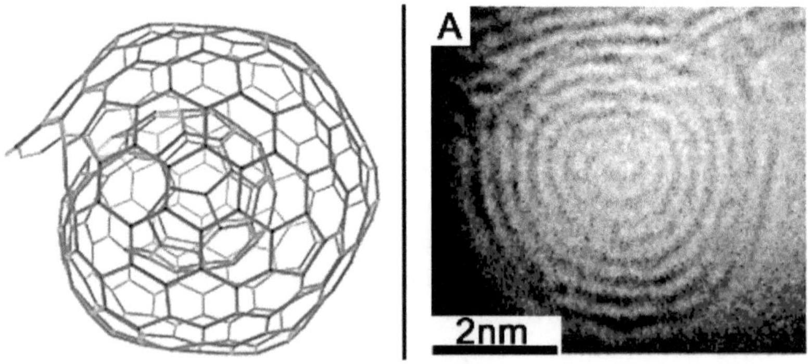

Figure 1.23. Spiroid structures (a) molecular model and (b) HRTEM of a spiroid created by electron bombardment of Toka Black #8500F commercial furnace black particles. Further irradiation converts this structure into a concentric shell multi-layered fullerene (images reproduced from ref. 192). See also Color Insert.

the layers (unlike the logarithmic spiral in the Nautilus structure). This is again consistent with the spiral forms observed in CNSs and is due to the van der Waals interaction between the layers.

Once again, analogously with nanotubes and nanoscrolls, the molecule can interchange between a spiroid and a multi-walled fullerene via the passage of dislocations.

Ozawa and coworkers show that, irrespective of the source carbon particle, such spiroids consistently form before transforming into carbon onions,[194] and thus propose this as the standard formation mechanism for spherical carbon onions.

1.13 Nanotori and Circular Nanotube Bundles

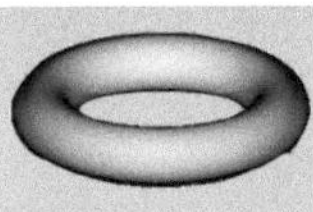

A carbon nanotorus is a single closed ring of carbon, e.g., a nanotube which is bent so that its axis remains a constant distance from a fixed central point, resulting in a single continuous surface with no dangling bonds. A circular nanotube bundle is made by bending a nanotube bundle around a central point, but in this case each individual tube does not form a closed loop (i.e., individual tubes have ends).

Circular structures have been observed in SEM and atomic force microscopy (AFM) images of laser-grown SWCNT[196,197] samples as well as in CVD-grown MWCNT.[198] In all cases, the observation is similar: rings of 300–500 nm diameter, where the thickness of the ring is 5–20 nm. The thickness of the ring matches the diameter of SWCNT ropes or the MWCNTs in each case and these circular structures are just a minority of the sample.

There is some controversy as to whether these structures are genuine nanotori (i.e., closed-loop nanotubes with no ends) or simply circular bundles (bundle of long nanotubes wrapped round into circles). Liu[196] originally concluded their structures were tori because no discontinuities were observed in electron microscopy. However, later experiments have observed incomplete

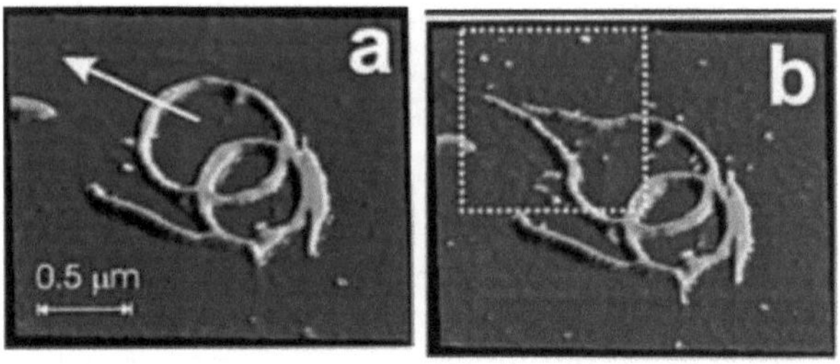

Figure 1.24. Circular nanotube bundles on hydrogenated Si(100) surface, imaged using the AFM, and after applying a vertical load of 30nN with the AFM tip to unfold the ring (taken from ref. 196).

circles and also overlappingrings, suggesting that these rings are in fact coiled bundles of nanotubes.[196,197] Furthermore, circles have been mechanically opened using AFM tips which seems unlikely if the structures were perfect tori[196] (see Fig. 1.24). Similar structures have also been observed by laser ablation of fullerenes samples.[199] In addition to the ring structures, Q-shaped structures have also been observed,[198] which also supports the idea of coiled nanotubes rather than genuine tori.

The nomenclature of this structure is not very consistent between authors. Nanotori have been denoted as fullerene "crop-circles,"[195] toroidal fullerenes,[195] nanohoops,[200] carbon-based toroids,[198] doughnut-shaped tubes,[198] and carbon nanotube rings. We would like to note that nanotori should only be used when the structure is perfectly closed (like the toriod geometrical solid). For those structure which are open and they are indeed coiled nanotubes where the ratio coil diameter to pitch is very large, and we suggest therefore the term *circular nanotube bundles.*

Small diameter nanotube tori have been a playground for theoretical structural modeling, but have not been observed experimentally to date. The topological thought experiment to obtain a nanotorus is to bend an open carbon nanotube and join the two ends, resulting in a doughnut shape. The overall morphology is the geometrical form of a nanotorus.

The interest is primarily because smaller diameter tori in principle require the addition of pentagons and heptagons to form a closed structure. Multiple theoretical models have been proposed,[201,202] and a good (if now somewhat dated) review of toroidal and nanocoil geometry is given in ref. 203. In particular, there have been a number of theoretical studies proposing small diameter toroidal structures formed from Haeckelite layers (pentagon, heptagon, andoptionally hexagonal periodic arrangements of carbon). László and Rassat showed that a rolled stripe of pentagon-heptagon pairs (possibly mixed with hexagons) results in a tube that spontaneously bends and can close into a torus.[204] This is discussed further in the section on Haeckelites below.

We note, however, that to date all experimental reports of nanotorii are of much larger diameter, where pentagons and heptagons need not be invoked to explain the structure. In addition, elastic theory studies suggest that a SWCNT torus of diameter $>$ 200 nm should be stable just by bending without pentagon or heptagon defects.[205]

1.14 Hybrid Nanoforms

A hybrid carbon nanoforms is constituted of two or more carbon nanoforms. A nanoform can be attached to the outer surface (//) or encapsulated inside (@) another.

There are a near infinite range of potential hybrid carbon nanoforms, but the majority observed experimentally to date can be classified either as one carbon nanoform which sits within another or one nanoform attached to the outer surface of another. Of these the most well known are probably "peapods," fullerenes within carbon nanotubes.

The naming convention developed by the fullerene community for describing a specific endohedral fullerene structure is of the form $x@y$, where species x lies inside species y, and this is used for other hybrid carbon nanomaterials, e.g., peapods can be described as C_{60}@SWCNT.[206] Brackets can be added where necessary, e.g., N atoms within C_{60}, which are themselves encapsulated in SWCNTs would be written as (N@C_{60})@SWCNT. The convention can be extended further to incorporate carbon nanoforms attached to the exterior surfaces by using a "//," so species *x* attached to the exterior of species *y* would be written *x*//*y*.

This convention allows description of even some of the most complicated carbon nanoforms. As an example, if nanohorns containing ferrocene were inserted into a phosphorus-doped SWCNT, which had porphyrine groups attached to its surface, this would be described as Porphyrine//(($Fe(C_5H_5)_2$)@CNH)@P-SWCNT.

We described below the primary hybrid forms that have been predicted or observed.

1.14.1 *Hybrid Forms Based on Filling (Peapods etc.)*

> The term peapod applies to nanotubes filled with fullerene molecules. This can be written as, e.g., C_{60}@SWCNT.

The hybrid material consisting of multiple fullerenes encapsulated inside a SWCNT is better known as a "peapod" since in the electron microscope it resembles a string of peas in a pod.[207] As well as completely filled tubes, when the carbon nanotube is filled with just one or two fullerenes it has been denoted as "bucky shuttle."[208]

The first identification of peapods was by HRTEM in 1998.[205] The images show tubes with diameters of 1.3–1.4 nm, where between the two lines of the HRTEM images circles of approximately 0.7 nm were observed (see Fig. 1.25a). The size of the circles matched that of C_{60} molecules, and the distance between circles centers was consistent with the distance between C_{60} centers in fcc C_{60}.

While nanotube filling with crystalline salts and oxides is typically either performed in the liquid phase,[216] for peapods the common route is through vacuum annealing of acid-treated nanotubes in the presence of fullerenes.[209] This works well since fullerenes are very stable with a relatively low sublimation temperature (~350°C).[210] Acid treatment (e.g., reflux in HNO_3 for 48 h followed by rinsing and neutralization) opens the nanotube ends. Vacuum annealing then facilitates the mobility of C_{60} which enters

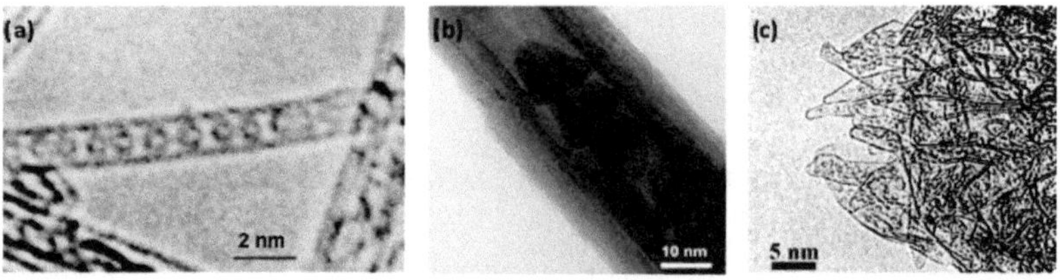

Figure 1.25. HRTEM of hybrid carbon forms based on the filling of one form with another.: (a) a peapod, C_{60} molecules encapsulated within a single-walled nanotube,[205] (b) multi-walled nanocones encapsulated within a multi-walled nanotube during synthesis,[214] and (c) fullerenes encapsulated within Dahlia-like carbon nanocones.[215]

the tubes. Temperatures of at least 325°C are needed to promote the mobility of the fullerenes.[208]

Different packing arrangements are observed depending on the nanotube diameter, and these can be accurately reproduced with simple models assuming the fullerenes to be hard spheres packing within a fixed cylinder.[211]

The electronic density of states of the nanotubes is perturbed by the encapsulated fullerenes, which give rise to a hybrid electronic state.[212] As well as pristine fullerenes, encapsulated fullerenes can also be previously treated and, e.g., $La_2@C_{80}$ has been successfully encapsulated inside carbon nanotubes.[213]

There is interest in peapods due to observations of improved bending modulus as compared to empty single-walled tubes by as much as 170%.[214] In addition peapods can be annealed, causing fusion of the interior fullerenes, which generates a secondary tube and is one route to formation of DWCNTs.[215]

There have also been reports in the literature of nanotube filling with other nanoforms, e.g., growth of MWCNTs which encapsulate small groups of stacked nanohorns, as in Fig. 1.25b (referred to in the publication as cone-type multi-shell in the hollow core of MWCNT).[216] As far as we are aware, this work has not been repeated.

Carbon nanocones can also be filled with other nanoforms such as C_{60} (C_{60}@SWNC). Oxidized laser-ablation synthesized nanohorns have been successfully filled, with fullerenes occupying up to 36% of the available hollow spaces[217] (see Fig. 1.25c).

For a good review from 2002 of nanotube filling, focusing on peapods, we refer the reader to ref. 218.

1.15 Hybrid Forms Based on Surface Interaction

The exterior walls of carbon nanoforms can be functionalized, not only with molecular groups but also with other carbon nanoforms. Notably SWCNTs with exterior walls functionalized with C_{60} (C_{60}//SWCNT) or short SWCNT sections (SWCNT//SWCNT) have been synthesized.[220] These were referred to as "nanobuds" by the authors (Fig. 1.26.a).

In arc-discharge production of Dahlia-type single-walled nanocones in a helium atmosphere, a considerable amount of fullerenes are also produced, and it has been observed that the fullerenes tend to be attached to the tip of carbon nanocones,[221] which has been explained through oxygen cross-linking.[222]

Finally by controlling CVD synthesis conditions, Trasobares *et al.* were able to produce MWCNTs with multi-layered graphitic sheets attached to their walls much like thorns on a rose stem[223] (referred to by the authors as "nanowings", see Fig. 1.26.b). These were proposed as interesting candidates for composite reinforcement due to assumed enhanced pull-out energies.

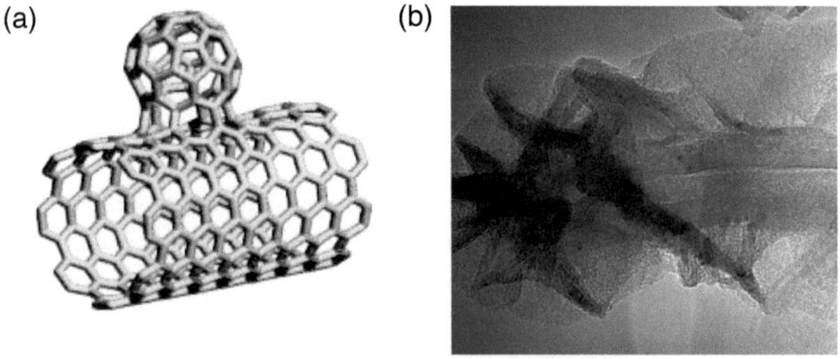

Figure 1.26. Hybrid forms produced through surface attachment, (a) fullerenes/short nanotube segments attached to nanotube surfaces ("nanobuds"), Wikipedia: http://en.wikipedia.org/wiki/File:Nanobud Computations70%25.jpg (b) "nanowings," segments of multi-layered graphite fused to nanotube walls. Susana Trasobares, Private Communication (2011). See also Color Insert.

1.16 Other Molecular Forms

As well as the fullerenes, once hydrogen termination is included there are a vast range of other carbon molecular forms. There are books devoted to the structure and properties of fused polycyclic aromatic hydrocarbons,[224] small "graphene-like" platelets of finite size, beginning with benzene (C_6H_6). We mention here only some special cases due to unusual topologies, and only discuss those for whom an experimental synthetic route has been devised; there are many more theoretically proposed structures in the literature.

Bowl-shaped aromatic hydrocarbons have been experimentally synthesized,[191] precursors to fullerenes or spiroids (Fig. 1.27.a). Various intermediate Buckminsterfullerene fragments including $C_{21}H_{12}$ (the elegant "sumanene"[225]), $C_{26}H_{12}$, $C_{28}H_{12}$, and $C_{36}H_{12}$ now have synthesis routes, which are summarized in ref. 191. $C_{60}H_{27}Cl_3$, a "propeller-shaped" molecule has also been produced synthetically, and this can be converted into C_{60} with 100% yield via flash vapor pyrolysis.[149]

Similarly, the small molecular equivalent of nanotubes have recently been synthesized: *cycloparaphenylenes*, or "carbon

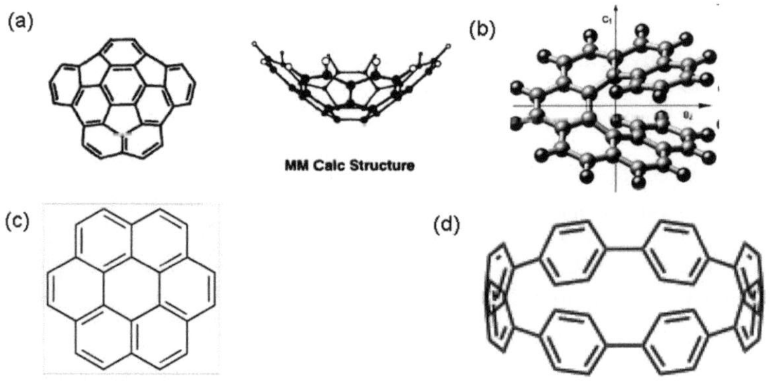

Figure 1.27. Various topologically unusual hydrogen-terminated carbon molecules, for which synthetic chemistry routes have been devised. (a) Carbon nanobowls such as $C_{32}H_{12}$ (ref. 191), (b) helicine structures $C_{30}H_{18}$, heptahelicene (ref. 229), (c) fused polycyclic aromatic hydrocarbons (coronene $C_{24}H_{12}$), (d) cycloparaphenylene ("carbon nanohoops") (ref. 224; Wikipedia: http://en.wikipedia.org/wiki/File:Cycloparaphenylene.PNG).

nanohoops" (see Fig. 1.27d). These are single ring structures of polymerized linked benzene, with radially oriented p-orbitals.[226] Different ring sizes have been synthesized ([9]-, [12]- and [18]cycloparaphenylene). Since these form a single ring of an arm-chair nanotube structure, as the authors speculate, the possibility of using these as templates for synthetic nanotube growth is "an intriguing prospect."

Finally we make special mention of *helicines*,[227] a special family of helical fused polycyclic aromatic hydrocarbons whose structure represents the core of a screw dislocation in graphite (Fig. 1.27.b). They are of particular interest since the screw direction can be clockwise or anti-clockwise, giving rise to chiral pairs of each isomer. An excellent early review of Helicenes and their chemistry from 1974 is ref. 228.

1.17 Non-Hexagon-Based SP^2 Carbon Nanoforms

While the majority of forms discussed above involve either hexagonal carbon layers, distorted hexagonal layers, or hexagonal layers containing periodic pentagons, other geometric alternatives exist. Notable proposed structures involving higher order polygons are the Schwarzites, and the Haeckelites.

1.17.1 *Schwarzites: Heptagon (and Above)-Hexagon Networks*

As a theoretical exercise, Terrones *et al.* proposed a closed fullerene structure using just hexagons and heptagons.[230] A large double-layered fullerene has its pentagonal corners removed and replaced by holes, the two layers connected via a ring of heptagons. However, in general heptagonal and higher order polygonal defects do not result in closed structures but instead result in "triple periodical minimal surfaces," continuous 3D surfaces with negative Gaussian curvature.[231–234] Schwarzites are one family of these, zeolite-type structures constructed from sp^2 graphitic planes containing heptagons and other higher order polygons.[228, 235] They are named

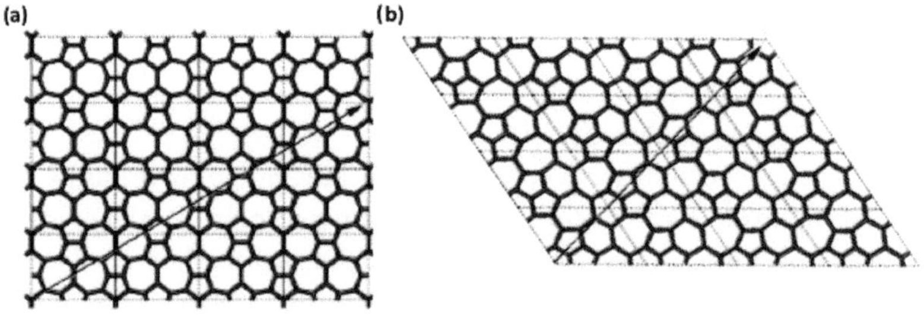

Figure 1.28. (a) Pentagon–heptagon, (b) pentagon–hexagon–heptagon Haeckelite structures (taken from ref. 240).

after the mathematician H. A. Schwarz, who in 1890 was the first to study such surfaces.[236]

Recent observations of spongy-carbon nanostructures seem to show strong resemblance to random Schwarzite networks.[237]

Since schwarzites are continuous extended surfaces resulting in porous "graphitic foams," rather than individual nanoobjects, we do not consider them further in this chapter.

1.17.2 *Haeckelites: Pentagon–(Hexagon)–Heptagon Networks*

A layered material based on ordered arrangement of pentagons, hexagons, and heptagons in a sheet was proposed with the name of Haeckelites.[228] The name was chosen in memory of the 19th century biologist Ernst Haeckel, whose beautiful drawings of radiolarians as viewed under the optical microscope exhibit geometric layers consisting of hexagons, pentagons, and heptagons. The authors proposed three types of structures: rectangular (only consisting in pentagons and heptagons — earlier work proposed the same structure under the name of pentaheptites[241]), hexagonal, and oblique. This new family of layered materials was proposed to be a similar family to graphene, and that the haecklite sheet could be rolled and therefore nanotubes of haecklite would be possible. Theoretical calculations suggested that they are more stable than C_{60} but less stable than "standard" hexagonal layers.[228]

It has also been suggested that Haeckelite sheets could be rolled up similarly to graphene. In this case, e.g., bands of heptagons will result in negative curvature while bands of pentagons will result in positive curvature, generating a variety of unusual structures such as periodic undulations and "string of pearls"-type structures, nanocoils, and nanotoroids.[242–244]

Coiled haeckelites have at least one advantage as a model for small-pitch coiled nanotubes over the "conventional" model evoking occasional periodic pentagons and heptagons in an otherwise hexagonal network, namely that the Haeckelite structure iscontinuous and does not need to invoke the periodic introduction of defects, which is difficult to explain experimentally.[245] Notably László and Rassat showed that a rolled stripe of pentagon–heptagon pairs (possibly mixed with hexagons) results in a tube that naturally bends, and can close into a torus.[246] However, periodic defects are not invoked in the purely hexagonal model for larger diameter coils, whose size corresponds closer to those observed experimentally (see section on helical nanotubes above).

The existence, or otherwise, of Haeckelites remains an open question. While Haeckelite structures have not been identified experimentally, distinguishing between these and conventional hexagonal graphene layers is not an easy task, and calculations suggest they are energetically close to conventional graphene. It has been suggested by Biró *et al.* that kinetic effects may in some cases favor Haeckelite formation, e.g., if it results in twisting which carries the carbon layer away from the catalyst particle.[245] If this occurs in low-temperature CVD, the carbon network may not be able to reconfigure itself to the ground state hexagonal network. In any case, it seems like that should such structures exist, they are most likely in coiled structures such as helical nanotubes.

1.18 Conclusions

The above catalogue of different nanoobjects shows clearly the fascinating and beautiful abundance of geometric variation that is possible with layered sp^2 carbon. Carbon has turned from being an apparently well-understood material less than 30 years ago, into a strange and complex element resulting in a multitude of forms, each with their own distinct properties.

For many of these structures, the key thing which differentiates one form from another are the precise conditions of synthesis. For example, CVD catalyzed synthesis conditions for "standard" MWCNTs can be adapted. Lower growth temperatures are associated with increasing yields of helical nanotubes, while introduction of impurities such as nitrogen or boron encourages bamboo, herringbone (and sometimes also helical) nanotube growth. Higher temperature growth, e.g., via arc-electric, leads to more graphitic structures but also introduces polygonal carbon impurities such as facetted carbon onions. The precise path taken by carbon atoms during synthesis is a complex dance taking place at high temperatures far from thermal equilibrium, and we are currently still a long way from being able to produce high-yield high-purity samples of different carbon nanoforms on demand.

Acknowledgments

NG would like to thank the Royal Society, STREP project BNC Tubes, NMP4-CT-2006-033350, and ERC Starting Grant ERC-2009-StG-240500 for funding. CPE and NG would like to thank COST Project MP0901 NanoTP. We would like to thank those who provided preprints or unpublished material that was used in this book chapter.

References

1. P. Levi, *The Periodic Table*, Penguin Classics, UK, New Ed edition (2000).
2. N. Grobert, *Materials Today* **9**(10), 64 (2006).
3. N. Grobert, *Materials Today* **10**(1–2), 28 (2007).
4. Y. Saito, *Carbon* **33**(7), 979 (1995).
5. H. Okuno, E. Grivei, F. Fabry, T. M. Gruenberger, J. Gonzalez-Aguilar, A. Palnichenko, L. Fulcheri, N. Probst, and J.-C. Charlier, *Carbon* **42**, 2543 (2002).
6. M. Muñoz-Navia, J. Dorantes-Davila, M. Terrones, T. Hayashi, Y. A. Kim, M. Endo, M. S. Dresselhaus, and H. Terrones, *Chem. Phys. Lett.* **407**, 327 (2005).
7. I. Suarez-Martinez, N. Grobert, and C. P. Ewels, *Carbon*, in press (2011).

8. E. H. L. Falcao and F. Wudl, *J. Chem. Technol. Biotechnol.* **82**, 524–531 (2007).
9. R. E. Peierls, *Ann. I. H. Poincare* **5**, 177–222 (1935).
10. L. D. Landau, *Phys. Z. Sowjetunion* **11**, 26–35 (1937).
11. K. S. Novoselov *et al.*, *Science* **306**, 666–669 (2004).
12. K.S. Novoselov, *et* al., *PNAS* **102**, 10451-10453 (2005).
13. A. K. Geim and K. S. Novoselov, *Nat. Mater.* **6**, 183–192 (2007).
14. H.-P. Boehm, R. Setton, and E. Stumpp, *Pure Appl. Chem.* **66**(9) 1893–1901 (1994). IUPAC Recommendations.
15. A. Fasolino, J. H. Los, and M. I. Katsnelson, *Nat. Mater.* **6**(11), 858–861 (2007).
16. S. Latil and L. Henrard, *Phys. Rev. Lett.* **97**, 036803 (2006).
17. K. S. Novoselov, A. K. Geim, S. V. Morozov, D. Jiang, Y. Zhang, S. V. Dubonos, *et al.*, *Science* **306**, 666 (2004).
18. Y. Hernandez, V. Nicolosi, M. Lotya, F. M. Blighe, Z. Sun, S. De, I. T. McGovern, B. Holland, M. Byrne, Y. K. Gun'ko, J. J. Boland, P. Niraj, G. Duesberg, S. Krishnamurthy, R. Goodhue, J. Hutchison, V. Scardaci, A. C. Ferrari, and J. N. Coleman, *Nat. Nanotechnol.* **3**(9), 563–568 (2008).
19. V. C. Tung, M. J. Allen, Y. Yang, and R. B. Kaner, *Nat. Nanotechnol.* **4**, 25–29 (2009).
20. C. Berger *et al. J. Phys. Chem.* **108**(19), 912 (2004).
21. C. Berger, Z. Song, X. Li, *et al.*, *Science* **312**, 1191–1196 (2006).
22. K. S. Kim, Y. Zhao, H. Jang, S. Y. Lee, J. M. Kim, K. S. Kim, J.-H. Ahn, P. Kim, J.-Y. Choi, and B. H. Hong, *Nature* **457**(7230), 706 (2009).
23. S. Latil, V. Meunier, and L. Henrard, *Phys. Rev. B* **76**, 201402(R) (2007).
24. Y. Wu, P. Qiao, T. Chong, and Z. Shen, *Adv. Mat.* **14**(1), 64–67 (2002).
25. A. Cresti, N. Nemec, B. Biel, G. Niebler, F. Triozon, G. Cuniberti, and S. Roche, *Nano Res.* **1**, 361–394 (2008).
26. M. Fujita, K. Wakabayashi, K. Nakada, and K. Kusakabe, *J. Phys. Soc. Jpn.* **65**, 1920–1923 (1996).
27. K. Nakada, M. Fujita, G. Dresselhaus, and M. S. Dresselhaus, *Phys. Rev. B* **54**, 17954–17961 (1996).
28. L. Tapasztó, G. Dobrik, P. Lambin, and L. P. Biró, *Nat. Nanotechnol* **3**, 397–401 (2008).
29. L. Jiao, L. Zhang, X. Wang, G. Diankov, and H. Dai, *Nature* **458**, 877–880 (2009).

30. E. W. S. Caetano, V. N. Freire, S. G. dos Santos, D. S. Galvao, and F. Sato, *J. Chem. Phys.* **128**, 164719 (2008).
31. M. I. Katsnelson, *Materials Today* **10**, 20 (2007).
32. Ph. Avouris and J. Chen, *Materials Today* **9**, 46 (2006).
33. Y. M. Lin, K. A. Jenkins, A. Valdes-Garcia, J. P. Small, D. B. Farmer, and Ph. Avouris, *Nanoletters* **9**(1), 422–426 (2009).
34. A. K. Geim and K. S. Novoselov, *Nat. Mater.* **6**(3), 169 (2007).
35. Y. G. Yang, C. M. Chen, Y. F. Wen, Q. H. Yang, and M. Z. Wang, *New Carbon Mater.* **23**(3), 193–200 (2008).
36. L.V. Radushlekiv and V.M. Lukyanovich, *Zurn. Fisic. Chim.* **26**, 88 (1952).
37. W. R. Davis, R. J. Slawson, and G. R. Rigby, *Nature* **171**, 756 (1953).
38. A. Oberlin, M. Endo, and T. Koyama, *J. Cryst. Growth* **32**(3), 335 (1976).
39. S. Iijima, *Nature* **354**, 56–58 (1991).
40. M. Monthioux, *Carbon* **44**, 1621 (2006).
41. P. J. F. Harris, Cambridge University Press, UK 2009. ISBN 978-9-521-82895-6.
42. N. Hamada, S. Sawada, and A. Oshiyama, *Phys. Rev. Lett.* **68**, 1579 (1992).
43. A FORTRAN computer program for generating nanotube carbon coordinates is available in the nanotube book: R. Saito, G. Dresselhaus, and M. S. Dresselhaus, *Physical Properties of Carbon Nanotubes*, Imperial College Press, UK. ISBN 978-1-86094-093-4.
44. J. W. Mintmire and C. T. White, *Carbon* **33**(7), 893–902 (1995).
45. M. Liu and J. M. Cowley, *Ultramicroscopy* **64**, 333–342 (1994).
46. M. Terrones, H. Terrones, F. Banhart, J.-C. Charlier, and P. M. Ajayan, *Science* **288**, 1226–1229 (2000).
47. M. Motta, A. Moisala, I. A. Kinloch, and A. H. Windle, *Adv. Mater.* **19**, 3721–3726 (2007).
48. Y. Wang, D. Maspoch, S. Zou, G. C. Schatz, R. E. Smalley, and C. A. Mirkin, *Proc. Natl. Acad. Sci. USA* **103**(7), 2026–2031 (2006).
49. A. G. Rinzler, *Nat. Nanotechnol* **1**, 17–18 (2006).
50. L. X. Zheng, M. J. O'Connell, S. K. Doom, X. Z. Liao, Y. H. Zhao, E. A. Akhadov, M. A. Hoffbauer, B. J. Roop, Q. Jia, R. Dye, D. E. Peterson, S. M. Huang, J. Liu, and Y. T. Zhu, *Nat. Mater.* **3**(10), 673–676 (2004).
51. G. Otieno, A. Koos, F. Dillon, A. Wallwork, N. Grobert, and R. Todd, *Carbon,* **48** (8), 2212–2217 (2009).

52. Z. Gu, H. Peng, R. H. Huage, R. E. Smalley, and J. L. Margrave, *Nano Lett.* **2**(9), 1009–1013 (2002).
53. Z. Y. Chen, K. Kobashi, U. Rauwald, R. Booker, H. Fan, W. F. Hwang, and J. M. Tour, *J. Am. Chem. Soc.* **128**(32), 10568 (2006).
54. X. Sun, S. Zaric, D. Daranciang, K. Welsher, Y. Lu, X. Li, and H. Dai, *J. Am. Chem. Soc.* **130**(20), 6551–6555 (2008).
55. X. Gui, J. Wei, K. Wang, A. Cao, H. Zhu, Y. Jia, Q. Shu, and D. Wu, *Adv. Mater.* **22**(5), 617–621 (2010).
56. R. H. Baughman, A. A. Zakhidov, and W. A. de Heer, *Science* **297**(5582), 787–792 (2002).
57. A regularly updated list of nanotube products can be found on the web at http://www.nanotechproject.org/inventories/consumer/search/
58. http://www.eikos.com/dispapplications.html
59. http://www.ameinfo.com/83600.html
60. A. K. Schaper, *Carbon Nanotubes and Related Structures*, Eds. V. Blank and B. Kulnitskiy, Research Signpost Press, Kerala, India, 2008, pp. 1–47. ISBN 978-81-308-0214-5.
61. I. Suareez-Martinez, G. Savini, A. Zobelli, and M. I. Heggie, *J. Nanosci. Nanotechnol.* **7**(10), 3417 (2007).
62. Y. Chen, J. Lu, and Z. Gao, *J. Phys. Chem. C Nanomater. Interfaces*, **111**, 1625–1630 (2007).
63. S. Berber and D. Tomanek, *Phys. Rev. B.* **69**, 233404 (2004).
64. R. Bacon, *J. App. Phys.* **31**(2), 283 (1960).
65. A. L. Chuvilin, V. I. Kuznetsov, and A. N. Obraztsov, Proc. ICEM-15, 1–6 September 2002, Durban, South Africs, Vol. 1, p. 245.
66. F. Y. Wu and H. M. Cheng, *J. Phys. D Appl. Phys.* **38**, 4302 (2005).
67. W. Ruland, A. K. Schaper, H. Hou, and A. Greiner, *Carbon* **41**(3) 423–427 (2003).
68. L. Qin, *Phys. Chem. Chem. Phys.* **9**, 31–48 (2007).
69. N. F. Yudanov, V. Okotrub, Y. V. Shubin, L. I. Yudanova, and L. G. Bulusheva, *Chem. Mater.* **14**, 1472 (2002).
70. J. G. Lavin, S. Subramoney, R. S. Ruoff, S. Berber, and D. Tomanek, *Carbon* **40**, 1123 (2002).
71. H. Shioyama and T. Akita, *Carbon*, **41**, 179–198 (2003).
72. L. M. Viculis, J. L. Mack, and R. B. Kaner, *Science* **299**, 1361 (2003).
73. Y. Hernandez, V. Nicolosi, M. Lotya, F. M. Bligue, Z. Y. Sun, *et al.*, *Nat. Nanotechnol.* **3**(9) 563 (2008).

74. C. Ducati, K. Koziol, S. Friedrichs, T. J. V. Yates, M. S. Shaffer, P. A. Midgley, and A. H. Windle, *Small* **2**(6), 774 (2006).

75. A. Alloul *et al.*, Abstract International Carbon Conference, Aberdeen (2006).

76. S. F. Braga, V. R. Coluci, R. H. Baughman, and D. S. Galvao, *Chem. Phys. Lett.* **441**(1–3), 78 (2007).

77. H. W. Kroto, *Nature* **329**, 529 (1987).

78. H. Terrones, *J. Math. Chem.* **15**, 143–156 (1994).

79. A. Krishman, E. Dujardin, M. M. Treacy, J. Hugdahl, S. Lynum, and T. W. Ebbensen, *Nature* **388**, 451 (1997).

80. S. Iijima, M. Yudasaka, R. Yamada, S. Bandow, K. Suenaga, F. Kokai, and K. Takahashi, *Chem. Phys. Lett.* **309**, 165–170 (1999).

81. P. J. F. Harris, S. C. Tsang, J. B. Claridge, and M. L. H. Green, *J. Chem. Soc. Faraday Trans.* **90**, 2799 (1994).

82. M. Pacheco, M. Monthioux, M. Razafinimanana, L. Donnadieu, H. Allouche, N. Caprais, and A. Gleizes, International Conference on Carbon Carbon'02, Beijing (China), Extend. Abstr. (CD-Rom), Oral #I014 (2002).

83. H. Wang, M. Chhowalla, N. Sano, S. Jia, and G. A. J. Amaratunga, *Nanotechnology* **15**, 546 (2004).

84. H. Takikawa, M. Ikeda, K. Hirahara, Y. Hibi, Y. Tao, P. A. Ruiz Jr., T. Sakakibara, S. Itoh, and S. Iijima, *Physica E Low Dimens. Syst. Nanostruct.* **323**, 277 (2002).

85. T. Yamaguchi, S. Bandow, and S. Iijima, *Chem. Phys. Lett.* **389**, 181 (2004).

86. S. Subramoney, R. S. Ruoff, D. C. Lorents, and R. Malhotra, *Nature* **366**, 637 (1993).

87. I. Suarez-Martinez, M. Monthioux, and C. P. Ewels, *J. Nanosci. Nanotechnol.* **9**, 6144 (2009).

88. W.-C. Xu, K. Takahashi, Y. Matsuo, Y. Hattori, M. Kumagai, S. Ishiyama, K. Kaneko, and S. Iijima, *Int. J. Hydrogen Energy* **32**, 2504–2512 (2007).

89. K. Ajima, T. Murakami, Y. Mizoguchi, K. Tsuchida, T. Ichihashi, S. Iijima, and M. Yudasaka, *ACS Nano* **2**(10), 2057–2064 (2008).

90. A. A. Koos, M. Dowling, K. Jurkschat, A. Crossley, and N. Grobert, *Carbon* **47**(1), 30–37 (2009).

91. X. C. Ma, E. G. Wang, W. Z. Zhou, D. A. Jefferson, J. Chen, S. Z. Deng, N. S. Xu, and J. Yuan, *Appl. Phys. Lett.* **75**(20), 3105–3107 (1999).

92. C. P. Ewels and M. Glerup, *J. Nanosci. Nanotechnol.* **5**(9) 1345–1363 (2005).
93. A. A. Koos, M. Dowling, K. Jurkschat, A. Crossley, and N. Grobert, *Carbon* **47**, 30–37 (2009).
94. X. Ma, E. G. Wang, W. Zhou, D. A. Jefferson, J. Chen, S. Z. Deng, N. S. Xu, and J. Yuan, *Appl. Phys. Lett.* **75**, 3105 (1999).
95. J. Bill and R. Riedel, *Mater. Res. Soc. Symp. Proc.* **271**, 839 (1992).
96. M. Glerup, M. Castignolles, M. Holzinger, G. Hug, A. Loiseau, and P. Bernier, *Chem. Commun.* **20**, 2542 (2003).
97. M. Castignolles, Thesis, Université Montpellier II, France (2004).
98. C. J. Lee, S. C. Lyu, H.-W Kim, J. H. Lee, and K. I. Cho, *Chem. Phys. Lett.*, **359**, 115 (2002).
99. H. J. Burch, S. A. Contera, M. R. R. de Planque, N. Grobert, and J. F. Ryan, *Nanotechnology* **19**(38), 384001 (2008).
100. S. Peng and K. Cho, *Nanoletters* **3**, 513 (2003).
101. D. Y. Zhong, G. Y. Zhang, S. Liu, E. G. Wang, Q. Wang, H. Li, and X. J. Huang, *Appl. Phys. Lett.* **79**, 3500 (2001).
102. M. Holzinger, J. Steinmetz, S. Roth, M. Glerup, and R. Graupner, *AIP Conf. Proc.* **786**, 211–214 (2005).
103. K. Y. Jiang, A. Eitan, L. S. Schadler, P. M. Ajayan, R. W. Siegel, N. Grobert, M. Mayne, M. Reyes-Reyes, H. Terrones, and M. Terrones, *Nano Lett.* **3**(3) 275–277 (2003).
104. M. Monthioux, L. Noé, L. Dussault, J.-C. Dupin, N. Latorre, T. Ubierto, E. Romeo, C. Royo, A. Monzón, and C. Guimon, *J. Mater. Chem.* **17**, 4611–4618 (2007).
105. M. Endo, Y. A. Kim, T. Hayashi, Y. Fukai, K. Oshida, M. Terrones, T. Yanagisawa, S. Higaki, and M. S. Dresselhaus, *Appl. Phys.* **80**(7) 1267 (2002).
106. M. Monthioux, L. Noe, L. Dussault, J.-C. Dupin, N. Latorre, T. Ubieto, E. Romeo, C. Royo, A. Monzon, and C. Guimon, *J. Mater. Chem.* **17**, 4611–4618 (2007).
107. B. Bouchet-Fabre, C. Castro, D. Ballutaud, T. Nociniewski, J.-N. Rouzaud, M. Pinault, M. Mayne-L'Hermite, and C. Reynaud, ex. Abstract, *Carbon* 2009 Biarritz (2009).
108. A. Bachmatiuk, E. Borowiak-Palen, R. Jedrzejewski, M. H. Rummeli, T. Gemming, T. Pichler, and R. J. Kalenczuk, *Phys. Status Solidi B*, **243**(13), 3226–3229 (2006).

109. J. W. Jang, C. E. Lee, C. I. Oh, and C. J. Lee, *J. Appl. Phys.* **98**(7), 074316 (2005).

110. M. Endo, Y. A. Kim, M. Ezaka, K. Osada, T. Yanagisawa, T. Hayashi, M. Terrones, and M. S. Dresselhaus, *Nano Lett.* **3**, 723 (2003).

111. T. Hasobe, S. Fukuzumi, and P. V. Kamat, *Angew. Chem. Int. Ed. Engl.* **45**, 755–759 (2006).

112. K. T. Lau, M. Lu, and D. Hui, *Composites Part B* **37**, 437–448 (2006).

113. N. Tang, W. Zhong, C. Au, and Y. Du, *Carbon Nanotubes and Related Structures*, Eds. Vladimir Blank and Boris Julnitskiy, Research Signport, Kerala, India, 2008, pp. 69–96. ISBN 978-81-308-0214-5.

114. S. Motojima, M. Kawaguchi, K. Nozaki, and H. Iwanaga, *Appl. Phys. Lett.* **56**, 321 (1990).

115. S. Itoh, S. Ihara, and J. Kitakami, *Phys. Rev. B* **48**, 5643–5647 (1993).

116. X. B. Zhang, X. F. Zhang, D. Bernaerts, G. Van Tendeloo, S. Amerlinckx, J. Van Landuyt, *et al.*, *Europhys. Lett.* **27**, 141–146 (1994).

117. S. Amelinckx, X. B. Zhang, D. Bernaerts, X. F. Zhang, V. Ivanov, and J. B. Nagy, *Science* **265**, 635 (1994).

118. M. Zhang, Y. Nakayama, and L. Pan, *J. Appl. Phys.* **39**, L1442 (2000).

119. V. K. Varadan and J. Xie, *Smart Mater. Struct.* **11**, 728 (2002).

120. N. Grobert, M. Terrones, S. Trasobares, K. Kordatos, H. Terrones, J. Olivares, J. P. Zhang, P. Redlich, W. K. Hsu, C. L. Reeves, D. J. Wallis, Y. Q. Zhu, J. P. Hare, A. J. Pidduck, H. W. Kroto, and D. R. M. Walton, *Appl. Phys. A* **70**(2), 175–183 (2000).

121. N. J. Tang, W. Zhong, A. Gedanken, and Y. W. Du, *J. Phys. Chem. B* **110**, 11772 (2006).

122. A. Szabo, A. Fonseca, J. B. Nagy, Ph. Lambin, and L. P. Biro, *Carbon* **43**(8), 1628–1633 (2005).

123. J. Xie, K. Mukhopadyay, J. Yadev, and V. K. Varadan, *Smart Mater. Struct.* **12**, 744 (2003).

124. M. Yudasaka, R. Kikuchi, Y. Ohki, and S. Yoshimura, *Carbon* **35**, 195 (1997).

125. S. Amelickx, X. B. Zhang, D. Bernaerts, X. F. Zhang, V. Ivanov, and J. B. Nagy, *Science* **265**, 635 (1994).

126. Ph. Lambin, A. Fonseca, J. P. Vigneron, J. B. Nagy, and A. A. Lucas, *Chem. Phys. Lett.* **245**, 85–89 (1995).

127. K. Akagi, R. Tamura, M. Tsukada, S. Itoh, and S. Ihara, *Phys. Rev. Lett.* **74**(12), 2307 (1995).

128. K. Akagi, R. Tamura, M. Tsukada, S. Itoh, and S. Ihara, *Phys. Rev. B* **53**(4), 2114 (1996).
129. D. Bernaerts, X. B. Zhang, X. F. Zhang, S. Amerlickx, G. Van Tendeloo, J. Van Landuyt, V. Ivanov, and J. B. Nagy, *Phil. Mag. A* **71**(3), 605 (1995).
130. C. N. Ralachandran and N. Sathyamurthy, *Curr. Sci.* **91**(11), 1503 (2006).
131. A. Volodin, M. Ahlskog, E. Seynaeve, C. V. Aesendonck, A. Fonseca, and J. B. Nagy, *Phys. Rev. Lett.* **84**, 3342 (2000).
132. G. A. Karp, A. Ya'akobovitz, M. David-Pur, Z. Ioffe, O. Cheshnovsky, S. Krylov, and Y. Hanein, *J. Micromech. Microeng.* **19**, 085021 (2009).
133. H. Okuno, E. Grivei, F. Fabry, T. M. Gruenberger, J. Gonzalez-Aguilar, A. Palnichenko, L. Fulcheri, N. Probst, and J.-C. Charlier, *Carbon* **42**(12–13), 2543 (2004).
134. V. D. Blank, I. G. Gorlova, J. L. Hutchison, N. A. Kiselev, A. B. Ormont, E. V. Poyakov, J. Sloan, D. N. Zakharov, and S. G. Zybtsev, *Carbon* **38**, 1217–1240 (2000).
135. X. Ma, E. G. Wang, R. D. Tilley, D. A. Jefferson, and W. Zhou, *Appl. Phys. Lett.* **77**(25), 4136–4138 (2000).
136. L. P. Biró, G. I. Márk, Z. E. Horváth, K. Kertész, J. Gyulai, J. B. Nagy, and Ph. Lambin, *Carbon* **42**, 2561 (2004).
137. X. Ma, E. G. Wang, R. D. Tilley, D. A. Jefferson, and W. Zhou, *Appl. Phys. Lett.* **77**(25), 4136 (2000).
138. J.-M. Ting and J. B. C. Lan, *Appl. Phys. Lett.* **75**, 3309 (1999).
139. H. W. Kroto, J. R. Heath, S. C. O'Brien, R. F. Curl, and R. E. Smalley, *Nature* **318**, 14 (1985).
140. H. W. Kroto, *Nature* **329**, 529 (1987).
141. L. A. Chernozatonskii, *Phys. Lett. A* **166**, 55–60 (1992).
142. F. Cozzi, W. H. Powell, and C. Thilgen, *Pure Appl. Chem.* **77**, 843–923 (2005).
143. P. W. Fowler and D. E. Manolopoulos, *An Atlas of Fullerenes*, Dover Publications, UK, 2007.
144. http://www.mathematik.uni-bielefeld.de/~CaGe/fullerenes.html
145. K. Hedberg, L. Hedberg, D. S. Bethune, C. A. Brown, H. C. Dorn, R. D. Johnson, and M. De Vries, *Science* **254**(5030), 410–412 (1991).
146. H. W. Kroto and K. McKay, *Nature*, **331**, 328–331 (1988).
147. C. P. Ewels, *Nanoletters* **6**(5), 890 (2006).

148. Y. Z. Tan, S. Y. Xie, R. B. Huang, and L. S. Zheng, *Nat. Chem.* **1**(6), 450–460 (2009).
149. Y. M. Solonin, *Powder Metall. Met. Ceram.* **40**(11–12), 618 (2001).
150. W. Kratchmer, L. D. Lamb, K. Fostiropoulos, and D. R. Huffman, *Nature* **347**, 354 (1990).
151. L. T. Scott, M. M. Boorum, B. J. McMahon, S. Hagen, J. Mack, J. Blank, H. Wegner, and A. de Meijere, *Science* **295** (5559), 1500 (2002).
152. *Fullerene-Related Materials*, Ed. S. Margadonna, Springer Academic Press, Germany, 2010. ISBN 978-1-4020-4458-8.
153. J. R. Heath, S. C. O'Brien, Q. Zhang, Y. Liu, R. F. Curl, H. W. Kroto, F. K. Tittel, and R. E. Smalley, *J. Am. Chem. Soc.*, **107**, 7779–7780 (1985).
154. M. Murata, Y. Murata, and K. Komatsu, *J. Am. Chem. Soc.* **128**(24) 8024–8033 (2006).
155. Z. Ge, Jc Duchamp, T. Cai, Hw Gibson, and Hc Dorn, *J. Am. Chem. Soc.* **127**(46), 16292–16298 (2005).
156. S. Margadonna and K. Prassides, *J. Solid State Chem.* **168**, 639 (2002).
157. M. R. Beasley, S. Datta, H. Kogelnik, H. Kroemer, and D.Monroe, Bell Labs Report (2002). http://www.alcatel-lucent.com/wps/DocumentStreamerServlet?LMSG_CABINET=Docs_and_Resource_Ctr&LMSG_CONTENT_FILE=Corp_Governance_Docs/researchreview.pdf
158. A. L. Ayzner, C. J. Tassone, S. H. Tolbert, and B. J. Schwartz, *J. Phys. Chem. C Nanomater. Interfaces* **113**(46), 20050 (2009).
159. M. Srivastava, M. Kumar, R. Singh, U. C. Agrawal, and M. O. Garg, *J. Sci. Ind. Res.* **68**(2), 93 (2009).
160. D. Suter and K. Lim, *Phys. Rev. A* **65**, 052309 (2002).
161. L. W. Tutt and T. F. Boggess, *Prog. Quantum Electron.* **17**(4), 299 (1993).
162. R. Bakry, R. M. Vallant, M. Najam-Ul-Haq, M. Rainer, Z. Szabo, C. W. Huck, and G. K. Bonn, *Int. J. Nanomedicine* **2**(4), 639 (2007).
163. L. L. Dugan, E. G. Lovett, K. L. Quick, J. Lotharius, T. T. Lin, and K. L. O'Malley, *Parkins. Rel. Disord.* **7**, 243 (2001).
164. A. H. Ringwood, N. Levi-Polyachenko, and D. L. Carroll, *Environ. Sci. Technol.* **43**, 7136 (2009).
165. V. D. Blank, S. G. Buga, N. R. Serebryanaya, G. A. Dubitsky, M. Yu. Popov, V. N. Denisov, A. N. Ivlev, and B. N. Mavrin, *Phys. Lett. A* **205**, 208 (1995).
166. V. D. Blank, S. G. Buga, G. A. Dubitsky, N. R. Serebryanaya, V. N. Denisov, A. N. Ivlev, B. N. Mavrin, and M. Yu. Popv, *Mol. Mat.* **7**, 251 (1996).

167. V. D. Blank, S. G. Buga, N. R. Serebryanaya, G. A. Dubitsky, R. H. Bagramov, M. Yu. Popov, V. M. Prokhorov, and S. A. Sulyanov, *Appl. Phys. A* **64**, 247 (1997).
168. S. Margadonna, D. Pontiroli, M. Belli, T. Shiroka, M. Ricco, M. Brunelli, *J. Am. Chem. Soc.* **126**, 15032 (2004).
169. D. Ugarte, *Carbon* **33**(7) 989–993 (1995).
170. S. Iijima, *J. Cryst. Growth* **50**(3), 675–683 (1980).
171. S. Iijima, *J. Phys. Chem.* **91**(13), 3466–3467 (1987).
172. D. Ugarte, *Nature* **359**, 707 (1992).
173. W. A. de Heer and D. Ugarte, *Chem. Phys. Lett.* **207** (4–6), 480 (1993).
174. B. Xu, *New Carbon Mater.* **23**(4), 289–301 (2008).
175. P. Fowler, *Chem. Phys. Lett.* **131**, 444 (1986).
176. D. J. Klein, W. A. Seitz, and T. G. Schmalz, *Nature* **323**, 703 (1986).
177. B. S. Xu and S. I. Tanaka, *Proc. Int. Conf.*, ICSE, Cambridge 355–360 (1997).
178. T. Li, G. Liu, X. Liu, *et al.*, *J. Trans. Mat. Heat Treatment* **26**(3), 28–30 (2005).
179. N. Sano, H. Akazawa, T. Kikuchi, *et al.*, *Carbon* **41**(11) 2159–2162 (2003).
180. B. Y. Li, B. Q. Wei, J. Liang, *et al.*, *Carbon* **37**(3), 493–497 (1999).
181. J. Y. Huang, H. Yasuda, and H. Mori, *Chem. Phys. Lett.* **303**(1–2) 130–134 (1999).
182. T. Cabioc'h, J. P. Riviere, and J. Delafond, *J. Mater. Sci.* **30**(19), 4787–4792 (1995).
183. N. Sano, H. Wang, I. Alexandrou, *et al.*, *J. Appl. Phys.* **92**(5), 2783–2788 (2002).
184. F. Banhart and P. M. Ajayan, *Adv. Mat.* **9**(3), 261 (1997).
185. I. V. Ponomareva and L. A. Chernozatonskii, *JETP Lett.* **79**(8), 375 (2004).
186. J. Hiraki, H. Mori, E. Taguchi, *et al.*, *Appl. Phys. Lett.* **86**(22), 223101 (2005).
187. M. Chhowalla, H. Wang, N. Sano, *et al.*, *Phys. Rev. Lett.* **90**(15), 155501–15504 (2003).
188. C. P. Chen, T. H. Chang, and T. F. Wang, *Ceram. Int.* **28**(8) 925–930 (2002).
189. K. W. Street, M. Marchetti, W. R. L. Vandar, *et al.*, *Tribol. Lett.* **16**(1–2), 143–149 (2004).

190. Y. L. Yao, X. M. Wang, J. J. Guo, *et al.*, *Mater. Lett.* **62**(16), 2524–2527 (2008).
191. L. Joly-Pottuz, N. Matsumoto, H. Kinoshita, *et al.*, *Tribol. Int.* **41**(2), 69–78 (2008).
192. H. W. Kroto, *Comput. Math. Appl.* **17**(1–3), 417 (1989).
193. M. D. Clayton and P. W. Rabideau, *Tetrahedron Lett.* **38**(5), 741–744 (1997).
194. M. Ozawa, H. Goto, M. Kusunoki, and E. Osawa, *J. Phys. Chem. B* **106**(29), 7135 (2002).
195. J. A. Simpson and E. S. C. Weiner, *Oxford English Dictionary*, 2nd ed., Clarendon Press, Oxford, 1989, Vol. 7, p. 113.
196. J. Liu, H. Dai, J. H. Hafner, D. T. Colbert, R. E. Smalley, S. J. Tans, and C. Dekker, *Nature* **385**, 780 (1997).
197. R. Martel, H. R. Shea, and Ph. Avouris, *J. Phys. Chem. B* **103**(36), 7551 (1999).
198. M. Ahlskog, E. Seynaeve, R. J. M. Vullers, C. Van Haesendonck, A. Fonseca, K. Hernadi, and J. B. Nagy, *Chem. Phys. Lett.* **300**(1–2), 202 (1999).
199. M. E. Lyn, J. He, and B. Koplitz, *Appl. Surf. Sci.* **246**, 44 (2005).
200. E. Yazgan, E. Taşci, O. B. Malcioğlu, Ş. Erkoç. c Yazgan, *J. Mol. Struct. (Theochem)* **681**, 231–234 (2004).
201. S. Ihara, S. Itoh, and J. Kitakami, *Phys. Rev. B* **47**, 12908–12911 (1993).
202. S. Itoh and S. Ihara, *Phys. Rev. B*, **48**, 8323–8328 (1993).
203. M. Terrones, W. K. Hsu, J. P. Hare, H. W. Kroto, H. Terrones, and D. R. M. Walton, *Phil. Trans. R. Soc. Lond. A* **354**, 2025–2054 (1996).
204. I. László and A. Rassat., *Int. J. Quantum. Chem.* **84**, 136 (2001).
205. V. Meunier, Ph. Lambin, and A. A. Lucas, *Phys. Rev. B* **57**, 14886 (1998).
206. B. Burteaux, A. Claye, B. W. Smith, M. Monthioux, D. E. Luzzi, and J. E. Fischer, *Chem. Phys. Lett.* **310**, 21 (1999).
207. B. W. Smith, M. Monthoiux, and D. E. Luzzi, *Nature* **354**, 56 (1998).
208. Y. K. Kwon, D. Tomanek, and S. IIjima, *Phys. Rev. Lett.* **82**(7), 1470 (1999).
209. K. Hirahara, K. Suenaga, S. Bandow, H. Kato, T. Okazaki, *et al.*, *Phys. Rev. Lett.* **85**, 5384 (2000).
210. B. W. Smith and D. E. Luzzi, *Chem. Phys. Lett.* **321**(1–2), 169 (2000).
211. M. Yoon, S. Berber, and D. Tomanek, *Phys. Rev. B* **71**, 155406 (2005).

212. D. J. Hornbaker, S.-J. Kahng, S. Misra, B. W. Smith, A. T. Johnson, E. J. Mele, D. E. Luzzi, and A. Yazdani, *Science* **295**, 828 (2002).

213. B. W. Smith, D. E. Luzzi, and Y. Achiba, *Chem. Phys. Lett.* **331**(2–4), 137 (2000).

214. P. Jaroenapibal, S. B. Chikkannanavar, D. E. Luzzi, and S. Evoy, *J. App. Phys.* **98**(4), 044301 (2005).

215. B. W. Smith, M. Monthioux, and D. E. Luzzi, *Chem. Phys. Lett.*, **315**, 31 (1999).

216. Y. A. Kim, T. Hayashi, K. Osawa, M. Endo, and M. S. Dresselhaus, *Chem. Phys. Lett.* **367**(5–6), 537 (2003).

217. R. Yuge, M. Yudasaka, J. Miyawaki, Y. Kubo, T. Ichihashi, H. Imai, E. Nakamura, H. Isobe, H. Yorimitsu, and S. Iijima, *J. Phys. Chem. B* **109**, 17861 (2005).

218. M. Monthioux, *Carbon* **40**, 1809 (2002).

219. A. G. Nasibulin, A. S. Anisimov, P. V. Pikhitsa, H. Jiang, D. P. Brown, M. Choi, and E. I. Kauppinen, *Chem. Phys. Lett.* **446**, 109 (2007).

220. A. G. Nasibulin, P. V. Pikhitsa, H. Jiang, D. Brown, *et al.*, *Nat. Nanotechnol.*, **2**, 156 (2007).

221. I. Suarez-Martinez, M. Monthioux, and C. P. Ewels, *J. Nanosci. Nanotechnol.* **9**, 6144 (2009).

222. I. Suarez-Martinez, C. P. Ewels, and M. Monthioux, in preparation (2011).

223. S. Trasobares, C. P. Ewels, J. Birrell, O. Stephan, B. Q. Wei, *et al. Adv. Mater.* **16**(7), 610–613 (2004).

224. R. G. Harvey, *Polycyclic Aromatic Hydrocarbons*, Wiley-VCH, 1997. ISBN 978-0471186083.

225. H. Sakurai, T. Daiko, and T. Hirao, *Science* **301**(5641), 1878 (2003).

226. R. Jasti, J. Bhattacharjee, J. B. Neaton, and C. R. Bertozzi, *J. Am. Chem. Soc.* **130**, 17646 (2008).

227. M. S. Newman and D. Lednicer, *J. Am. Chem. Soc.* **78**, 4765 (1956).

228. R. H. Martin, *Angew. Chem. Int. Ed.* **13**, 649 (1974).

229. K.-H. Ernst, Y. Kuster, R. Fasel, C. F. McFadden, and U. Ellerbeck, *Surf. Sci.* **530**(3), 195 (2003).

230. H. Terrones and M. Terrones, *Phys. Rev. Lett.* **84**, 1716 (2000).

231. D. Vanderbilt and J. Tersoff, *Phys. Rev. Lett.* **68**, 511 (1992).

232. A. L. Mackay and H. Terrones, *Nature* (London) **355**, 762 (1991).

233. T. Lenosky, X. Gonze, M. Teter and V. Elser, *Nature* (London) **355**, 333 (1992).
234. S. J. Townsend, T. Lenosky, D. A. Muller, C. S. Nichols, and V. Elser, *Phys. Rev. Lett.* **69**, 921 (1992).
235. T. Lenosky, X. Gonze, M. Teter, and V. Elser, *Nature* **355**, 333 (1992).
236. H. A. Schwarz, *Gesammelte Mathematische Abhandlugen*, Springer, Berlin, 1890.
237. E. Barborini, P. Piseri, P. Milani, G. Benedek, C. Ducati, and J. Robertson, *Appl. Phys. Lett.* **81**(18), 3359 (2002).
238. S. J. Townsend, T. J. Lenosky, D. A. Muller, C. S. Nichols, and V. Elser, *Phys. Rev. Lett.* **69**(6), 921 (1992).
239. H. Terrones, M. Terrones, F. L. Urias, J. A. Rodriguez Manzo, and A. L. Mackay, *Phil. Trans. R. Soc. Lond. A* **362**, 2039–2063 (2004).
240. H. Terrones, M. Terrones, E. Hernandez, N. Grobert, J.-C. Charlier, and P. M. Ajayan, *Phys. Rev. Lett.* **84**(8), 1716 (2000).
241. V. H. Crespi, L. X. Benedict, M. L. Cohen, and S. G. Louie, *Phys. Rev. B* **53**(20), 13303 (1996).
242. L. P. Biro, G. I. Mark, A. A. Koos, J. B. Nagy, and Ph. Lambin, *Phys. Rev. B* **66**, 165405 (2002).
243. Ph. Lambin, G. I. Mark, and L. P. Biro, *Phys. Rev. B* **67**, 205413 (2003).
244. Ph. Lambin and L. P. Biro, *New J. Phys.* **5**, 141.1–141.14 (2003).
245. L. Biró, G. I. Márk, A. A. Koós, J. B. Nagy, and Ph. Lambin, *Phys. Rev. B* **66**, 165405 (2002).
246. I. László and A. Rassat. *Int. J. Quantum. Chem.* **84**, 136 (2001).

Chapter 2

Surfaces and Thin Films of Fullerenes

Roberto Macovez[1*] and Petra Rudolf[2]

[1] *Grup de Caracterització de Materials, Departament de Física i Enginyeria Nuclear, Universitat Politècnica de Catalunya, Av. Diagonal 647, 08028 Barcelona, Spain*

[2] *Zernike Institute for Advanced Materials, University of Groningen, Nijenborgh 4, 9747AG Groningen, the Netherlands*

[*] *Previously at ICFO — Institut de Ciencies Fotoniques, Mediterranean Technology Park, Av. Canal Olímpic, 08860 Castelldefels (Barcelona), Spain.*

We review the basic properties of fullerene thin films, focusing on issues such as morphology, electronic structure, conduction and optical properties, and phase transitions. After discussing the preparation methods of fullerene films, we describe some of the most significant experimental results obtained on these systems by optical and electron spectroscopy, scanning probe microscopy, and electrical measurements. Throughout the chapter, we compare several different materials ranging from pristine fullerite, compounds with alkali, alkaline earth and rare earth elements, fullerene polymers, as well as pristine and intercalated endofullerenes. The emphasis is on the aspects related to the impact of surfaces and interfaces on electronic and structural features, on

This chapter is dedicated to the memory of Paul A. Bruhwiler (1961–2010), a dear friend and colleague who made very important contributions to fullerene science.

Advances in Carbon Nanomaterials: Science and Applications
Edited by Nikos Tagmatarchis

ISBN 978-981-426-78-78 (Hardcover), 978-981-426-78-85 (eBook)
www.panstanford.com

the dependence of physical properties upon film thickness (from mono- to multilayer to thick films), and on the comparison of thin-film and surface characteristics with corresponding bulk properties.

2.1 Introduction

With their extremely rich variety of behaviors in the solid state, fullerenes constitute a unique playground to investigate the fundamental properties of molecular condensed matter. The simple chemical formula and highly symmetric structure of the fullerene molecules, together with their ability to support different oxidation states allowing the formation of charge-transfer compounds within a wide range of stoichiometries, are all features that make these molecules the prototypical building block of organic molecular solids.

The main characteristic of fullerene systems, common to all molecular condensed matter, is their heavy molecular imprint. All fundamental physical properties of fullerene solids, from cohesive forces to electronic states and phonon excitations, from the conduction and dielectric behavior to the magnetic response, are a direct emanation of molecular features. A prominent manifestation of the molecular character is the high degree of localization of electrons on individual molecules in condensed fullerene phases. The corollaries of this are multiple: one is the formation of narrow electronic bands (from corresponding molecular orbitals,) in which electron correlation effects are usually important; also, integer molecular oxidation states are strongly favored, a fact which has consequences for compound formation and metallic behavior, as well as for the properties of systems with non-integer electron filling such as the surface of some C_{60} compounds.

Electronic localization is also accompanied by a strong coupling to intramolecular phonon modes, manifest for example in a pronounced Jahn–Teller effect which plays a fundamental role for the magnetic and conduction properties (especially superconductivity) of fullerene solids. In this scenario, the primary challenge from a fundamental perspective is to understand how collective solid-state properties such as metallicity, superconductivity, and magnetism emerge from the molecular degrees of freedom. Fullerene materials are also archetypal systems to investigate the impact of π conjugation and molecular orientation dynamics and ordering on solid-state properties.

The thin-film form of fullerenes is the most suited for several types of studies as well as for most device applications. C_{60} films are readily obtained on flat surfaces, where the quasi-spherical shape of the molecule favors the formation close-packed structures via the growth of planar hexagonal layers stacked upon one another. The typical growth method of C_{60} films is by vapor deposition in vacuum or controlled atmosphere (generally N_2 or Ar). The choice of an inert environment is dictated by the instability of C_{60} when exposed to air and light: due to the relatively large (on the atomic scale) interstitial voids between the molecules in pristine fullerite, molecular oxygen readily diffuses into it[1] and subsequent illumination by light triggers photochemical reactions leading to C–O binding and disruption of the fullerene cages.[2,3] Therefore, the characterization of fullerene thin films often requires *in situ* measurements on samples freshly deposited in vacuum or controlled atmosphere.

Obvious choices of characterization tools in such experimental conditions are electron spectroscopies and scanning tunneling microscopy (STM) and spectroscopy, which due to the finite mean free path of electrons in solids are inherently surface sensitive. The effect is even more pronounced in fullerene systems, where the electron attenuation length is of the order of the intermolecular spacing for a wide range of electron kinetic energies.[4–6] The use of electron-based techniques constrains the choice of substrates to conducting and semiconducting ones and the film thickness to few molecular layers to avoid charging effects, and it might be wondered whether such limitations and extreme surface sensitivity actually restrict the amount of information that can be obtained by these methods on the intrinsic properties of fullerene solids and thin films. In fact, as it will be shown in this chapter, it turns out that the characteristic features of fullerene-based systems are already present in ultrathin films (a few molecular layers), and, remarkably, some of the most beautiful experimental results on alkali fullerides (e.g., by STM and angle-resolved photoemission spectroscopy) have been obtained on mono- and multilayer films.

Moreover, compounds with exotic stoichiometries can be obtained in thin-film form by controlled evaporation of the intercalant species, at chemical compositions for which bulk growth methods yield instead mixed-phase samples. Complementary information to that obtained with electron-based techniques can be acquired from conduction measurements on thin-film transistor devices as well as by surface-sensitive nonlinear optical techniques such as second harmonic generation.

2.2 Preparation of Fullerene Thin Films

Well-ordered face-centered cubic (*fcc*) polycrystalline C_{60} films have been successfully grown by thermal vapor deposition on several substrates with weak surface bonding, such as GaAs, GaN, GeS, mica, MoS_2, VSe_2, ZrO_2, alkali and alkaline earth (*AE*) halides, as well as on highly oriented pyrolytic graphite and metals such as Au, Ag, and Cu (see refs. 7 and 8 and references therein). On strongly binding substrates such as Si surfaces with open dangling bonds, C_{60} growth results in amorphous films.[9]

The optimal substrate temperature is in the range 450–475 K, i.e., just below the desorption temperature of C_{60} multilayers, which lies in the range of 500–575 K. Good quality films were demonstrated also at higher substrate temperature (575 K) with very high deposition rates.[10,11] Film growth by supersonic molecular beam[12–14] and ionized cluster beam[15–17] deposition was also reported. All these methods aim at achieving a high surface mobility of the fullerene molecules during growth to obtain a high crystalline quality. Order in the film can be improved by a choice of substrate which allows epitaxial growth, as in the case of the single-crystalline Ag(111) and Au(111) surfaces.

Films of higher fullerenes such as C_{70} as well as of endohedral fullerenes such as $M@C_{82}$ (M = Y, Dy, La, Gd, Tm, Sc) can be similarly grown by thermal deposition.[18–20] Films of alkali endohedrals may also be obtained by ion bombardment.[21] Higher fullerenes with a quasi-spherical shape form *fcc* structures in the condensed phase. C_{70} also forms a close-packed lattice, but depending on the growth method and conditions both *fcc* and hexagonal close-packed structures are reported.[22,23] Thin films of C_{60} derivatives such as PCBM, some C_{60} compounds,[24,25] and some endohedrals[26] may be obtained by solution growth, self-assembly, or Langmuir–Blodgett techniques.[8,27,28] Such films are more stable in air/light and also allow for other types of characterization and easier application in electronic and photovoltaic devices. Thin films of endofullerenes processed from solution generally contain large fractions of solvent molecules, which need to be eliminated after solution casting by annealing at high temperatures if the intrinsic properties of the endofullerene condensed phase are to be probed.[29]

Fullerenes and their derivatives are good electron acceptors, and charge-transfer salts (fullerides) are easily obtained with electron-donor elements. When C_{60} or C_{70} films are intercalated with alkali (*AE*) and rare earth (*RE*) elements, stable charge-transfer compounds form for well-defined integer stoichiometries, due to the possibility of accommodating only an integer number of electrons on each molecule. Since the highest occupied molecular orbital (HOMO) of the C_{60} molecule is totally filled, the extra electrons donated by the intercalant fill the electronic states derived from the lowest unoccupied molecular orbital (LUMO), which is threefold degenerate and hence may accommodate up to six electrons (the LUMO of C_{70} is instead only twice degenerate). At higher electron filling the electronic states derived from higher orbitals (usually denoted as LUMO+1, LUMO+2, and so on) start to be occupied. The thin-film form of C_{60} salts is generally obtained via evaporation of the electron-donor species on top of a previously deposited well-ordered pristine film (intercalation of amorphous films results instead in inhomogeneous film with phase separation). Either a very controlled deposition method of the intercalant or a vacuum annealing[30] is usually necessary to obtain well-ordered phase-pure films.

Mixtures of C_{60} with other elements which form solids with high cohesive energies[31] do not usually yield compounds or solid-state solutions: deposition of Au, Cr, or Si on top of C_{60}, e.g., results in the formation of nanocrystals embedded in the fullerene matrix or at its surface.[32–35] Some of the *d*-shell transition metals, such as Pd, Pt, and Fe, intermix to some extent with C_{60}, but the obtained phases show poor crystallinity and are thermally unstable.[36,37] Nb and Ti form C_{60} compounds only in thin-film form,[38] while other transition metals like V, Co, and Au show no sign of compound formation. The nature of cohesive forces in these transition metal-fullerene systems is unclear, although many studies suggest an important contribution of covalent bonding. For example, Ti and La evaporated on top of C_{60} films display a tendency to form single atomic layers at the surface, which reflects the hybridization of metal d and fullerene π orbitals as in bulk metal carbides.[32] In Pd_xC_{60} phases, it has been suggested that the Pd atoms bridge C_{60}, molecules forming polymer-like structures in one, two, or three dimensions depending on the composition.[39]

Another route to tune the properties of C_{60} films is by irradiation with light in an inert environment. Exposure of pristine C_{60} films in vacuum or controlled atmosphere to intense visible or UV light, during[40] or after[41] deposition, results in photopolymerization where some of the "double" bonds which constitute the π electronic structure of the molecule break and intermolecular σ bonds are formed between next neighbors, usually arranged in one-dimensional chains or two-dimensional (2D) networks. C_{60} deposition under irradiation yields phase-pure polymerized films displaying an orthorhombic structure of parallel polymer chains.[40] Polymeric phases are also observed in some C_{60} compounds with alkali and *AE* elements, where the polymer bonds form spontaneously upon charge transfer (without light irradiation). These phases consist either of parallel polymer chains[42–44] or of parallel planes of 2D networks,[45–47] which have a different geometry with distinct bonding motifs depending on the compound.

C_{70}-based systems display some of the features observed in C_{60} solids (effect of orbital degeneracy, polymerization), but have been less studied than the latter, especially in thin-film form. In the following, we will mainly deal with C_{60}-based systems, focusing on the properties of pristine, photopolymerized, and intercalated C_{60} films, and discuss also endohedral systems.

2.3 Monolayer Systems

Single-layer fullerene films on crystalline surfaces are highly ordered quasi-2D systems which for their peculiar character constitute a class on its own right. The binding of the fullerene monolayer to the substrate, both in the case of metallic surfaces or Si wafers, is much stronger than the intermolecular van der Waals cohesive forces that keep together the fullerene molecules,[48] which allows annealing to high temperatures (typically 850–1050 K) without desorption of the monolayer, and induces orientational ordering up to temperatures which are much higher than in the bulk form (see Section 2.4.2). On the other hand, C_{60} monolayers can also be grown on very weakly bonding substrates such as self-assembled alkyl-thiol monolayers, which interact very weakly with fullerene and allow rotational motions at low temperature.[49]

As interatomic distances in an inorganic substrate are usually much shorter than the molecular diameter or the intermolecular spacing, the substrate–adsorbate bonding usually depends on the molecular orientation, and epitaxial growth, when it occurs, entails a monolayer periodicity over several substrate unit cells. The epitaxial properties of the interface can induce a lower symmetry than that expected for a close-packed single layer, which both for C_{60} and C_{70} is the hexagonal symmetry of a (111) plane of the corresponding bulk crystal structure. Also when the hexagonal symmetry is retained, non-equivalent surface adsorption sites or intermolecular interactions may give rise to distinct molecular orientations, resulting in a larger effective monolayer periodicity (see below).

STM studies on C_{60} and C_{70} monolayers on metals are able to distinguish intramolecular features even at room temperature, which implies that molecular orientations are more or less fixed also at temperatures where bulk phases display rotational freedom (see Section 2.4.2). In C_{70} monolayers, the molecules are usually oriented with their long axis perpendicular to the surface plane.[18,50] In contrast, C_{60} monolayers on self-assembled alkyl-thiols display orientational ordering only at very low temperatures: an STM study has shown that at room temperature the C_{60} molecules are free to rotate and move to different locations on the self-assembled substrate (the molecules display a smooth hemispherical protrusion in STM images), at 77 K they are still capable of rotation around a fixed axis (C_{60} molecules appear as hemispheres, tilted donuts, or asymmetric dumbbells), and only at 5 K it is possible to discern their internal fine structure.[51] Interestingly, three types of molecular ordering are observed at low temperatures, each corresponding to a distinct local minimum in the theoretical potential energy surface for a perfect 2D (free standing) C_{60} layer,[51] which indicates that the C_{60} monolayer on alkyl-thiols is representative of a truly 2D fullerene system.

In monolayer C_{60} films deposited on metal substrates the tunneling intensity displays characteristic inhomogeneities from molecule to molecule, which may be periodic or aperiodic depending on the substrate and either static or dynamic (on the experiment's timescale of seconds).[52–54] This difference in STM contrast originates in the different molecular orientation and/or different bonding to the substrate, and it is not observed in the second or higher layer in multilayer films. Some authors suggest that the inhomogeneous tunneling intensity might also reflect a inhomogeneous charge distribution[53] (in monolayers grown on metallic substrates, the high electron affinity of C_{60} leads to a charge transfer from the metal of up to several electrons per molecule, depending on the substrate). STM contrast inhomogeneities analogous to those of C_{60} monolayers have been reported for the potassium-doped K_3C_{60} and K_5C_{60} monolayers (the latter phase was only identified in ultrathin fulleride films and does not exist as bulk phase).[55] Here too, the contrast is lost at the second molecular layer. While in some studies only two intensity levels are observed, with a fraction of the molecules being brighter than the rest, recent high-resolution characterizations of C_{60} monolayers on metals have shown that orientational ordering may lead to more complicated patterns in STM images (see ref. 56 and references therein). These patterns seem to be associated with an adsorbate-induced reconstruction of the metallic substrate.[57–60]

The orientation-specific binding, together with the higher annealing temperature, results in a high degree of ordering which enhances intermolecular interactions. One of the most spectacular findings on such systems is the direct observation by STM of a static Jahn–Teller distortion in the K_4C_{60} single-layer film.[55,61] Figure 2.1 shows bias-dependent STM topographs of a K_4C_{60} monolayer. The left picture is an image of the filled molecular orbitals, in which each molecule appears bisected by a single nodal line; the panel to the right shows instead the empty states, which are characterized

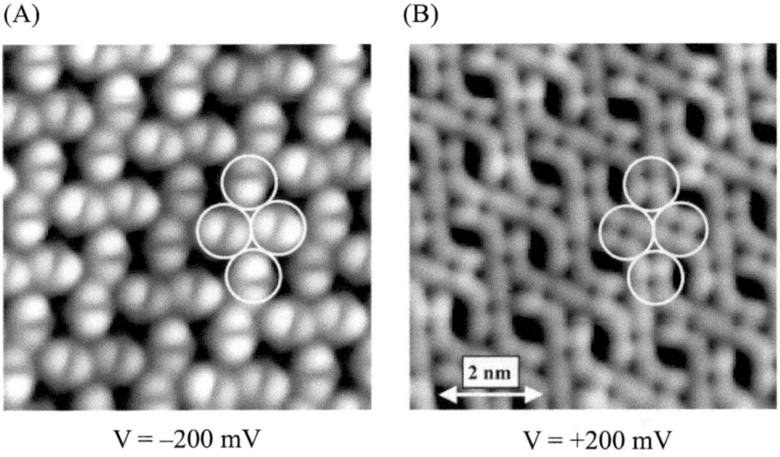

Figure 2.1. Energy-dependent STM topographs of the filled (a) and empty (b) electronic states of the same region of a K_4C_{60} monolayer (7×7 nm^2). Single molecules are marked by circles (courtesy of Prof. M. F. Crommie).

by an additional nodal plane rotated by 90° with respect to the node observed in the filled-state image. The characteristic nodal structure above the Fermi level was observed in the bias range +0.1 to +0.6 V, while the filled-state image did not change over the bias range −0.1 to −0.7 V. The measured local electronic density of states (DOS) is in agreement with the expectations for a C_{60} LUMO orbital split into two degenerate filled Jahn–Teller levels and one non-degenerate empty sublevel, corresponding to a filling of four electrons per molecule.

Another milestone achievement on monolayer systems is the experimental determination of the band dispersion of C_{60} and K_3C_{60} monolayers by angle-resolved photoelectron spectroscopy.[62–65] As already mentioned, due to the strong substrate–molecule interaction it would be hasty to consider such systems as representative of the thin-film form of fullerides (let alone the bulk form). For example, the inhomogeneous contrast and static Jahn–Teller distortion in K-doped monolayers are not observed at all in the second or higher layer in multilayer films, and orientational ordering in thicker films resembles more closely bulk orientational transitions than those of single-layer systems (see also Section 2.4.2).

In monolayer and ultrathin films grown on metal surfaces, the presence of the conducting substrate also has an important affect on the electronic properties. Obvious examples include charge transfer or strong chemical binding, which deeply impact the electronic landscape. However, also in systems with weaker bonding, the presence of the metallic interface induces specific electronic states in the molecular film which stem from the image potential felt by charges in the proximity of the metal surface (image states). These electronic states have been observed in many monolayer and double-layer fullerene films[66–68] and are dispersive in the plane of the interface (following the in-layer periodicity) and quantized in the perpendicular direction. While molecular polarization in multilayer films is effective in screening interfacial charges originating from the electron transfer/charge redistribution at the metal surface, leading thus to delocalized free-electron-like image states, in monolayer films electron scattering upon the lattice of interfacial dipoles results in an increased effective mass of the image states.[66]

2.4 Properties of Multilayer and Thick C_{60} Films

2.4.1 *Electronic States*

Electron spectroscopies (photoemission, inverse photoemission, X-ray absorption, electron energy loss spectroscopy, and scanning tunneling spectroscopy) have been extensively applied to the study of the occupied and unoccupied DOS in fullerenes (in the case of scanning tunneling spectroscopy, the local DOS is measured). An example is given in Fig. 2.2, which shows the valence-band photoemission and inverse photoemission spectra of a crystalline C_{60} thin film on Au(110).[69] Each peak corresponds to an electronic band derived from a molecular orbital, and its position reflects the energy of that band relative to the vacuum level. The peaks in the photoemission spectrum arise respectively, starting from the gap region around −5 eV and going to the left, from the HOMO, HOMO−1, HOMO–2 levels, and so on, while the features to the right of the band gap in the inverse photoemission spectrum arise from the LUMO, LUMO+1 states, and so on.

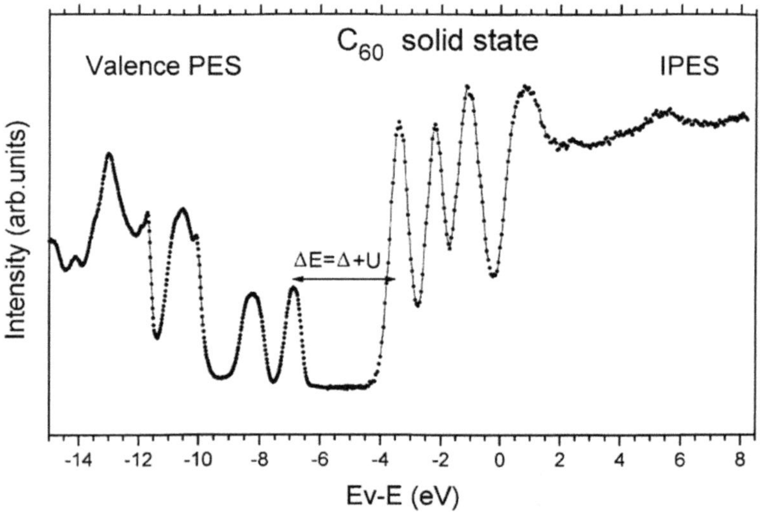

Figure 2.2. Valence-band photoemission and inverse photoemission spectra of a pristine C_{60} film. Reprinted from ref. 69.

It is important to note that in electronic spectra, as well as in tunneling spectra, the width of the features does not reflect that of the corresponding electronic bands, which are much narrower and display only weak dispersion. The large width of the spectral features is mostly due to Franck-Condon broadening (i.e., to phonon satellites) with a contribution of band dispersion effects. Evidence for such effects is provided by several electron spectroscopy studies on C_{60} films, both for occupied and empty states.[64,70,71]

As visible in Fig. 2.2, the energy separation ΔE between the HOMO-derived band in electron removal and the LUMO-derived states in electron addition is 3.5 eV (peak to peak). This separation is equal to the sum of the band gap and the screened intramolecular electron repulsion (or correlation energy) U. The latter quantity has been determined independently by comparing the self-convolution of the photoemission spectrum with the Auger spectrum.[72] The measured value is 1.4 ± 0.2 eV, which gives a band gap of slightly above 2 eV, in agreement with theoretical calculations and other experimental estimates (see below). It should be noted that the value of U at the film surface is higher than the bulk value[73] because

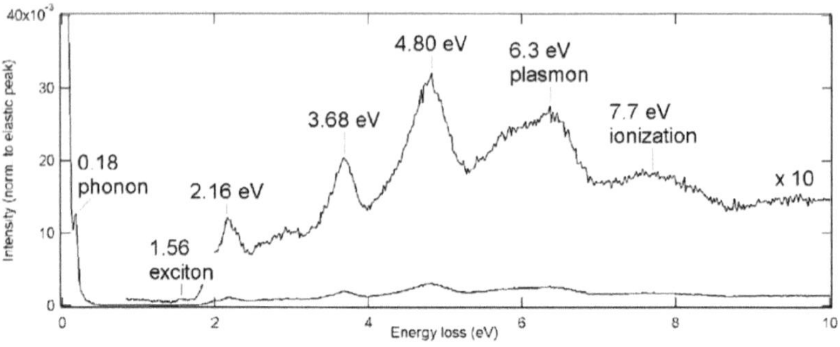

Figure 2.3. High-resolution electron energy loss spectrum of the electronic excitations of a thick C_{60} multilayer on Ag(111).

of the poorer screening due to the lower molecular coordination,[74] so that the value obtained in an experimental measurement of the correlation energy with electron spectroscopy techniques actually depends on the probing depth, which is generally rather low. Experimental estimates of the relevant electronic energies also exist for some alkali fullerides (Section 2.5.1).

An experimental determination of the bandgap in C_{60} films can also be obtained by high-resolution electron energy loss spectroscopy. Figure 2.3 shows a typical spectrum acquired on a multilayer sample. The first prominent loss feature is observed at 2.2 eV, which can be taken as the experimental determination of the bandgap in pristine C_{60} films. The peak at 0.18 eV in the tail of the elastic peak originates from the excitation of a high-frequency vibrational mode of the C_{60} molecule (around 176 meV). A distinctive feature of fullerene solids is indeed the presence of stiff intramolecular phonons at energies which are comparable with the electronic bandwidth. The relatively strong coupling of LUMO electrons to these high-frequency phonons gives rise to hybrid vibronic states and places fullerenes outside the range of validity of the adiabatic electron–phonon coupling regime.

The weak feature at 1.56 eV corresponds to the lowest energy molecular excitonic triplet state (see below).[75] Several peaks can be observed in the spectrum of Fig. 2.3 above the first excitonic feature. Beside the peaks at 2.2, 3.7, and 4.8 eV, which stem from interband one-electron transitions, a broad feature can be observed around

6 eV which corresponds to the excitation of collective oscillations (plasmons) of the π electron cloud. Also higher energy excitations exist (not visible in the range of Fig. 2.3), which are assigned to interband transitions between σ orbitals and to a mixed plasmon involving both π and σ electrons. The energy of the broad feature at 7.7 eV is in close agreement with the vertical ionization potential of C_{60} on gold.[76]

Besides with electron energy loss spectroscopy, excitonic states in C_{60} films have been probed by second harmonic and sum-frequency generation spectroscopy,[77,78] photoluminescence,[79] as well as excited state photoemission spectroscopy.[80] Quadratic nonlinear optical techniques are able to probe exciton states selectively and with high spectral resolution, and are virtually insensitive to interband transitions above the conductivity gap. The sensitivity to excitonic states has been attributed to the lower coherence of electron and hole states with respect to excitons.[77] The mutual Coulomb attraction between a LUMO electron and a hole sitting in the HOMO orbital gives rise to four distinct Frenkel exciton singlets, respectively of $^1T_{1g}$, $^1T_{2g}$, ^1G_g, and ^1H_g symmetry.[81] All four are electric dipole forbidden, but two of them, the magnetic dipole-allowed $^1T_{1g}$ and the electric quadrupole-allowed ^1H_g, can be observed by second harmonic and sum-frequency generation, as shown in Fig. 2.4.

The peak at 1.83 eV in Fig. 2.4 corresponds to the $^1T_{1g}$ state.[82] The resonance at 1.86 eV is assigned to the mixing of the $^1T_{1g}$ singlet with the nearly degenerate ^1G_g singlet, while that at 2.02 eV stems from the vibronic mixing of the $^1T_{1g}$ exciton with the high-frequency phonon at 176 meV.[83] The peak at 2.3 eV corresponds instead to the ^1H_g exciton. The large difference in second harmonic intensity between the ^1H_g and the $^1T_{1g}$ states is due to the fact that the electric quadrupole-induced susceptibility is only of the order of 10% of the magnetic dipole-induced one.[77] The large magnetic dipole and electric quadrupole contribution to second harmonic generation hinder the surface sensitivity of this technique, which is only attained for centrosymmetric media if the electric dipole approximation holds. It is nonetheless possible to distinguish surface and bulk contributions performing second harmonic generation experiments on films of different thickness.[84]

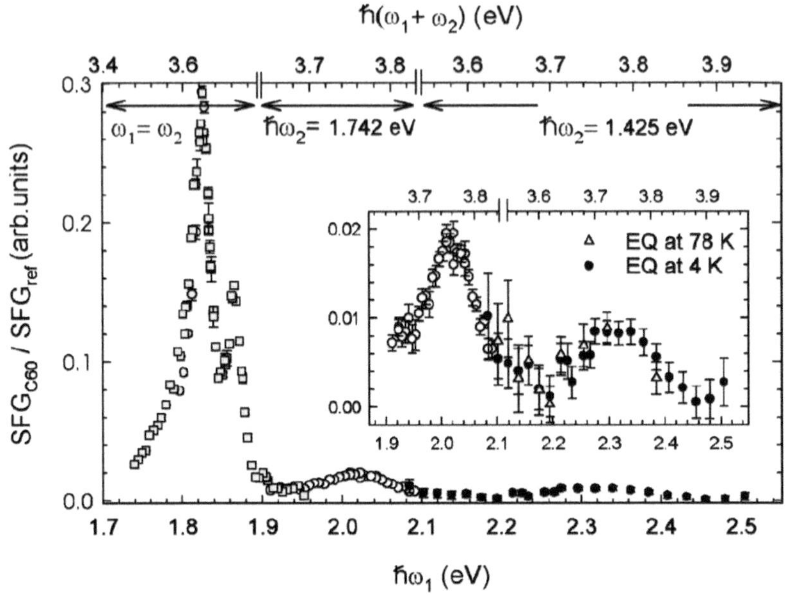

Figure 2.4. Second harmonic (gray) and sum-frequency (black and white) spectra of a C_{60} film at 78 K, as a function of the fundamental (infrared) frequency (lower *x* axis). The inset shows a close-up view of the higher energy features. For each experiment, the corresponding second harmonic or sum-frequency energies are indicated in the upper *x* axis. Reprinted from ref. 77.

All the spectral features visible in Fig. 2.4 arise form singlet exciton states. Triplet states have a lower energy due to exchange interactions, and the lowest energy triplet states in C_{60} films can be probed with several techniques including electron energy loss,[75] as shown above, and photoluminescence.[79] Pump-probe excited state photoemission spectroscopy allows detecting both singlet and triplet exciton states, while at the same time providing a means of distinguishing between them due to their different lifetimes.[80]

An interesting feature of the electronic structure of thin C_{60} films, which has been discussed in two-photon photoemission studies, is the effect of electron confinement in the direction normal to the film surface. The epitaxial C_{60} multilayer on Au(111) is in fact reported to behave as a quantum well system for the (somewhat more delocalized) electronic states derived from the LUMO+2 and LUMO+3 molecular orbitals, that are nearly degenerate in energy. The resulting quantum well states have a nearly free-electron-like dispersion in the plane, and are characterized by a progressive splitting into an increasing number of sublevels with increasing film thickness, which is due to wave function confinement inside the film boundaries.[85]

2.4.2 *Molecular Orientations and Surface Morphology*

Fullerene molecules display interesting orientational dynamics and ordering in solid phases. In bulk fullerite at 300 K the C_{60} molecules rotate very rapidly (they are in fact more labile than in solution), resulting in a lattice with effective *fcc* symmetry.[86,87] Below 260 K there is a first-order phase transition to a simple cubic (*sc*) phase with orientational order (the icosohedral point group symmetry of the C_{60} molecule is in fact incompatible with an orientationally ordered *fcc* phase), in which the C_{60} molecules continue to "ratchet" from one preferred orientation to another. This motion is finally frozen out on crossing a glass transition at 90 K, which leaves 85% of the molecules in one orientation and the remaining 15% in another orientation of slightly higher energy.[88] Analogous transitions and phases are observed in thin C_{60} films and at their surfaces, where the truncation of the lattice introduces non-equivalent surface sites.[89,90] An STM study of the surface of a multilayer C_{60} film showed the presence of two distinct molecular orientations at the surface of the *sc* phase. A 2×2 superlattice was reported, where one molecule in each unit cell is takes up the minority orientation while the other three are in the majority orientation.[91]

Figure 2.5 shows the temperature dependence of the width of the C 1*s* photoemission peak of a C_{60} film,[92] after subtraction of a smooth curve which represents the Gaussian phonon broadening of the core-level spectrum and fits rather well the experimental data at low and high temperatures (i.e., away from all ordering transition). The inset shows individual C 1*s* spectra acquired at various temperatures. Four different regimes may be identified, separated by three critical temperatures. The highest transition temperature corresponds to the bulk *fcc* to *sc* transition. Orientational ordering at the surface occurs instead in two steps. While the bulk rotations

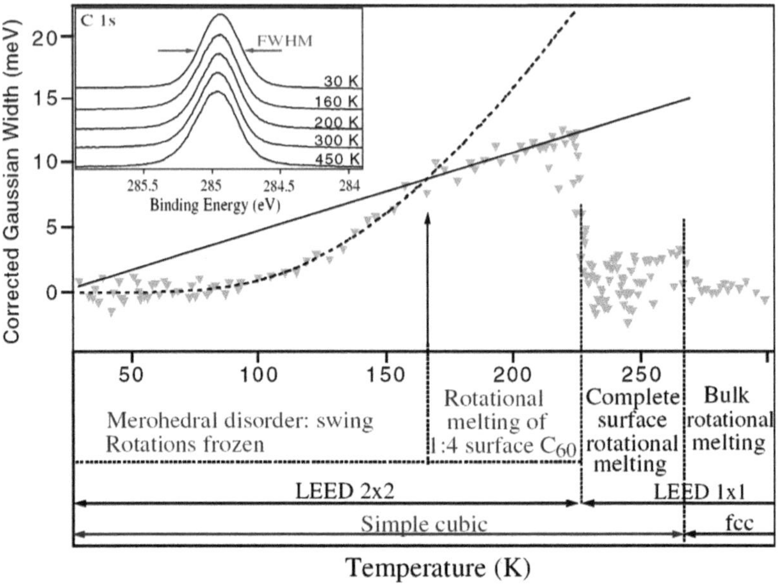

Figure 2.5. Gaussian width of the C 1*s* photoemission core level acquired on a multilayer C_{60} film, after subtraction of a smooth curve (see text). Solid and dashed curves are guide to the eye. The inset shows individual C 1*s* spectra acquired at different temperatures. The phase diagram of the different orientational phases at the (111) surface of C_{60} crystals is also shown in the bottom part of the figure, with the corresponding low-energy electron diffraction (LEED) pattern. Reprinted from ref. 92.

are already frozen, the surface molecules remain free to rotate down to 230 K. At this temperature, three out of four molecules (the ones which assume the majority orientation in the low-temperature phase) stop rotating. The rotation of the fourth molecule is only frozen at 160 K (one hundred degrees lower that the bulk ones), when complete orientational ordering sets it. As visible in Fig. 2.5, the temperature variation of the C 1*s* width closely reflects these four regimes. This interpretation is further corroborated by the low-energy electron diffraction patterns at distinct temperatures (see caption of Fig. 2.5 and ref. 93) by and electron energy loss spectra.[89]

Molecular motions and disorder have an effect on the intensity of second harmonic generation from C_{60} films. For example, the feature at 1.86 eV in the low temperature second harmonic

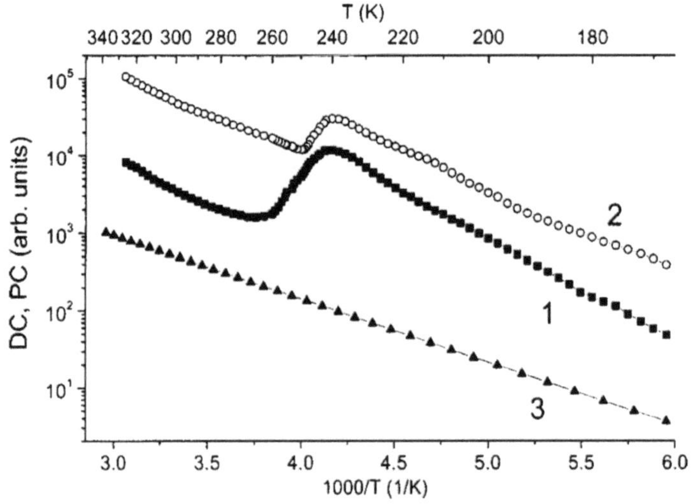

Figure 2.6. Temperature-dependent conductivity (curve 1) and photoconductivity (curve 2) of a C_{60} film. A large anomaly is observed across the *fcc* to *sc* transition around 260 K, which is instead absent in the conductivity of C_{60} films exposed to oxygen (curve 3). Reprinted from ref. 95.

spectrum (Fig. 2.4) was not detected at room temperature, and the intensity of the second harmonic resonance resulting from the $^1T_{1g}$ exciton state at 1.83 eV was observed to decrease dramatically as the sample temperature was raised from 200 to 260 K.[77,94] The incoherent motion of the C_{60} molecules in the orientationally disordered phase leads to an induced nonlinear polarization with no correlations between neighboring molecules, thus resulting in destructive interference and quenching of the quadratic nonlinear optical response.[94]

Orientational ordering also has a deep impact on the electronic properties of C_{60} films. Figure 2.6 shows the dependence of the conductivity (curve labeled with 1) and photoconductivity (curve 2) of a C_{60} film across the *fcc* to *sc* transition at 260 K.[95] When molecular rotations freeze, the conductivity raises by more than one order of magnitude, implying that orientational order significantly enhances hopping between the molecules. Conversely, the conductivity of C_{60} films previously exposed to molecular oxygen (curve 3) does not show any abrupt change at this temperature.

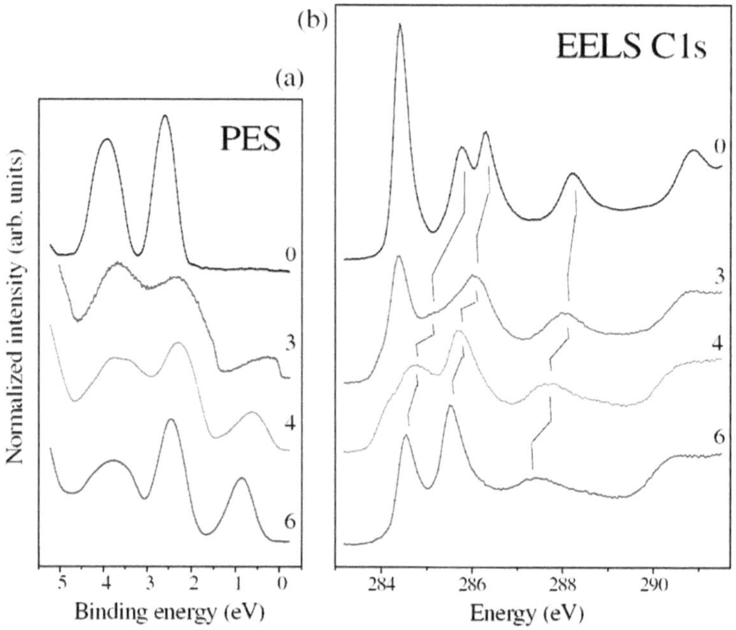

Figure 2.8. Photoemission (a) and C 1*s* electron energy loss (b) spectra of phase-pure K_xC_{60} films for various concentrations. The upper spectra in both panels are acquired on a pristine C_{60} film ($x = 0$), and the peaks visible in these spectra arise, from left to right, from the HOMO-1- and HOMO-derived states in (a) and from the LUMO, LUMO+1, and so on in (b) (see Fig. 2.2 for comparison). As the K content increases, the LUMO-derived band starts to be filled, giving rise to a new photoemission feature above the HOMO-derived states. At the same time a broadening and a change in the energy and relative intensity of the other features is observed, but the labeling according to the C_{60} molecular orbitals can still be applied. The triply degenerate LUMO-derived band is only partially occupied in K_3C_{60} and K_4C_{60}, while it is totally filled in K_6C_{60}. Reprinted from ref. 98. See also Color Insert.

becomes less important until it disappears for $x = 6$, while the empty states at higher energy gradually shift towards the Fermi level (i.e., to lower energy loss).

Both in (a) and (b), the spectra show significant broadening due to phonon-gain and phonon-loss satellites, as may be expected in a system with strong electron–phonon coupling. Despite this broadening, band dispersion effects can be observed in thick films also, as reported in angle-resolved photoemission experiments on potassium-intercalated C_{60} films.[99]

Comparison of K-intercalated C_{60} multilayers with different potassium content shows that the resistance of K_xC_{60} is strongly dependent on the stoichiometry, displaying a minimum for $x = 3$.[100–102] This is in agreement with the phase diagram of the K_xC_{60} solid phase, for which the $x = 0, 4, 6$ compounds are insulating and the $x = 1$ stoichiometry (see below) is only weakly conducting,[103] while the A_3C_{60} salts (A = K, Rb, Cs) are metallic and even superconducting at remarkably high critical temperatures[104–106] (20–40 K). An STM study of ultrathin K_xC_{60} films ($x = 3, 4, 5$)[107] has highlighted the dependence of the correlation energy U, as obtained from the energy separation between the leading features with positive and negative bias, versus alkali content and film thickness. This study has shown that the screening of Coulomb interactions in thin K_3C_{60} films goes beyond the simple molecular polarization screening which characterizes pristine C_{60} and actually involves the contribution of itinerant charge carriers.

K_3C_{60} thin films are not only conducting but also become superconducting at critical temperatures similar to those of bulk samples. Superconductivity has been observed in ordered thick films,[108] where a detrimental effect of disorder on superconducting properties is reported,[100] as well as in ordered multilayers on semiconducting substrates for thicknesses as low as 2.4 molecular layers.[109] This low value is emblematic of the abrupt character of interfaces of fullerene films, with the bulk behavior recovered in the space of two or three molecular layers.

We have seen in Section 2.3 that the interface between a fullerene film and the substrate exhibits characteristic electronic features which are far from being trivial. The other extremity of fullerene films, namely their free vacuum surface, similarly displays peculiar electronic properties. One prominent example are the surfaces of AC_{60} and A_3C_{60} films (A = K, Rb, Cs). The AC_{60} stoichiometry displays a very rich phase diagram as a function of temperature and thermal treatment. Similar to pristine fullerite, two distinct cubic phases exist in the bulk compound: a *fcc* structure of rapidly spinning molecules,[110] thermodynamically stable above 400 K, and a metastable *sc* phase,[111] obtained by fast cooling

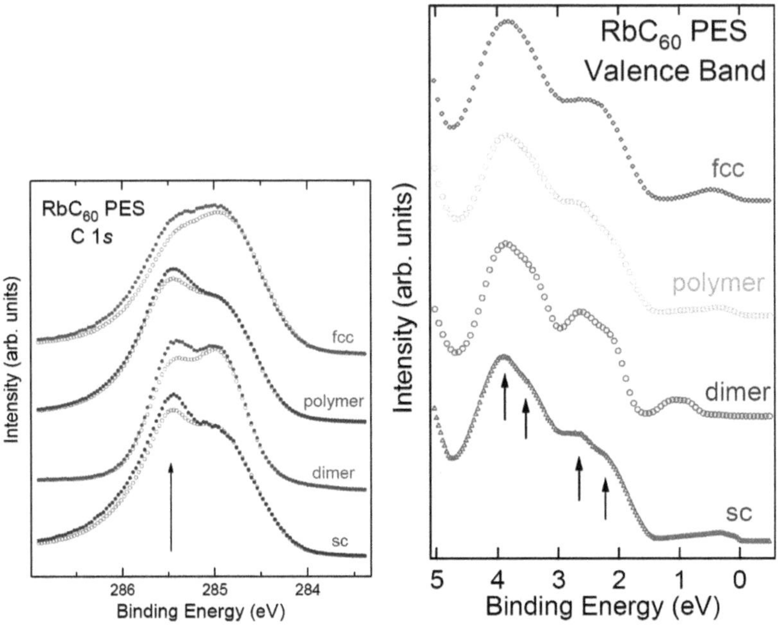

Figure 2.9. (Left) C 1*s* photoemission spectra of the four phases of RbC_{60}, acquired at normal (empty circles) and grazing (filled circles) photoelectron emission. In all spectra the presence of (at least) two components is visible (see for comparison the C 1*s* spectrum of C_{60} at different temperatures, shown in the inset of Fig. 2.5). The highest binding energy component (neutral C_{60} molecules) has a higher relative intensity in grazing than in normal emission, signaling the presence of neutral molecules at the film surface. (Right) Valence-band normal-emission photoemission spectra of the four phases of RbC_{60}. Comparison with the valence-band spectrum of C_{60} films (Figs. 2.2 and 2.8a) reveals the presence of two molecular charge states (see arrows). Adapted from refs. 115 and 116. See also Color Insert.

the *fcc* phase to below 100 K, which differs from the latter due to orientational order. Two more phases are observed: upon annealing to 200 K, the *sc* structure transforms irreversibly into a metastable phase of $(C_{60})_2^{(2-)}$dimers,[112,113] which can also be obtained by fast cooling the *fcc* phase to below room temperature. A weakly conducting phase of polymer chains[42,43] is instead thermodynamically stable below 400 K.

Figure 2.9 shows the photoemission spectra of the C 1*s* level (left panel) and frontier valence-band states (right panel) of a RbC_{60}

thin film in all four crystallographic phases. A double-component structure is clearly visible in the C 1*s* spectra of all RbC_{60} phases, in contrast to the C 1*s* spectrum of pristine C_{60} films where only one component is detected (inset of Fig. 2.5). The valence-band spectra reveal the presence of two non-equivalent molecular states at the film surface (see arrows in the right panel of Fig. 2.9). A similar behavior is observed at the K_3C_{60} surface, where three non-equivalent molecular contributions can be discerned in the photoemission spectra,[4] and where the comparison with X-ray emission spectra[114] clearly demonstrates the surface nature of the phenomenon. For RbC_{60},[115,116] the comparison of the valence-band spectral lineshape with that of pristine C_{60} (see Figs. 2.2 and 2.8a) and with theoretical calculations for the $(C_{60})_2^{2-}$ dimer,[117] as well as the relative intensity of the valence-band features, indicates that the two components arise from neutral and charged surface molecules. As visible in Fig. 2.9, the neutral C_{60} molecules indeed contribute a higher relative C 1*s* signal (arrow in left panel) if a more surface-sensitive experiment is performed collecting photoelectrons at grazing emission.

A similar interpretation in terms of distinct molecular oxidation states (instead of a single one as observed in the bulk) holds for the K_3C_{60} case.[4] The presence of several charge states corresponds in both cases to a reduction by 50% of the electron density in the C_{60} termination layer of the film, and is indicative of the occurrence of a surface charge reconstruction. Reconstructions are often observed at the free surface of polar solids (fullerene salts being an obvious example) where they usually involve the displacement of the surface atoms or molecules (so-called structural reconstruction). In some cases, however, a surface reconstruction can consist only in a redistribution of the charge density near the surface (charge reconstruction). A 50% reduction of the surface electronic charge corresponds indeed to the expectation for a charge-reconstructed $C_{60}(111)$ termination layer of a AC_{60} or A_3C_{60} film.[101,118]

A clear Fermi edge is detected in the low-temperature valence-band photoemission spectra of *sc* RbC_{60} (see Fig. 2.10) and A_3C_{60} (A = K, Rb) films, which is indicative of metallic character.[116] The *sc* phase of RbC_{60} is indeed more conducting than the *fcc* phase of the same compound, presumably due to the beneficial effect of orientational order (see Section 2.4.2). In the photoemission spectra of RbC_{60} thin films in the *sc* phase, evidence is found for the presence of not just two but actually three distinct charge states at the surface, the third one corresponding to doubly charged anions. The presence of doubly charged molecules was also reported in the bulk *sc* phase of the twin CsC_{60} compound. The observation of distinct charge states and a sharp Fermi edge in the spectra of K_3C_{60} and *sc* RbC_{60} suggests that in both compound the (surface) metallicity is accompanied by fluctuations in the oxidation state of the molecules. There is evidence that similar molecular charge fluctuations occur in most bulk fullerides,[119–121] and it has been argued that they are a key feature of fulleride superconductors[122] which favors the local pairing of electrons through Jahn–Teller electron–phonon coupling. The impact of electron correlation on the occurrence of charge fluctuations and thus metallicity and superconductivity is not clear. In contrast with the expectation that strong repulsion between electrons on the same molecules should hinder local charge fluctuations, theoretical studies have shown in fact that correlation effects may, in the presence of Jahn–Teller coupling, result in an effective enhancement of the local pairing.[123,124] A fully developed theory of fullerene metallicity and superconductivity is still lacking.

Another peculiar feature of the free surface of fullerene solids with respect to their bulk properties is the different character and critical temperature of phase transitions at the surface. An example is the orientational ordering transition at the surface of C_{60} films, which was discussed in Section 2.4.2. Another one is teh case of RbC_60, where the transformation from the *sc* phase to the dimer phase is irreversible in the bulk,[112,113] while it is a fully reversible phase transition at the film surface,[116] as shown in Fig. 2.10. Panel (a) displays the temperature evolution of the frontier electronic states during the quench from the *fcc* to the dimer phase and as the temperature is further lowered. The feature closest to the Fermi level (E_F) in the spectrum of the *fcc* phase arises from the partial filling of the band derived from the LUMO of the C_{60} molecule due to charge transfer from Rb. The transition to the insulating dimer phase is accompanied by the opening of a gap at E_F in the DOS and by the rise of two new features around 1 eV (spectra acquired at 230 and 170 K), which stem from the highest filled molecular orbitals

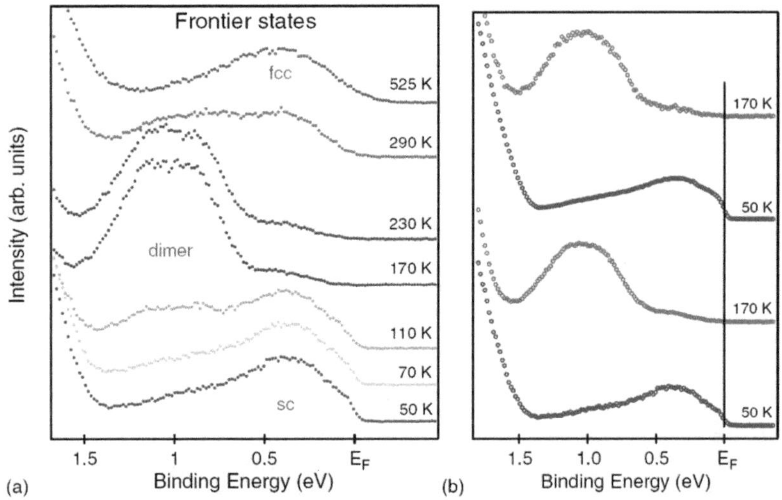

Figure 2.10. High-resolution photoemission spectra of the frontier states of a RbC_{60} film. (a) Sequence of spectra acquired during the quench from the *fcc* to the *sc* phase at a rate of 50 K per minute, evidencing the temperature dependence of the frontier states. The dimer phase is obtained at intermediate temperatures during the quench. (b) Spectra obtained while cycling the sample temperature between 170 and 50 K, which show the reversible character of the *sc*-to-dimer phase transition at the film surface. Reprinted from ref. 116. See also Color Insert.

of the charged $(C_{60})_2^{2-}$ dimer. As the temperature is lowered below 135 K, the dimerized film undergoes a transition to the conducting *sc* phase, which is characterized by a sharp Fermi edge. Panel (b) shows the effect of repeated annealing and cooling through the *sc*-to-dimer transition, which shows its reversible character. The reversibility of the *sc*-to-dimer phase transformation at the film surface might be related to a higher degree of rotational freedom of the C_{60} monomers at the film surface (such as observed at the surface of pristine fullerite, see Section 2.4.2).

Another transition which displays a modified behavior at surfaces is the metal–insulator transition in the polymer phase of the AC_{60} compounds, which is reported around 50 K in the bulk polymer.[125,126] At the surface of RbC_{60} films, this transition takes place at much higher temperature (90 K),[127] which has been attributed to the poorer screening of electron correlation[124] at the surface. Analogous differences between the bulk and surface critical behavior are common also to inorganic strongly correlated systems.

In the case of Li and Na fullerides, the phase diagram is very different from that of the larger alkalis. No stable compound seems to exist below a stoichiometry of four alkali atoms per fullerene, for which a 2D-polymer phase forms in bulk samples.[45,47,128] Lower alkali content results presumably in mixed-phase samples, both in bulk samples and thin films.[128–131] A reproducible increase in the n-type semiconductor-like conductivity by several orders of magnitude has been reported for C_{60} films intercalated by Li and Na, accompanied by a decrease in activation energy.[132,133]

Since Li easily diffuses through the fullerene matrix, the ionic conduction properties of Li fullerides have also attracted some attention.[134] Charge transfer from Li is generally incomplete and compounds with very high stoichiometry can be obtained.[135] The lowest stable Li_4C_{60} stoichiometry has been demonstrated also in thin-film form.[136] The structure of Li_4C_{60} can be described as a set of rectangular 2D-polymer planes stacked onto each other along the $< 100 >$ direction of the pristine C_{60} crystal. Since the termination plane of a pristine C_{60} film is perpendicular to the $< 111 >$ direction, when the Li_4C_{60} phase forms upon intercalation of C_{60} films with Li one of the polymerization directions lies in the surface plane, so that the triangular surface symmetry is distorted into a quasi-hexagonal symmetry with contraction of the unit cell along the polymerization direction. This shows up in the low-energy diffraction pattern of Li_4C_{60} films[136] (Fig. 2.11), where three equivalent surface domains rotated by 60° can be discerned, corresponding to the three possible directions of polymerization in the hexagonal surface plane.

Evidence for a significant mobility of the Li ions near the film surface is reported at room temperature,[136] consistent with the ionic conduction properties[134] of bulk Li_4C_{60}.

2.5.2 *Thin Films of AE and RE Fullerides*

Compounds of C_{60} with *AE* elements have attracted a lot of attention in the first years of solid-state fullerene research. Thin-film studies of AE_xC_{60} fullerides (AE = Ca, Sr, Ba) have shown that while charge transfer is complete up to a stoichiometry of $x = 3$, corresponding

Figure 2.11. Low-energy electron diffraction pattern of a Li_4C_{60} film (top) compared to that of a pristine C_{60} film (bottom). Reprinted from ref. 136. See also Color Insert.

to the complete filling of the C_{60} LUMO-derived band, for higher stoichiometries the LUMO+1-derived states hybridize with the *AE* *s* shell (*s*-*d* hybrid orbital in the case of Sr and Ba), leading to an only partial further transfer of charge.[137–139] While thin films of stoichiometry less or equal to 3 are insulating, evidence is found for metallicity at higher intercalation levels.[137,139] This is in agreement with bulk studies, which report a superconducting ground state for AE_xC_{60} (AE = Ca, Sr, Ba) with x around 4 or 5, though at lower temperatures than in alkali fullerides.[140–142] The similarity between the LUMO- and LUMO+1-derived states (threefold degeneracy, coupling to the same phonon modes, and an electron filling similar to that of A_3C_{60} superconductors) suggests that the mechanism of superconductivity is the same as in alkali fullerides, although hybridization with the intercalant's electronic levels complicates the scenario.

The surface of the Ca_5C_{60} compound was investigated in an early STM study which has provided submolecular details of the surface morphology.[143] A very specific tunneling pattern is observed (Fig. 2.12), which reflects both the periodicity of the surface layer and the arrangement of Ca atoms. The proposed structure is that of a *fcc* lattice with multiple Ca intercalation into octahedral interstices (four Ca ions per octahedral void) and occupation of one out of two tetrahedral interstices. The fact that Ca ions also contribute a tunneling intensity is ascribed to a remnant charge density in Ca 4*s* states. At higher Ca concentration, a Ca-induced STM contrast is observed with a minority of the molecules exhibiting a lower intensity. The periodicity of the induced contrast indicates the formation of superlattices similar to those encountered in monolayer systems (see Section 2.3).

In the case of Mg, no stable phase appears to exist at low stoichiometry, while a 2D-polymer phase structurally similar to that of Li_4C_{60} or Na_4C_{60} is reported in bulk Mg_xC_{60} for x around 5, which appears to be metallic.[144] Very few thin-film studies exist on Mg fullerides, and ordered phase-pure samples have been achieved only recently.[145] These films display enhanced conductivity at room temperature already at low Mg intercalation, and partial evidence for polymerization in the film is reported at higher Mg content.[145]

The last family of fullerene salts, which has attracted much interest in recent years, is that of the rare-earth fullerides, RE_xC_{60} and RE_xC_{70}. Bulk studies have reported the existence of several stable RE_xC_{60} stoichiometries for $x = 2.75$, 3, and 6, with interesting physical properties besides superconductivity, such as strong magnetism, mixed valency and valency transition of the *RE* cations, and giant magnetoresistance.[146–149] These properties arise from the presence of the localized magnetic moments of the *RE* ions, which interact with the π electron system of the fullerene molecules. The $x = 2.75$ phase of Sm and Yb fullerides is particularly interesting. The unusual stoichiometry reflects the fact that the large size of the *RE* ions induces a distortion of the crystal structure in which some of the

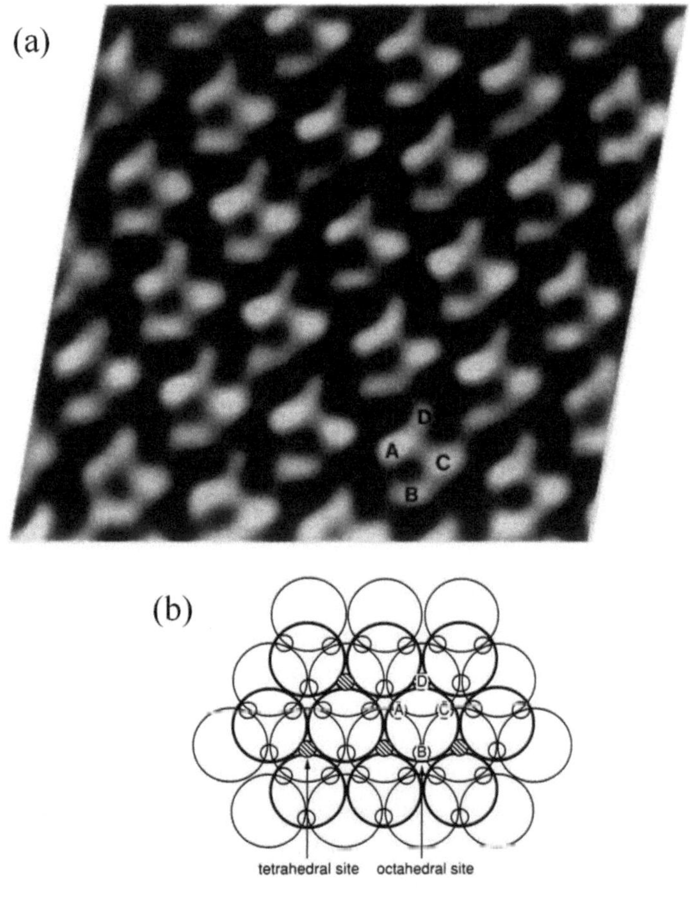

Figure 2.12. (a) Curvature-enhanced STM topograph of the surface of Ca_5C_{60} displaying the hexagonal symmetry of the (111) termination. A Ca-induced fine structure is clearly visible. (b) Schematic of the surface morphology: top view of the first two C_{60} layers along with Ca ions. Small open (resp. shaded) circles indicate Ca ions in the multiply filled octahedral (resp. singly filled tetrahedral) sites (empty tetrahedral sites are not shown). Reprinted from ref. 143.

interstitial voids remain unoccupied, forming an ordered lattice of *RE* vacancies. The low-temperature ground state of bulk $Sm_{2.75}C_{60}$ and $Yb_{2.75}C_{60}$ is a Kondo-like state which exhibits mixed *RE* valency, with an average cationic charge between +2 and +3. As the temperature is raised to above 30 K a contraction of the crystal lattice is observed, which is accompanied by a *RE* valency transition towards a purely divalent state that remains stable up to high temperatures.[146,147]

For the $x = 2.75$ stoichiometry, a divalent *RE* state implies an average formal molecular oxidation state of -5.5. An X-ray absorption study on $Yb_{2.75}C_{60}$ has indeed found evidence for distinct molecular charge states at room temperature, as well as for a strong distortion of the molecular anions induced by strong local Madelung potential gradients.[150] Photoemission studies on Sm_xC_{60} thin films with x near 2.75 reports a divalent character of the Sm ions at all temperatures, with no evidence for the formation of a Kondo state at cryogenic temperature.[151] The discrepancy between bulk and thin-film studies might be due to the polymorphism of *RE* fullerides and to the difficulty of obtaining the periodic arrangement of *RE* vacancies which characterizes the bulk $x = 2.75$ compounds in the thin-film form.

Superconductivity is reported in bulk *RE* fullerides with higher Sm or Yb content, in correspondence to a partial filling of the LUMO+1-derived band as in superconducting *AE* fullerides.[152] Thin Yb_xC_{60} films with x near 5 are reported to be metallic.[151] but no study has yet found superconductivity at low temperature. Superconducting *RE* fullerides are especially interesting as they offer the possibility of studying the interplay of magnetism and superconductivity in the same phase.

C_{60} films intercalated with Eu show divalent character of the cations for all stoichiometries,[153] in contrast with bulk studies which disagree on the valency of the Eu ions,[154–159] both in the paramagnetic Eu_3C_{60} compound and in the interesting Eu_6C_{60} fulleride,[155,157] which exhibits ferromagnetism and giant magnetoresistance. No studies are yet available on the magnetic properties of the thin-film form of *RE* fullerides.

2.6 Thin Films of Endohedral Fullerenes

The hollow nature of the fullerene molecule allows encapsulation of single atoms, molecules, and even metal-carbon and metal-nitride clusters inside the carbon cage.[160] The presence of encapsulated species, besides stabilizing certain fullerene structures that are otherwise unstable, introduces new degrees of freedom which give rise to new physical properties. Even when the parent fullerene is centrosymmetric, the inversion symmetry is lost in the endohedral derivative, and this is true also in monometallofullerenes, where the encapsulated atom usually occupies a noncentrosymmetric position. The reduced symmetry entails a lifting or reduction of the degeneracy of the molecular orbitals. Charge transfer (even if only partial) in the endohedral complex is thus accompanied by the formation of a net electric dipole moment, which affects intermolecular interactions, phase transitions and dielectric properties,[161,162] as well as thin-film growth. For example, $Y@C_{82}$ molecules on crystalline surfaces have a strong tendency to form dimers and larger clusters, due to the attraction of the positive Y ions to neighboring negatively charged fullerene cages.[163,164] This interaction does not prevent the formation of well-ordered $Y@C_{82}$ monolayers at higher coverages.[163]

If the encapsulated unit contains transition metal ions, a net magnetic dipole moment is also present. Magnetization studies of Er endofullerenes by means of soft X-ray magnetic circular dichroism have shed light on the types of magnetic interactions that are present in these systems.[29] If electron transfer to the carbon cage results in a closed-shell configuration, the encapsulated metal spin is isolated from that of neighboring endofullerenes by the diamagnetic cage, hence leading to paramagnetic behavior. If the charged fullerene cage presents instead an open shell configuration, the partially filled π frontier orbital carries a net spin moment which couples antiferromagnetically to the spin of the encapsulated lanthanoid ion. Moreover, the spin moment of the cage can couple through electron exchange to nearby spins, leading to antiferromagnetic-like intramolecular interactions.[165] The comparison of the X-ray circular dichroism spectra of $Er_2C_2@C_{82}$ with those of $ErYC_2@C_{82}$, in which a magnetic Er^{3+} ion is replaced by a diamagnetic Y^{3+} ion, shows that the intramolecular magnetic interaction between the two metal ions trapped inside the C_{82} cage does not contribute substantially to the magnetic response of the dimetallocarbide-endofullerene, as both direct exchange as superexchange mediated by the C_2 unit and the by close-shell $(C_{82})^{6-}$ cage are very weak.[29] It would be interesting to extend this study to investigate magnetic interactions in the absence of the separating C_2 units or/and when the encapsulating cage has an open shell configuration.

Electric dipole and magnetic interactions between endofullerenes molecules in solid phases enrich the phase diagram of these systems, leading for example to a fascinating interplay of the dielectric and magnetic response with the molecular orientational dynamics and the confined motion of the encapsulated species.[161,166] As will be discussed in the following in connection with STM studies, also the conduction properties of endofullerenes are somehow connected with these internal degrees of freedom. It should be noted that to obtain information on intrinsic properties of endofullerene thin films, these should be grown by vapor condensation, or else annealing at high temperature should be performed to get rid of residual solvent molecules if the film is processed from solution.[29]

The reduction of symmetry in endofullerenes with respect to the parent fullerene molecules also affects their optical properties. The UV-vis-NIR absorption spectrum of endofullerenes can be used to extract information on the size, symmetry, and oxidation state of the carbon cage.[167] In the alkali endohedral Li@C_{60}, the reduction of symmetry with respect to the pristine C_{60} cage enables electric dipole contributions to the first hyperpolarizability, which boost the nonlinear optical response of the endohedral fullerene by one or two orders of magnitude, as confirmed experimentally by second harmonic generation on C_{60} films containing 30% of Li@C_{60}.[168] Second harmonic investigations on Li@C_{60} films with 95% purity have shown that also endofullerene films undergo photopolymerization when irradiated with UV or visible light, similarly to pristine C_{60} films, and have moreover allowed probing the dynamics of photopolymerization.[169] While second harmonic generation from pristine C_{60} films vanishes for normal laser incidence, as expected for an isotropic film, this is not the case for films containing 30% of Li@C_{60} prepared by Li bombardment. In these samples, the minimum of second harmonic generation occurs for incidence angles in the range 10–20°, while a non-vanishing second harmonic signal is detected at normal incidence, indicative of anisotropic molecular orientation in the film.[168] Given the ability of Li@C_{60} to photopolymerize[169] and the observed formation of a 2D-polymeric phase in Li-doped C_{60} films,[136] it is possible that the observed anisotropy is related to the formation of polymeric bonds due to the reaction of the fullerene molecules with implanted (exohedral) Li atoms.

Synchrotron-based photoelectron spectroscopy investigations of endofullerene films have proven extremely useful to obtain information on the degree of charge transfer and hybridization between the endohedral species and the surrounding cages.[170–172] These studies have for example evidenced that the amount of transferred charge and the extent of hybridization vary significantly among endofullerene species even when the encapsulated atom is the same. In the case of Sc, e.g., the ion is close to monovalent in Sc_3@C_{82} and in Sc_2@C_{66}, while its valency is intermediate between +2 and +3 in Sc_2@C_{84}, where strong hybridization of Sc and fullerene levels takes place. Electron spectroscopy studies have also been employed to investigate the properties of endofullerenes as a function of chemical composition. In the trimetal nitride endofullerenes M_3N@C_{80} (M = Dy, Sc, Tm), for instance, the effective metal valency depends on the size of the metal ion as well as on the orbital overlap between the encapsulated ions and the fullerene cage.[173]

The family of the lanthanide monometallic endofullerenes LN@C_{82} (LN = lanthanoid element) constitutes an interesting and simple system to carry out comparative studies on distinct endofullerene species with the same cage structure. Their electronic structure may be written formally as M^{x+}@C_{82}^{x-}, where x is between 2 and 3. Roughly speaking, Sm, Eu, Tm, and Yb are divalent inside the C_{82} cage, while the other lanthanides are trivalent.[160] The $4f$ configurations of the endohedral lanthanide ion are, formally, $4f^0$ in La@C_{82}, $4f^{13}$ in Tm@C_{82} and $4f^7$ in Gd@C_{82}, which correspond, assuming complete charge transfer from the lanthanide $5d$ and $6s$ levels, to the formal valencies La^{3+}, Tm^{2+}, and Gd^{3+}, respectively.[19]

Gd is indeed found to be trivalent, albeit with a small hybridization between its valence levels and the fullerene orbitals. Tm is purely ionic and divalent, while for La@C_{82} there is a clear indication of hybridization which results in the partial occupancy of the La 5d-shell.[19] Figure 2.13 shows the result of a resonant photoemission characterization of La@C_{82} thin films.[174] In panel (a) several photoemission spectra are shown, one acquired at a photon energy away from any absorption feature of the film (196 eV), and two acquired at higher photon energies, namely 840 and 848 eV. Despite the similar photon energy, the corresponding photoemission spectra look dramatically different. The 840 eV photon energy is just below the La $3d_{3/2}$ absorption edge, and with this photon energy, as well as with hν = 196 eV, only the normal photoemission process can occur. For the slightly higher photon energy of 848 eV, however, a new channel for electron emission is available, corresponding to the Auger decay of the La core hole following absorption from the La $3d_{3/2}$ level to the empty valence DOS (resonant photoemission), which is in principle mainly derived from the π orbitals of the fullerene cage. The comparison of the 848 eV and 840 eV spectra shows that the resonant photoemission process dominates the former. To highlight this contribution, the two spectra have been subtracted from one another in panel (b). This difference spectrum is an image of the La character of the valence-band DOS.

The inset of Fig. 2.13b shows the valence-band photoemission spectrum of the La@C_{82} film acquired with He I radiation, together with a fit of this spectrum with nine components representing as many molecular orbitals. The two (non-degenerate) frontier states of La@C_{82}, displayed with a dotted line, are empty in pristine C_{82} and become occupied in La@C_{82} due to electron transfer from the La atom. The difference spectrum in Fig. 2.13b could also be fitted with nine components at the same binding energies but different relative intensities. This fit gives an estimate of the La character of each valence-band orbital of the La@C_{82} molecule. It is evident that the states which display the largest enhancement in resonant photoemission (and hence the largest La character) are not the two frontier states, but rather the next two at slightly lower binding energy, plotted with a continuous line, which correspond to the frontier orbitals of pristine (empty) C_{82}. These states display, besides a C-derived π character, an important contribution of La character. From the intensity of these two components it may be

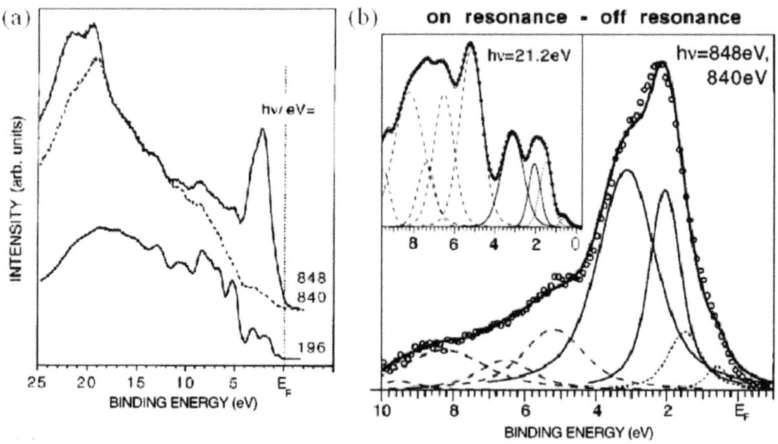

Figure 2.13. Resonant photoemission study of La@C_{82} thin films. (a) Spectra of La@C_{82} acquired at different photon energies, away from any absorption level (196 eV), and just below (840 eV) and above (848 eV) the La 3*d* 3/2 adsorption threshold. (b) Difference spectrum obtained subtracting the 840 eV spectrum from the 848 eV one. A fit of the difference spectrum with nine components (corresponding to as many orbitals) is shown, together with a similar fit of the valence-band spectrum acquired with 21.2 eV photon energy with the same number of components at the same binding energies (inset). Reprinted from ref. 174.

inferred that roughly one third of an electron remains on the 5*d* orbital of the encapsulated La ion.[174]

Interestingly, the valency of the species encapsulated in the fullerene cage is generally remarkably robust, indicating that the endofullerene complex forms a tightly bound, super-atom-like unit. For example, the divalent character of Tm in Tm@C_{82} resists even to air exposure, and intercalation of potassium into monometallo-endofullerene films to form $K_x(M@C_{82})$ thin films (M = Tm, Gd, Y) does not affect the valency of the encapsulated ion.[19,20] The same is true also in K-intercalated thin films of endofullerenes containing lanthano-nitride complexes, such as $M_3N@C_{80}$ (M = Sc, Tm).[175] The stability of the endohedral valency is also observed in monolayer Ce@C_{82} films on metallic substrates in which the substrate–adsorbate bonding can be tuned by annealing.[176] The robustness of the encapsulated ions' valency even when the net charge (and hence also the spin moment) on the cage is varied suggests that it may be possible to tune the magnetic properties of endofullerene films by controlled intercalation of electron-donor species or by charge injection.

STM studies on thin films of endofullerenes report the formation of hexagonal layers without any feature in the DOS that hints to the presence of encapsulated ions, hence confirming their endohedral nature.[177,178] Scanning tunneling spectroscopy characterizations of isolated endofullerenes on crystalline surfaces have allowed imaging their internal structure and local electronic DOS (see e.g. ref. 179). In monometallofullerenes, the encapsulated ion occupies a noncentrosymmetric position and is in some cases capable of thermal motion along the inner wall of the cage.[180] These characterizations have also allowed a direct visualization of the metallic or semiconducting character of single endofullerene molecules. It was found for instance that the $Ce@C_{60}$ molecule is semiconducting,[181] as expected since the noncentrosymmetric position of the Ce ion lifts the orbital degeneracy, and the charge transfer of four electrons fills completely the two (non-degenerate) frontier molecular orbitals. On the contrary, single $La@C_{60}$ molecules, in which the formal lanthanide charge is 3+ yielding a partially occupied frontier molecular orbital, display a metallic DOS. The metallic character appears to be somehow linked with the vibrational degree of freedom of the encapsulated ion.[181]

Thin films of gapped endohedral fullerenes display interesting semiconductor properties. The n-type room temperature of $Li@C_60$ films is higher by four orders of magnitude with respect to pristine C_60.[182] Electronic transport in thin films of $Dy@C_{82}$ and $La@C_{82}$ similarly shows n-type semiconducting behavior, with energy gap values of 0.2 and 0.3 eV, respectively, as estimated from the observed temperature dependence of the conductivity.[183] Although n-channel field effect transistors based on $Dy@C_{82}$ (see refs. 184 and 185) and $La@C_{82}$ thin films[186] display electron mobility values which are considerably lower than in C_{60}-based devices (a fact that has been attributed to the low crystallinity[185] of the endofullerene thin films), promising results have been reported for C_{60} field effect transistors in which the electrode/fullerene interfaces were modified with $La@C_{82}$.[187] Besides enhancing the carrier density, the functionalization of the gold electrodes with La@C_{82} is so effective in reducing the trapping levels at the interface between the electrode and the C_{60} thin film that transistor operation was observed without any annealing processes and even after the fabricated devices were exposed to air, in sharp contrast with conventional C_{60} devices.[187] Endofullerenes have been also demonstrated to be highly beneficial in improving the performance of fullerene photovoltaic devices. A recent study[188] on spin-coated films of a soluble derivative of Lu_3N@C_{80} has shown that endohedral encapsulation allows tuning the energy position of the frontier electronic levels of the fullerenes and hence their photophysical activity. With these modifications, the theoretical efficiency limit for fullerene photovoltaics[189] has been boosted to above 10%,[188] thus in principle enabling further improvement of the (already relatively high) performance of fullerene-based solar cell devices.[190]

2.7 Conclusions and Outlook

In this chapter we have explored the fundamental properties of thin films and interfaces of fullerenes. After a survey of available growth procedures of ordered monolayer and thicker films, we have given an overview of the most important properties of the thin-film form of several C_{60}-derived solids, focusing on each of the three large families of C_{60} compounds, namely alkali, *AE*, and *RE* fullerides. The leitmotif of our discussion has been twofold: on one hand, we have dealt with the film morphology on the molecular scale, in particular with issues related to molecular orientations and covalent intermolecular bonding; on the other hand, we have focused on electronic features, from band structure to vibronic coupling, to excitons, to linear and nonlinear optical response, to conduction properties, to superconductivity and magnetism.

We have shown that interfacial (2D) systems such as surfaces of C_{60} solids and fullerene monolayers display a very wide range of behaviors depending on the relative strength of the in-plane interactions versus the out-of-plane bonding. The relative importance of out-of-plane interactions decreases going from the strong substrate–adsorbate binding on clean metal surfaces, to the large electric field gradients at the surface of ionic C_{60} compounds, to the more balanced situation encountered at the surface of pristine fullerite, to the quasi-free-standing character of C_{60} monolayers on self-assembled alkyl-thiol monolayers where the substrate–adsorbate interaction is extremely weak. The thermal dynamics of the molecules in these systems is directly correlated with the strength of the involved interactions. STM studies on monolayer systems have allowed a direct visualization of both orientational order and disorder, as well as of molecular distortions and details of the atomic structure of the fullerene molecules. Electron spectroscopy studies of C_{60} and RbC_{60} surfaces have enabled monitoring the effect of the lower coordination upon the characteristics of phase transitions.

We have reviewed the electronic and magnetic properties of endofullerene thin films, focusing on the degree of charge transfer and hybridization, and on the impact of electric and magnetic dipole moment on the properties of condensed phases. Electron spectroscopy and STM investigations on monolayers of endohedral metallofullerenes have unraveled the endofullerene's electronic structure and demonstrated the insensitivity of the metal valency with respect to the bonding strength to the substrate. Besides for applications related to their magnetic properties, thin-film and interfacial endofullerene systems hold potential for optoelectronic devices.

The field of fullerenes constitutes a vast area of research, and even restricting it to the experimental studies on crystalline thin films and surfaces, it is impossible to do justice to all researchers and lines of investigation that have been or are being pursued. This chapter provides nonetheless a panoramic view of the basic features of these systems, while at the same time dwelling in more detail on the issues that we have considered most relevant or with which are more familiar. While many of the features of fullerene systems are well understood, others are still the subject of debate inside the scientific community. An issue which remains at least partially open, despite the large number of studies addressing it, is the precise mechanisms of charge conduction and especially superconductivity in fullerene materials. The difficulty of the problem is intimately related on one hand to the large number of molecular degrees of freedom involved in conduction processes, and on the other to the inherent complexity of modeling electron correlation effects. Another open and interesting line of research, which ought to be developed further, concerns the magnetic properties of fullerene thin films, in particular those obtained by endohedral or exohedral intercalation with lanthanide elements. X-ray and electron-based spectroscopy tools will certainly prove very useful for this task thanks to their elemental specificity.

Acknowledgments

The authors are grateful to Dr. Andrea Goldoni for critical reading of this manuscript.

References

1. R. A. Assink, J. Schirber, D. Loy, B. Morosin, and G. A. Carlson, *J. Mater. Res.* **7**, 2136 (1992).
2. A. Hamed, Y. Y. Sun, Y. K. Tao, R. L. Meng, and P. H. Hor, *Phys. Rev.* **B47**, 10873 (1993).
3. E. A. Katz, D. Faiman, B. Mishori, Y. Shapira, A. I. Shames, S. Shtutina, and S. Goren, *J. Appl. Phys.* **84**, 3333 (1998).
4. J. Schiessling, L. Kjeldgaard, T. Käämbre, I. Marenne, J. N. O'Shea, J. Schnadt, C. J. Glover, M. Nagasono, D. Nordlund, M. G. Garnier, L. Qian, J.-E. Rubensson, P. Rudolf, N. Mårtensson, J. Nordgren, and P. A. Brühwiler, *Phys. Rev.* **B71**, 165420 (2005).
5. A. Goldoni, L. Sangaletti, F. Parmigiani, G. Comelli, and G. Paolucci, *Phys. Rev. Lett.* **87**, 076401 (2001).
6. H.-N. Li, X.-X. Wang, and W.-F. Ding, *J. Electron Spectros. Relat. Phenomena* **153**, 96 (2006).
7. E. A. Katz, in *Nanostructured Materials for Solar Energy Conversion*, Ed. T. Soga, Elsevier B.V., Amsterdam, 2006, p. 361.
8. C. A. Mirkin and W. B. Caldwell, *Tetrahedron* **52**, 5113 (1996).
9. A. F. Hebard, O. Zhou, Q. Zhong, R. M. Fleming, and R. C. Haddon, *Thin Solid Films* **257**, 147 (1995).

10. E. A. Katz, D. Faiman, S. Shtutina, and A. Isakina, *Thin Solid Films* **368**, 49 (2000).
11. E. A. Katz, D. Faiman, S. Shtutina, A. Isakina, K. Yagotintsev, and K. Iakoubovskii, *Solid State Phenomena* **80**, 15 (2001).
12. A. L. Shakhmin, M. A. Khodorkovskii, S. V. Murashov, T. O. Artamonova, and Y. A. Golod, *Tech. Phys. Lett.* **27**, 87 (2001).
13. M. A. Khodorkovskii, S. V. Murashov, T. O. Artamonova, Y. A. Golod, A. L. Shakhmin, V. L. Varentsov, and L. P. Rakcheeva, *Tech. Phys.* **48**, 523 (2003).
14. M. A. Khodorkovskii, S. V. Murashov, T. O. Artamonova, A. L. Shakhmin, and A. A. Belyaeva, *Tech. Phys. Lett.* **30**, 129 (2004).
15. H. Gao, Z. Xue, and Sh. Pang, *J. Phys.* **D29**, 1868 (1996).
16. Z. M. Ren, X. X. Xiong, Y. C. Du, Z. F. Ying, F. M. Li, and L. Y. Chen, *J. Appl. Phys.* **77**, 4142 (1995).
17. X. Zou, S. Zhu, J. Xie, and J. Feng, *J. Cryst. Growth* **200**, 441 (1999).
18. X.-D. Wang, V. Yu. Yurov, T. Hashizume, H. Shinohara, and T. Sakurai, *Phys. Rev.* **B49**, 14746 (1994).
19. M. S. Golden, T, Pichler, and P. Rudolf, in *Fullerene-Based Materials:Structures and Properties*, Ed. K. Prassides, *Structure and Bonding* vol. 109, Springer Verlag, Germany, 2004, p. 201.
20. C. Ton-That, A. G. Shard, V. R. Dhanak, H. Shinohara, J. S. Bendall, and M. E. Welland, *Phys. Rev.* **B73**, 205406 (2006).
21. R. Tellgmann, N. Krawez, S. H. Lin, I. V. Hertel, and E. E. B. Campbell, *Nature* **382**, 407 (1996).
22. G. B. M. Vaughan, P. A. Heiey, D. E. Luzzi, D. A. Ricketts-Foot, A. R. McGhie, J. E. Fischer, Y.-W. Hui, A. L. Smith, D. E. Cox, W. J. Romanow, B. H. Allen, N. Coustel, J. P. McCauley Jr., and A. B. Smith III, *Science* **29**, 1350 (1991).
23. W. B. Zhao, Z. D. Zhang, Z. Y. Ye, J. L. Zhang, C. Y. Li, D. L. Yin, Z. N. Gu, X. H. Zhou, and Z. X. Jin, *Solid State Commun.* **85**, 311 (1993).
24. J. Stinchombe, A. Penicaud, P. Bhyrappa, P. D. W. Boyd, and C. A. Reed, *J. Am. Chem. Soc.* **115**, 5212 (1993).
25. A. J. Schell-Sorokin, F. Mehran, G. R. Eaton, S. S. Eaton, A. Viehbeck, T. R. O'Toole, and C. A. Brown, *Chem. Phys. Lett.* **195**, 225 (1992).
26. H. Huang and S. Yang, *J. Organomet. Chem.* **599**, 42 (2000).
27. N. C. Maliszewskyj, P. A. Heiney, D. R. Jones, R. M. Strongin, M. A. Cichy, and A. B. Smith III, *Langmuir* **9**, 1439 (1993).

28. E. D. Mishina, T. V. Misuryaev, A. A. Nikulin, V. R. Novak, Th. Rasing, and O. A. Aktsipetrov, *J. Opt. Soc. Am.* **B16**, 1692 (1999).
29. H. Okimoto, R. Kitaura, T. Nakamura, Y. Ito, Y. Kitamura, T. Akachi, D. Ogawa, N. Imazu, Y. Kato, Y. Asada, T. Sugai, H. Osawa, T. Matsushita, T. Muro, and H. Shinohara, *J. Phys. Chem.* **C112**, 6103 (2008).
30. D. M. Poirier, *Appl. Phys. Lett.* **64**, 1356 (1994).
31. G. K. Wertheim and D. N. Buchanan, *Solid State Commun* **88**, 97 (1993).
32. T. R. Ohno, Y. Chen, S. E. Harvey, G. H. Kroll, P. J. Benning, J. H. Weaver, L. P. F. Chibante, and R. E. Smalley, *Phys. Rev.* **B47**, 2389 (1993).
33. A. Devenyi, R. Manaila, A. Belu-Marian, D. Macovei, M. Manciu, E. M. Popescu, M. Tanase, D. Fratiloiu, N. D. Mihai, P. B. Barna, J. Labar, G. Safran, A. Kovacs, and T. Braun, *Thin Solid Films* **335**, 258 (1998).
34. P. Reinke and P. Oelhafen, *Phys. Rev.* **B71**, 045420 (2005).
35. H. Liu and P. Reinke, *Surf. Sci.* **601**, 3149 (2007).
36. H. Nagashima, H. Yamaguchi, Y. Kato, Y. Saito, M. Haga, and K. Itoh, *Chem. Lett.*, 2153 (1993).
37. P. Byszewsky, E. Kowalska, and R. J. Diduszko, *Therm. Anal.* **45**, 1205 (1995).
38. A. Talyzin, H. Högberg, and U. Jansson, *Thin Solid Films* **405**, 42 (2002).
39. J. M. Cowley, M. Q. Liu, B. L. Ramakrishna, T. S. Peace, A. K. Wertsching, and M. R. Pena, *Carbon* **32**, 746 (1994).
40. S. G. Stepanian, V. A. Karachevtsev, A. M. Plokhotnichenko, L. Adamowicz, and A. M. Rao, *J. Phys. Chem.* **B110**, 15769 (2006).
41. A. M. Rao, P. Zhou, K.-A. Wang, G. T. Hager, J. M. Holden, Y. Wang, W. T. Lee, X. X. Bi, P. C. Eklund, D. S. Cornett, M. A. Duncan, and I. J. Amster, *Science* **259**, 955 (1993).
42. P. W. Stephens, G. Bortel, G. Faigel, M. Tegze, A. Janossy, S. Peckker, G. Oszlanyi, and L. Forró, *Nature* **370**, 636 (1994).
43. O. Chauvet, G. Oszlanyi, L. Forró, P. W. Stephens, M. Tegze, G. Faigel, and A. Janossy, *Phys. Rev. Lett.* **72**, 2721 (1994).
44. G. M. Bendele, P. W. Stephens, K. Prassides, K. Vavekis, K. Kordatos, and K. Tanigaki, *Phys. Rev. Lett.* **80**, 736 (1998).
45. G. Oszlanyi, G. Baumgartner, G. Faigel, and L. Forró, *Phys. Rev. Lett.* **78**, 4438 (1997).
46. F. Borondics, G. Oszlanyi, G. Faigel, and S. Pekker, *Solid State Commun.* **127**, 311 (2003).

47. S. Margadonna, D. Pontiroli, M. Belli, T. Shiroka, M. Riccò, and M. Brunelli, *J. Am. Chem. Soc.* **126**, 15032 (2004).
48. S. Saito and A. Oshiyama, *Phys. Rev. Lett.* **66**, 2637 (1991).
49. S. D. Evans and A. Ulman, *Chem. Phys. Lett.* **170**, 462 (1990).
50. A. Goldoni, C. Cepek, R. Larciprete, L. Sangaletti, S. Pagliara, L. Floreano, R. Gotter, A. Verdini, A. Morgante, and Y. Luo, *J. Chem. Phys.* **116**, 7685 (2002).
51. L.-F. Yuan, J. Yang, H. Wang, C. Zeng, Q. Li, B. Wang, J. G. Hou, Q. Zhu, and D. M. Chen, *J. Am. Chem. Soc.* **125**, 169 (2003).
52. E. I. Altman and R. J. Colton, *Phys. Rev.* **B48**, 18244 (1993).
53. E. Giudice, E. Magnano, S. Rusponi, C. Boragno, and U. Valbusa, *Surf. Sci.* **405**, L561 (1998).
54. M. K. J. Johansson, A. J. Maxwell, S. M. Gray, P. A. Bruhwiler, D. C. Mancini, L. S. O. Johansson, and N. Mårtensson, *Phys. Rev.* **B54**, 13472 (1996).
55. Y. Wang, R. Yamachika, A. Wachowiak, M. Grobis, K. H. Khoo, D.-H. Lee, S. G. Louie, and M. F. Crommie, *Phys. Rev. Lett.* **99**, 086402 (2007).
56. G. Schull and R. Berndt, *Phys. Rev. Lett.* **99**, 226105 (2007).
57. J. K. Gimzewski, S. Modesti, and R. R. Schlittler, *Phys. Rev. Lett.* **72**, 1036 (1994).
58. M. Grobis, X. Lu, and M. F. Crommie, *Phys. Rev.* **B66**, 161408 (2002).
59. R. Felici, M. Pedio, F. Borgatti, S. Iannotta, M. Capozi, G. Ciullo, and A. Stierle, *Nat. Mater.* **4**, 688 (2005).
60. H. I. Li, K. Pussi, K. J. Hanna, L.-L. Wang, D. D. Johnson, H.-P. Cheng, H. Shin, S. Curtarolo, W. Moritz, J. A. Smerdon, R. McGrath, and R. D. Diehl, *Phys. Rev. Lett.* **103**, 056101 (2009).
61. A. Wachowiak, R. Yamachika, K. H. Khoo, Y. Wang, M. Grobis, D.-H. Lee, S. G. Louie, and M. F. Crommie, *Science* **310**, 468 (2005).
62. W. L. Yang, V. Brouet, X. J. Zhou, H. J. Choi, S. G. Louie, M. L. Cohen, S. A. Kellar, P. V. Bogdanov, A. Lanzara, A. Goldoni, F. Parmigiani, Z. Hussain, and Z.-X. Shen, *Science* **300**, 303 (2003).
63. V. Brouet, W. L. Yang, X. J. Zhou, H. J. Choi, S. G. Louie, M. L. Cohen, A. Goldoni, F. Parmigiani, Z. Hussain, and Z.-X. Shen, *Phys. Rev. Lett.* **93**, 197601 (2004).
64. J.-M. Themlin, S. Bouzidi, F. Coletti, J.-M. Debever, G. Gensterblum, L.-M. Yu, J.-J. Pireaux, and P. A. Thiry, *Phys. Rev.* **B46**, 15602 (1992).
65. K.-D. Tsuei, J.-Y. Yuh, C.-T. Tzeng, R.-Y. Chu, S.-C. Chung, and K.-L. Tsang, *Phys. Rev.* **B56**, 15412 (1997).

66. G. Dutton, J. Pu, D. G. Truhlar, and X.-Y. Zhu, *J. Chem. Phys.* **118**, 4337 (2003).

67. K.-D. Tsuei and P. D. Johnson, *Solid State Commun.* **101**, 337 (1997).

68. G. Dutton and X.-Y. Zhu, *J. Phys. Chem.* **B106**, 5975 (2002).

69. P. Rudolf, M. S. Golden, and P. A. Brühwiler, *J. Electron Spectros. Relat. Phenomena* **100**, 409 (1999).

70. P. J. Benning, C. G. Olson, and D. W. Lynch, *Phys. Rev.* **B50**, 11239 (1994).

71. G. Gensterblum, J.-J. Pireaux, P. A. Thiry, R. Caudano, T. Buslaps, R. L. Johnson, G. Le Lay, V. Aristov, R. Günther, A. Taleb-Ibrahimi, G. Indlekofer, and Y. Petroff, *Phys. Rev.* **B48**, 14756 (1993).

72. P. A. Brühwiler, A. J. Maxwell, A. Nilsson, N. Mårtensson, and O. Gunnarsson, *Phys. Rev.* **B48**, 18296 (1993).

73. C. T. Tzeng, K. D. Tsuei, H. M. Cheng, and R. Y. Chu, *J. Phys. Condens. Matter* **19**, 176009 (2007).

74. V. P. Antropov, O. Gunnarsson, and O. Jepsen, *Phys. Rev.* **B46**, 13647 (1992).

75. C. Cepek, A. Goldoni, S. Modesti, F. Negri, G. Orlandi, and F. Zerbetto, *Chem. Phys. Lett.* **250**, 537 (1996).

76. D. L. Lichtenberger, K. W. Nebesny, C. D. Ray, D. R. Huffman, and L. D. Lamb, *Chem. Phys. Lett.* **176**, 203 (1991).

77. A.-M. Janner, H. T. Jonkman, and G. A. Sawatzky, *Phys. Rev.* **B63**, 085111 (2001).

78. K. Kuhnke, M. Epple, and K. Kern, *Chem. Phys. Lett.* **294**, 241 (1998).

79. Y. Wang, J. M. Holden, A. M. Rao, P. C. Eklund, U. D. Venkateswaran, D. Eastwood, R. L. Lidberg, G. Dresselhaus, and M. S. Dresselhaus, *Phys. Rev.* **B51**, 4547 (1995).

80. J. P. Long, S. J. Chase, and M. N. Kabler, *Phys. Rev.* **B64**, 205415 (2001).

81. F. Negri, G. Orlandi, and F. Zerbetto, *Chem. Phys. Lett.* **144**, 31 (1988).

82. B. Koopmans, A.-M. Janner, H. T. Jonkman, G. A. Sawatzky, and F. van der Woude, *Phys. Rev. Lett.* **71**, 3569 (1993).

83. R. Eder, A.-M. Janner, and G. A. Sawatzky, *Phys. Rev.* **B53**, 12786 (1996).

84. D. Wilk, D. Johannsmann, C. Stanners, and Y. R. Shen, *Phys. Rev.* **B51**, 10057 (1995).

85. X.-Y. Zhu, G. Dutton, D. P. Quinn, C. D. Lindstrom, N. E. Schultz, and D. G. Truhlar, *Phys. Rev.* **B74**, 241401 (2006).

86. R. Tycko, G. Dabbagh, R. M. Fleming, R. C. Haddon, A. V. Makhija, and S. M. Zahurak, *Phys. Rev. Lett.* **67**, 1886 (1991).

87. R. M. Fleming, A. R. Kortan, B. Hessen, T. Siegrist, F. A. Thiel, P. Marsh, R. C. Haddon, R. Tycko, G. Dabbagh, M. L. Kaplan, and A. M. Mujsce, *Phys. Rev.* **B44**, 888 (1991).
88. W. I. F. David, R. M. Ibberson, J. C. Matthewman, K. Prassides, T. J. S. Dennis, J. P. Hare, H. W. Kroto, R. Taylor, and D. R. M. Walton, *Nature* **353**, 147 (1991).
89. A. Goldoni, C. Cepek, and S. Modesti, *Phys. Rev.* **B54**, 2890 (1996).
90. C. Laforge, D. Passerone, A. B. Harris, P. Lambin, and E. Tosatti, *Phys. Rev. Lett.* **87**, 085503 (2001).
91. H. Wang, C. Zeng, B. Wang, J. G. Hou, Q. Li, and J. Yang, *Phys. Rev.* **B63**, 085417 (2001).
92. A. Goldoni, C. Cepek, R. Larciprete, L. Sangaletti, S. Pagliara, G. Paolucci, and M. Sancrotti, *Phys. Rev. Lett.* **88**, 196102 (2002).
93. P. J. Benning, F. Stepniak, and J. H. Weaver, *Phys. Rev.* **B48**, 9086 (1993).
94. A.-M. Janner, R. Eder, B. Koopmans, H. T. Jonkman, and G. A. Sawatzky, *Phys. Rev.* **B52**, 17158 (1995).
95. E. A. Katz, D. Faiman, K. Iakoubovskii, A. Isakina, K. A. Yagotintsev, M. A. Strzhemechny, and I. Balberg, *J. Appl. Phys.* **93**, 3401 (2003).
96. A. Hassanien, J. Gasperic, J. Demsar, I. Musevic, and D. Mihailovic, *Appl. Phys. Lett.* **70**, 417 (1997).
97. K.-H. Han, D. Spemann, R. Höhne, A. Setzer, T. Makarova, P. Esquinazi, and T. Butz, *Carbon* **41**, 785 (2003).
98. M. S. Golden, M. Knupfer, J. Fink, J. F. Armbruster, T. R. Cummins, H. A. Romberg, M. Roth, M. Sing, M. Schmidt, and E. Sohmen, *J. Phys. Condens. Matter* **7**, 8219 (1995).
99. A. Goldoni, L. Petaccia, G. Zampieri, S. Lizzit, C. Cepek, E. Gayone, J. Wells, and Ph. Hofmann, *Phys. Rev.* **B74**, 045414 (2006).
100. S. K. Watson, K. Allen, D. W. Denlinger, and F. Hellman, *Phys. Rev.* **B55**, 3866 (1997).
101. R. Hesper, L. H. Tjeng, A. Heeres, and G. A. Sawatzky, *Phys. Rev.* **B62**, 16046 (2000).
102. G. Kochanski, A. Hebard, R. Haddon, and A. Fiory, *Science* **255**, 184 (1992).
103. M. S. Dresselhaus, G. Dresselhaus, and P. C. Eklund, in *Science of Fullerenes and Carbon Nanotubes*, Academic, San Diego, 1996.
104. O. Gunnarsson, *Rev. Mod. Phys.* **69**, 575 (1997).
105. K. Prassides, *Curr. Opin. Solid State Mater. Sci.* **2**, 433 (1997).

106. M. J. Rosseinsky, *Chem. Mater.* **10**, 2665 (1998).

107. Y. Wang, R. Yamachika, A. Wachowiak, M. Grobis, and M. F. Crommie, *Nat. Mater.* **7**, 194 (2008).

108. T. T. M. Palstra, R. C. Haddon, A. F. Hebard, and J. Zaanen, *Phys. Rev. Lett.* **68**, 1054 (1992).

109. S. Rogge, A.W. Dunn, T. Melin, C. Dekker, and L.J. Geerligs, *Carbon* **38**, 1647 (2000).

110. R. Tycko, G. Dabbagh, D. W. Murphy, Q. Zhu, and J. E. Fischer, *Phys. Rev.* **B48**, 9097 (1993).

111. M. Kosaka, K. Tanigaki, T. Tanaka, T. Atake, A. Lappas, and K. Prassides, *Phys. Rev.* **B51**, 12018 (1995).

112. G. Oszlányi, G. Bortel, G. Faigel, M. Tegze, L. Gránásy, S. Pekker, P. W. Stephens, G. Bendele, R. Dinnebier, G. Mihály, A. Jánossy, O. Chauvet, and L. Forró, *Phys. Rev.* **B51**, 12228 (1995).

113. Q. Zhu, D. E. Cox, and J. E. Fischer, *Phys. Rev.* **B51**, 3966 (1995).

114. T. Käämbre, J. Schiessling, L. Kjeldgaard, L. Qian, I. Marenne, J. N. O'Shea, J. Schnadt, D. Nordlund, C. J. Glover, J.-E. Rubensson, P. Rudolf, N. Mårtensson, J. Nordgren, and P. A. Brühwiler, *Phys. Rev.* **B75**, 195432 (2007).

115. R. Macovez, P. Rudolf, I. Marenne, L. Kjeldgaard, P. A. Brühwiler, T. Pichler, P. Vilmercati, R. Larciprete, L. Petaccia, G. Bertoni, and A. Goldoni, *Phys. Rev.* **B75**, 195424 (2007).

116. R. Macovez, A. Goldoni, L. Petaccia, I. Marenne, P. A. Brühwiler, and P. Rudolf, *Phys. Rev. Lett.* **101**, 236403 (2008).

117. T. Pichler, M. Knupfer, M. S. Golden, S. Haffner, R. Friedlein, J. Fink, W. Andreoni, A. Curioni, M. Keshavarez, K. C. Bellavia-Lund, A. Sastre, J. C. Hummelen, and F. Wudl, *Phys. Rev. Lett.* **78**, 4249 (1997).

118. J. G. Fripiat, A. A. Lucas, J. M. Andre, and E. G. Derouane, *Chem. Phys.* **21**, 101 (1977).

119. V. Brouet, H. Alloul, T.-N. Le, S. Garaj, and L. Forró, *Phys. Rev. Lett.* **86**, 4680 (2001).

120. V. Brouet, H. Alloul, S. Garaj, and L. Forró, *Phys. Rev.* **B66**, 155122 (2002).

121. V. Brouet, H. Alloul, S. Garaj, and L. Forró, *Phys. Rev.* **B66**, 155124 (2002).

122. T. Chida, S. Suzuki, and K. Nakao, *J. Physical Soc. Japan* **70**, 317 (2001).

123. J. E. Han, O. Gunnarsson, and V. H. Crespi, *Phys. Rev. Lett.* **90**, 167006 (2003).
124. M. Capone, M. Fabrizio, C. Castellani, and E. Tosatti, *Science* **296**, 2634 (2002).
125. F. Bommeli, L. Degiorgi, P. Wachter, Ö. Legeza, A. Jánossy, G. Oszlányi, O. Chauvet, and L. Forró, *Phys. Rev.* **B51**, 14794 (1995).
126. K. Mizoguchi, A. Sasano, H. Sakamoto, M. Kosaka, K. Tanigaki, T. Tanaka, and T. Atake, *Synth. Met.* **103**, 2395 (1999).
127. R. Macovez, M. R. C. Hunt, J. Shan, A. Goldoni, T. Pichler, M. Pedio, C. Castellarin-Cudia, J. Schiessling, L. Venema, and P. Rudolf, *New J. Phys.* **11**, 023035 (2009).
128. M. Riccò, T. Shiroka, M. Belli, D. Pontiroli, M. Pagliari, G. Ruani, D. Palles, S. Margadonna, and M. Tomaselli, *Phys. Rev.* **B72**, 155437 (2005).
129. G. Klupp, P. Matus, D. Quintavalle, L. F. Kiss, É. Kováts, N. M. Nemes, K. Kamarás, S. Pekker, and A. Jánossy, *Phys. Rev.* **B74**, 195402 (2006).
130. J. Schiessling, I. Marenne, L. Kjeldgaard, and P. Rudolf, *Surf. Sci.* **601**, 3933 (2007).
131. R. A. Dilanyan, S. S. Khasanov, S. I. Bredikhin, A. F. Gurov, V. V. Kveder, Y. A. Osip'yan, and A. I. Shalynin, *JETP* **93**, 1239 (2001).
132. Y. Chen, F. Stepniak, J. H. Weaver, L. P. F. Chibante, and R. E. Smalley, *Phys. Rev.* **B45**, 8845 (1992).
133. R. P. Gupta and M. Gupta, *Physica* **C219**, 21 (1994).
134. M. Riccò, M. Belli, M. Mazzani, D. Pontiroli, D. Quintavalle, A. Janossy, and G. Csanyi, *Phys. Rev. Lett.* **102**, 145901 (2009).
135. L. Cristofolini, M. Riccò, and R. De Renzi, *Phys. Rev.* **B59**, 8343 (1999).
136. R. Macovez, R. Savage, L. Venema, J. Schiessling, K. Kamaras, and P. Rudolf, *J. Chem. Phys.* **C112**, 2988 (2008).
137. Y. Chen, D. M. Poirier, M. B. Jost, C. Gu, T. R. Ohno, J. L. Martins, J. H. Weaver, L. P. F. Chibante, and R. E. Smalley, *Phys. Rev.* **B46**, 7961 (1992).
138. H. Romberg, M. Roth, and J. Fink, *Phys. Rev.* **B49**, 1427 (1994).
139. Th. Schedel-Niedrig, M. C. Böhm, H. Werner, J. Schulte, and R. Schlögl, *Phys. Rev.* **B55**, 13542 (1997).
140. C. M. Brown, S. Taga, B. Gogia, K. Kordatos, S. Margadonna, K. Prassides, Y. Iwasa, K. Tanigaki, A. N. Fitch, and P. Pattison, *Phys. Rev. Lett.* **83**, 2258 (1999).
141. A. R. Kortan, N. Kopylov, E. Ozdas, A. P. Ramirez, R. M. Fleming, and R. C. Haddon, *Chem. Phys. Lett.* **223**, 501 (1994).

142. A. R. Kortan, N. Kopylov, S. Glarum, E. M. Gyorgy, A. P. Ramirez, R. M. Fleming, F. A. Thiel, P. L. Trevor, and R. C. Haddon, *Nature* **355**, 529 (1992).

143. Y. Z. Li, J. C. Patrin, M. Chander, J. H. Weaver, L. P. F. Chibante, and R. E. Smalley, *Phys. Rev.* **B46**, 12914 (1992).

144. D. Quintavalle, F. Borondics, G. Klupp, A. Baserga, F. Simon, A. Jánossy, K. Kamarás, and S. Pekker, *Phys. Rev.* **77**, 155431 (2008).

145. N. Kojima, M. Natori, H. Suzuki, and M. Yamaguchi, *J. Phys. Conf. Ser.* **159**, 012018 (2009).

146. J. Arvanitidis, K. Papagelis, S. Margadonna, K. Prassides, and A. N. Fitch, *Nature* **425**, 599 (2003).

147. S. Margadonna, J. Arvanitidis, K. Papagelis, and K. Prassides, *Chem. Mater.* **17**, 4474–4478 (2005).

148. K. Ishii, A. Fujiwara, H. Suematsu, and Y. Kubozono, *Phys. Rev.* **B65**, 134431 (2002).

149. Y. Ksari-Habiles, D. Claves, G. Chouteau, Ph. Touzain, Cl. Jeandey, J. L. Oddou, and A. Stepanov, *J. Phys. Chem. Solids* **58**, 1771 (1997).

150. P. H. Citrin, E. Özdas, S. Schuppler, A. R. Kortan, and K. B. Lyons, *Phys. Rev.* **B56**, 5213 (1997).

151. S. He, M. Nakatake, M. Arita, X. Cui, and S. Qiao, *Appl. Phys. Lett.* **91**, 143103 (2007).

152. M. Akada, T. Hirai, J. Takeuchi, T. Yamamoto, R. Kumashiro, and K. Tanigaki, *Phys. Rev.* **B73**, 094509 (2006).

153. X.-X. Wang, H.-N. Li, and Y.-B. Xu, *Solid State Commun.* **147**, 436 (2008).

154. I. Margiolaki, S. Margadonna, K. Prassides, T. Hansen, K. Ishii, and H. Suematsu, *J. Am. Chem. Soc.* **124**, 11288 (2002).

155. K. Ishii, A. Fujiwara, H. Suematsu, and Y. Kubozono, *Phys. Rev.* **B65**, 134431 (2002).

156. D. Claves, Y. Ksari-Habiles, G. Chouteau, and Ph. Touzain, *Solid State Commun.* **106**, 431 (1998).

157. Y. Ksari-Habiles, D. Claves, G. Chouteau, Ph. Touzain, Cl. Jeandey, J. L. Oddou, and A. Stepanov, *J. Phys. Chem. Solids* **58**, 1771 (1997).

158. A. Vértes, Z. Klencsár, E. Kuzmann, L. Forró, G. Oszlányi, and S. Pekker, *J. Phys. Chem. Solids* **61**, 2013 (2000).

159. T. Shiroka, M. Riccò, O. Ligabue, F. Bolzoni, G. Concas, and G. Spano, *J. Magn. Magn. Mater.* **272–276**, 544 (2004).

160. H. Shinohara, *Rep. Prog. Phys.* **63**, 843 (2000).

161. C. J. Nuttall, Y. Hayashi, K. Yamazaki, T. Mitani, and Y. Iwasa, *Adv. Mater.* **14**, 293 (2002).
162. T. Watanuki, A. Fujiwara, K. Ishii, Y. Matuoka, H. Suematsu, K. Ohwada, H. Nakao, Y. Fujii, T. Kodama, K. Kikuchi, and Y. Achiba, *Mol. Cryst. Liq. Cryst.* **340**, 639 (2000).
163. H. Shinohara, M. Inakuma, M. Kishida, S. Yamazaki, T. Hashizume, and T. Sakurai, *J. Phys. Chem.* **99**, 13769 (1995).
164. T. Sakurai, X.-D. Wang, Q. K. Xue, Y. Hasegawa, T. Hashizume, and H. Shinohara, *Prog. Surf. Sci.* **51**, 263 (1996).
165. K. Furukawa, S. Okubo, H. Kato, H. Shinohara, and T. Kato, *J. Phys. Chem.* **A107**, 10933 (2003).
166. Y. Ito, W. Fujita, T. Okazaki, T. Sugai, K. Awaga, E. Nishibori, M. Takata, M. Sakata, and H. Shinohara, *Chem. Phys. Chem.* **8**, 1019 (2007).
167. T. Inoue, T. Tomiyama, T. Sugai, T. Okazaki, T. Suematsu, N. Fujii, H. Utsumi, K. Nojima, and H. J. Shinohara, *J. Phys. Chem* **B108**, 7573 (2004).
168. E. E. B. Campbell, M. Fanti, I. V. Hertel, R. Mitzner, and F. Zerbetto, *Chem. Phys. Lett.* **288**, 131 (1998).
169. C. Stanciu, R. Ehlich, and I. V. Hertel, *Appl. Phys.* **A79**, 515 (2004).
170. M. Takata, E. Nishibori, M. Sakata, M. Inakuma, E. Yamamoto, and H. Shinohara, *Phys. Rev. Lett.* **83**, 2214 (1999).
171. C. R. Wang, T. Kai, T. Tomiyama, T. Yoshida, Y. Kobayashi, E. Nishibori, M. Takata, and H. Shinohara, *Nature* **408**, 426 (2000).
172. T. Pichler, Z. Hu, C. Grazioli, S. Legner, M. Knupfer, M. S. Golden, J. Fink, F. M. F. de Groot, M. R. C. Hunt, P. Rudolf, R. Follath, C. Jung, L. Kjeldgaard, P. A. Brühwiler, M. Inakuma, and H. Shinohara, *Phys. Rev.* **B62**, 13196 (2000).
173. H. Shiozawa, H. Rauf, T. Pichler, D. Grimm, X. Liu, M. Knupfer, M. Kalbac, S. Yang, L. Dunsch, B. Büchner, and D. Batchelor, *Phys. Rev.* **B72**, 195409 (2005).
174. B. Kessler, A. Bringer, S. Cramm, C. Schlebusch, W. Eberhardt, S. Suzuki, Y. Achiba, F. Esch, M. Barnaba, and D. Cocco, *Phys. Rev. Lett.* 79, 2289 (1997).
175. X. Liu, M. Krause, J. Wong, T. Pichler, L. Dunsch, and M. Knupfer, *Phys. Rev.* **B72**, 085407 (2005).
176. K. Schulte, L. Wang, P. J. Moriarty, J. Purton, S. Patel, H. Shinohara, M. Kanai, and T. J. S. Dennis, *Phys. Rev.* **B71**, 115437 (2005).

177. C. Ton-That, A. G. Shard, S. Egger, A. Taninaka, H. Shinohara, and M. E. Welland, *Surf. Sci.* **522**, L15 (2003).

178. X.-D. Wang, Q. K. Xue, T. Hashizume, H. Shinohara, Y. Nishina, and T. Sakurai, *Phys. Rev.* **B48**, 15492 (1993).

179. K. Wang, J. Zhao, S. Yang, L. Chen, Q. Li, B. Wang, S. Yang, J. Yang, J. G. Hou, and Q. Zhu, *Phys. Rev. Lett.* **91**, 185504 (2003).

180. E. Nishibori, M. Takata, M. Sakata, H. Tanaka, M. Hasegawa, and H. Shinohara, *Chem. Phys. Lett.* **330**, 497 (2000).

181. R. Klingeler, G. Kann, I. Wirth, S. Eisebitt, P. S. Bechthold, M. Neeb, and W. Eberhardt, *J. Chem. Phys.* **115**, 7215 (2001).

182. V. N. Popok, I. I. Azarko, A. V. Gromov, M. Jönsson, A. Lassesson, and E. E. B. Campbell, *Solid State Commun.* 133, 499–503 (2005).

183. Y. Kubozono, Y. Takabayashi, K. Shibata, T. Kanbara, S. Fujiki, S. Kashino, A. Fujiwara, and S. Emura, *Phys. Rev. B* 67, 115410 (2003).

184. T. Nishikawa, S.-I. Kobayashi, T. Nakanowatari, T. Mitani, T. Shimoda, Y. Kubozono, G. Yamamoto, H. Ishii, M. Niwano, and Y. Iwasa, *J. Appl. Phys.* **97**, 104509 (2005).

185. T. Kanbara, K. Shibata, S. Fujiki,Y. Kubozono, S. Kashino, T. Urisu, M. Sakai, A. Fujiwara, R. Kumashiro, and K. Tanigaki, *Chem. Phys. Lett.* **379**, 223 (2003).

186. S. Kobayashi, S. Mori, S. Iida, T. Takenobu, Y. Taguchi, A. Fujiwara, A. Taninaka, H. Shinohara, and Y. Iwasa, *J. Am. Chem. Soc.* **125**, 8116 (2003).

187. N. Hiroshiba, K. Tanigaki, R. Kumashiro, H. Ohashi, T. Wakahara, and T. Akasaka, *Chem. Phys. Lett.* **400**, 235 (2004).

188. R. B. Ross, C. M. Cardona, D. M. Guldi, S. Gayathri Sankaranarayanan, M. O. Reese, N. Kopidakis, J. Peet, B. Walker, G. C. Bazan, E. Van Keuren, B. C. Holloway, and M. Drees, *Nat. Mater.* **8**, 208 (2009).

189. L. J. A. Koster, V. D. Mihailetchi, and P. W. M. Blom, *Appl. Phys. Lett.* **88**, 0935111 (2006).

190. S. E. Shaheen, C. J. Brabec, N. S. Sariciftci, F. Padinger, T. Fromherz, and J. C. Hummelen, *Appl. Phys. Lett.* **78**, 841 (2001).

Chapter 3

High-Resolution Transmission Electron Microscopy Imaging of Carbon Nanostructures

Kazu Suenaga, Yuta Sato, Zheng Liu, Masanori Koshino, and Chuanhong Jin

National Institute of Advanced Industrial Science and Technology (AIST)
AIST, Central 5, Tsukuba 305-8565 Japan
suenaga-kazu@aist.go.jp

Here we show how a high-resolution transmission electron microscopy can be applied to characterize the carbon nanostructures. Direct imaging of the hexagonal network of carbon nanotube enables us to determine the chiral index and to visualize the topological defects, such as pentagons and heptagons. Individual molecular imaging has also become possible, and atomic structure of fullerene molecules (C_{60} and C_{80}) has been successfully identified at a single-molecular basis. Some recent progress for *in situ* observation of the carbon nanotube/fullerene growth and the defect dynamics is also presented.

Advances in Carbon Nanomaterials: Science and Applications
Edited by Nikos Tagmatarchis

ISBN 978-981-426-78-78 (Hardcover), 978-981-426-78-85 (eBook)
www.panstanford.com

3.1 Introduction

The diversified properties of carbon nanostructures (nanotubes, graphenes, fullerenes, and their derivatives) are related to their polymorphic arrangement of carbon atoms. Therefore, the direct observation of carbon network is of great consequence in both scientific and technological viewpoints to predict the physical and chemical properties of carbon nanostructures. To identify the local configuration of pentagons and hexagons in carbon nanostructures, an electron microscope with higher spatial resolution and higher sensitivity is definitively required. Since the spatial resolution of the conventional transmission electron microscope (TEM) is limited by the spherical aberration coefficient (C_S) of its objective lens and the wave length (λ) of the incident electron beam, the C_s must be minimized to achieve the best performance because the reduction of the λ is detrimental to the carbon-based materials due to the higher knock-on probability. Lowering accelerating voltage is also beneficial to achieve the high sensitivity necessary to visualize individual carbon atoms. The spatial resolution of 0.14 nm (a typical C–C bond length) obtained at a moderate accelerating voltage can offer us a great advantage because we can realize the visualization of carbon atomic arrangement without massive electron irradiation damage. We will show here some examples for the atomic-level characterizations of carbon nanostructures by high-resolution transmission electron microscope (HR-TEM).

3.2 Experimental

A HR-TEM (JEOL-2010F) equipped with a post-specimen aberration corrector (CEOS) was operated at a moderate accelerating voltage of 120 kV (Fig. 3.1). The C_s was set to 0.5–10 μm in this work. The HR-TEM images were obtained under a slightly under-focus condition ($\Delta f = -2$ to -7 nm) where a point resolution better than 0.14 nm was achieved at 120 kV. A CCD camera (Gatan 894) was used for the digital recording of the HR-TEM images. A typical exposure time is 0.5–1.0 s for each frame, and some of the frames are superimposed after drift correction to enhance the contrast if necessary. In a

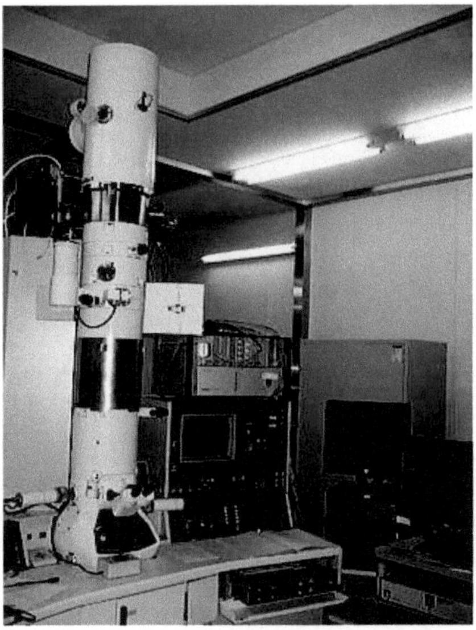

Figure 3.1. HR-TEM (JEM-2010F) equipped with an aberration corrector (CEOS) and a piezo-driven stage (Nanofactory) operated at 120 kV. The spatial resolution is better than 0.14 nm (typical C–C bond length). See also Color Insert.

typical high-dose condition (~100,000 electrons/nm^2), the contrast of single carbon atoms can be well isolated with a signal-to-noise (SN) ratio >3, which guarantees us a confidence level of 80% for single carbon atom detection. A piezo driven stage with mobile electrode (Nanofactory) was used for *in situ* experiment of the carbon nanostructure growth.

3.3 Visualization of Atomic Defects in Carbon Nanotubes

The physical properties of carbon nanotube are strongly dependent on its chirality as well as atomic defects. The chiral index (*n*,*m*) for individual single-walled carbon nanotubes (SWNTs) can be determined by either electron diffraction or HR-TEM.[1,2] Especially the discrimination of metallic and semiconducting SWNTs is quite important.[3] A great advantage of HR-TEM lies in its capability to determine the atomic defects as well.[4] Such defect structures of

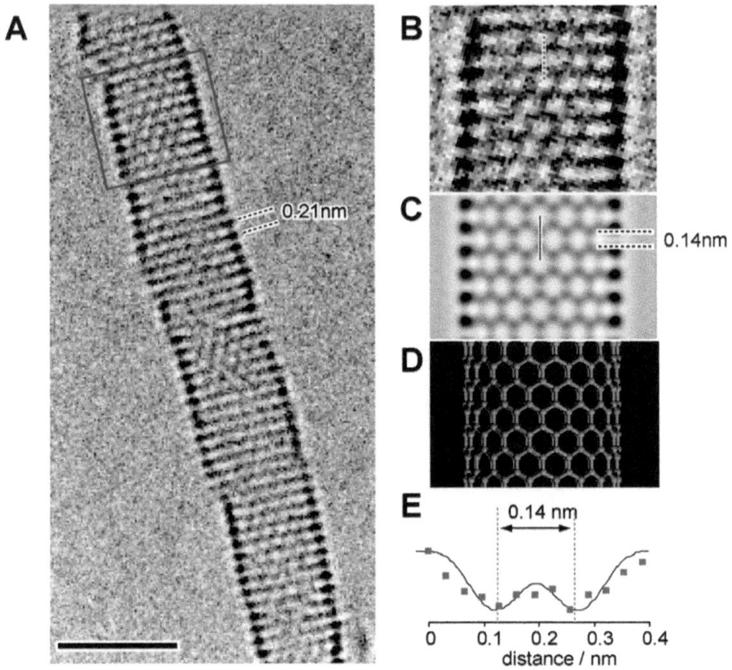

Figure 3.2. (a) HR-TEM image of SWNT taken at 120 kV with a CEOS image corrector. The chiral index was assigned as (18, 0). (b) An enlarged image from the rectangle in (a). (c) Simulated image for (18, 0) SWNT and its atomic model (d). (e) Contrast profiles from indicated lines in (b) and (c), showing a typical C–C bond length (~0.14 nm) can be clearly resolved. Scale bar = 2 nm.

carbon materials have long been of great scientific and technological importance especially for nuclear research. Although single vacancies, topological defects, interstitials, and their combination were theoretically predicted, no experimental evidence for these defects can be provided until they are directly identified. To visualize faithfully the atomic structures of carbon nanotubes, a high spatial resolution (~0.14 nm) is indispensable to resolve a typical C–C bond in the carbon networks. Figure 3.2a shows a HR-TEM image of a SWNT. One can easily see the zig-zag chains contrast (0.21 nm apart) all over the SWNT. Especially in the region of a red rectangle the hexagonal structures of the carbon network are clearly recognized

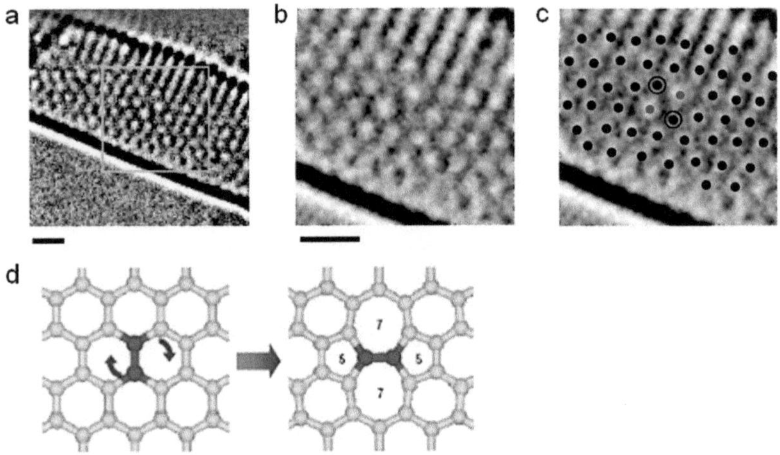

Figure 3.3. (a)–(c) A pentagon–heptagon pair defect found on a SWNT after a heat treatment at 2000 K. The defect is a proof of the Stone-Wales transformation due to the C–C bond rotation (d). Scale bar = 0.5 nm.

(Fig. 3.2b). Note that the hexagonal structure is only partly visible because the local distortion and/or inclination of the tube to the incident electron beam can largely critically affect the imaging conditions. By comparing the HR-TEM images with the image simulation and the structural model (Fig. 3.2c,d), the examined SWNT is proved to have a zig-zag structure with the index of (18,0) and is slightly rotated around the tube axis (~2°). A contrast line profile along the two neighboring carbon atoms is shown in the Fig. 3.2e. The red dotted curve obtained from the line profile (experiment) in Fig. 3.2b is fitted with the blue line profile (model) in Fig. 3.2c. Both profiles are identical and clearly show two minima corresponding to the carbon–carbon distance (0.14 nm), proving that the individual carbon atoms in the hexagon network have been faithfully imaged.

Non-hexagonal rings such as pentagons or heptagons can be regarded as topological defects within the carbon network. Especially a C–C bond rotation has been expected by a theory (known as the Stone-Wales transformation) and was supposed to lead to the pentagon–heptagon pair defect. Figure 3.3 shows a HR-TEM image of the pentagon–heptagon pair defect of SWNT after a heat treatment at 2000 K.[5] A fast Fourier transform (FFT) analysis has been

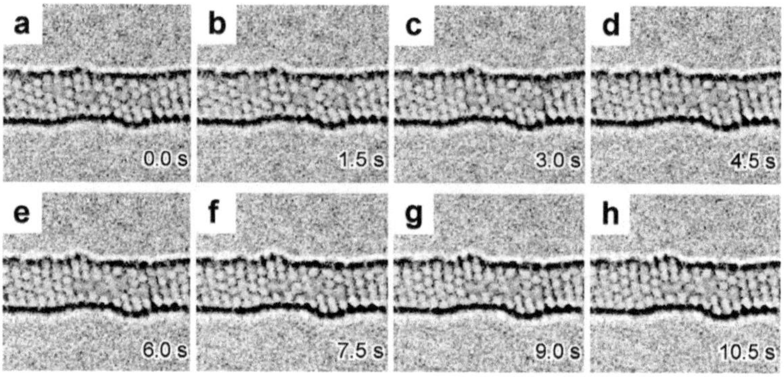

Figure 3.4. A series of HR-TEM images showing the active topological defects. The hexagons are indicated by green whereas the pentagons and heptagons are indicated by blue and red. The regions are heavily deformed. Note that another layer has been eliminated by FFT analysis and may have some interference on the images. See also Color Insert.

performed to eliminate one of the two layers overlapped for the SWNT. Plastic deformation of carbon nanotube indeed relies on the mobility of these topological defects. If any topological defect can migrate along the nanotube, this indeed means that the nanotube exhibits plasticity. The first experimental evidence for the active topological defects has been demonstrated by *in situ* HR-TEM.[5] Figure 3.4 shows a series of HR-TEM images of a SWNT. Here the hexagons are indicated by green, whereas the pentagons and heptagons are indicated by blue and red, respectively. Although the structure on the other layer may have affected these HR-TEM images after the FFT analysis, these topological defects are indeed active and do migrate along the SWNT during the observation. This is the first atomistic proof that SWNT can exhibit an authentic plastic deformation which should rely on the active topological defects.[6]

The other types of atomic defects rather than the topological defects have also been investigated seriously. *In situ* HR-TEM at elevated temperatures has shown the growth and migration of vacancies in carbon networks and given a rough estimation of the activation energy barrier for individual vacancies as ~2.2 eV.[7] Thermal relaxation of the Frenkel-type of defects (interstitial and vacancy pair) between the two layers of double-walled carbon nanotubes has been also investigated *in situ*.[8] The critical temperature for the annihilation of the Frenkel defects was found around 450–500 K, which is very close to the annealing temperature for releasing the Wigner energy at 473 K. Therefore, one can eventually conclude the Frenkel pair defects in graphite as the Wigner source which has been a well-known problem for half a century.

3.4 Imaging of Fullerenes and Their Derivatives

Another important usage of the HR-TEM with a moderate accelerating voltage is to visualize individual molecular structures. Organic molecules are known to suffer the irradiation damage due to the incident electrons and have been believed difficult to be imaged by HR-TEM. A common discussion about the difficulty in molecular imaging by HR-TEM often relies on the extremely small critical dose (typically several hundreds to thousands of electrons per nm^2 for protein specimens), with which any HR-TEM cannot attain an enough SN ratio to isolate the contrast of molecules. Such a discussion is valid for molecular crystal analysis because the critical dose is generally measured by the decrease of electron diffraction intensity. We should note that the major damage procedure in molecular crystal is attributed to the "cross-linking" of the adjacent molecules, which means that a broken bond due to the inelastic scattering will make a new bond to the adjacent molecules. Molecules in crystal will be heavily deformed due to the cross-link, which should lead to the decrease of diffraction intensities.

Damage process of isolated molecules should be completely different from that of molecular crystals. Even if the radical bonds are created due to inelastic scattering, there should be no adjacent molecules nearby to inter-link. The broken bonds can be instantly recovered unless any possibility to make other bonds. Consequently, no massive structural deformation could be observed on the isolated molecule except the knock-on displacements.

To observe the isolated molecules by HR-TEM, the SWNTs have been used as a specimen cell.[9] The inner surface of SWNTs is completely inert and is therefore very much suitable to hold the molecules inside because the broken bonds cannot easily make new bonds with the SWNT. By putting the damage-sensitive molecules inside the SWNTs, we have eventually succeeded to visualize the individual molecules in motion.[10–12] It is well known that the *cis/trans* isomerization of the retinal chromophores triggers biological activity in rhodopsins. Also their conformation change is crucial for animal's vision. The isomerization and conformation changes of single chains of carbon have been imaged for the first time.[12] Figure 3.5 shows an example for HR-TEM imaging of the functionalized fullerenes. The retinal chromophores attached to the C_{60} fullerenes are clearly visualized. The retinal chromophores consist of the conjugated carbon atomic chains ($\ldots -C = C-C = C - \ldots$) and have *cis/trans* isomers. The methyl groups as well as the cyclohexene are visible. Note that we need as much as 100,000 electrons/nm^2 to isolate the contrast of single retinal chromophores.

Isomer assignments of fullerene molecules have also been performed.[13,14] A C_{80} fullerene molecule consists of 80 carbon atoms, consequently 12 pentagons and 30 hexagons close the cage. Here we have chosen the C_{80} molecule with the D_{5d} symmetry among seven isomers. The D_{5d}-C_{80} molecules were encapsulated in SWNTs and observed by HR-TEM. Figure 3.6 shows a series of HR-TEM images in which a C_{80} molecule shows a rotational and translational movement inside the SWNT. At $t = 0$ s, a pair of pentagons are overlapped in projection (colored in orange). In the next frame at $t = 45$ s, the molecule shows the four bright spots corresponding to the pyrene-like tetracyclic components on the both sides (colored in orange); therefore, one of the five mirror planes of the D_{5d} symmetry is projected. In the last frame at $t = 79$ s, a pair of two dark lines appear and are attributed to the zig-zag chains of the anthracene-like tricyclic component (colored in orange). The individual molecules can be monitored as such during the rotational movements in SWNTs, so that the orientation changes can be investigated at each frame. In such a case, structural analysis and isomer identification are more reliably possible for the specific molecule. See ref. 13 for detailed analysis.

One of the disadvantages for the use of SWNT as a specimen cell is that the HR-TEM contrast of SWNT walls often disturbs the molecular images and makes it difficult to analyze an

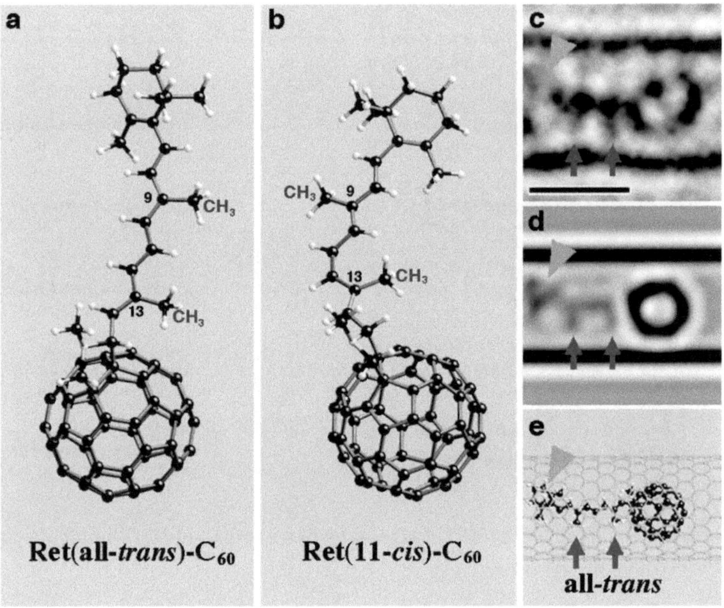

Figure 3.5. Atomic models of the all-trans (a) and 11-cis (b) retinal chromophores attached to C_{60} molecules. The carbon–carbon bonds around atom 11 are shown in red and the two methyl groups are highlighted in blue letters. In the trans form they point in the same direction, whereas in the cis form they do not. (c) An HR-TEM image of a Ret-C_{60} molecule inside a SWNT, showing fine structures that correlate well with a simulation (d) and a best-fit model (e). General agreements of discontinuous contrast, corresponding to the methyl groups (red arrows) and cyclohexene (green arrow), can be found, suggesting that the image in (c) is of the all-trans isomer. Scale bar = 1 nm. See also Color Insert.

individual molecular structure. In such a case one could try to fix the molecules outside the SWNTs so that the contrast of SWNT does not interfere with the molecular images. Figure 3.7 shows the HR-TEM of C_{60} fullerene molecules outside the SWNTs. Each molecule has been fixed to the SWNTs by using a functional group of pyrrolidine (C_{60}-C_3NH_7) as an anchor and does exhibit some intramolecular features. To corroborate the observed HR-TEM images, the image simulations for the C_{60} fullerene molecules were systematically made in more than 30 different orientations considering the I_h symmetry

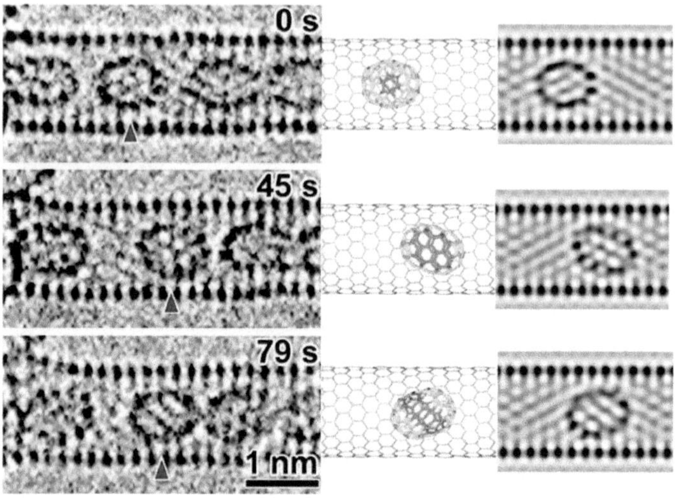

Figure 3.6. A series of HR-TEM images of a same C_{80} (D_{5d}) molecule (indicated by red arrows).

(only 16 types are shown in Fig. 3.8). Comparing the HR-TEM images with the image simulations, the molecular orientation can be reasonably assigned for some of the experimental images. The molecule in Fig. 3.7a shows a six-membered ring contrast inside which is quite close to the simulated image of Fig. 3.8(I) corresponding to the C_{60} molecule aligned parallel to the six-fold symmetry axis. Similarly the image in Fig. 3.7b corresponds to the simulation in Fig. 3.8(II), in which two pentagons are overlapped in projection and thus give rise to a small circle contrast in the middle of the C_{60} fullerene. The image in Fig. 3.7c is closely equivalent to the simulated image in Fig. 3.8(III). It is interesting to note that the molecule observed in Fig. 3.7d shows roughly 10 dark spots around the fullerene cage and therefore may correspond again to the simulation in Fig. 3.8(II) although this molecule exhibits a large deformation. The image in Fig. 3.7e again corresponds to the Fig. 3.8(I) in spite of a slight misorientation. Besides all the above, we were unable to convincingly identify the other three molecules in Fig. 3.7f–h for any orientation in simulated images. It strongly suggests that the observed molecules could have suffered a considerable deformation.[14]

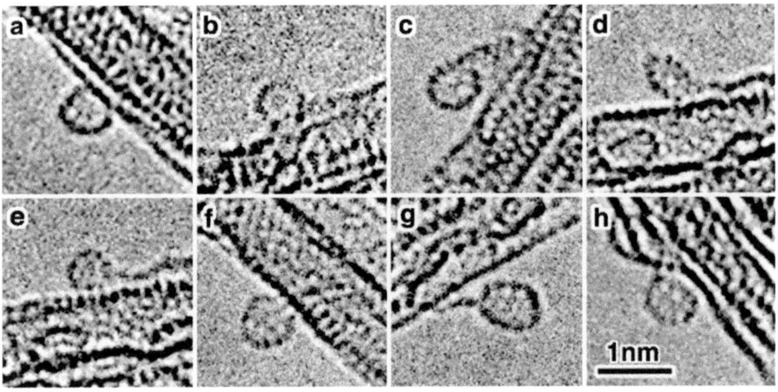

Figure 3.7. (a)–(h) HR-TEM images of fullerene molecules. C_{60}-C_3NH_7 derivatives are attached to the surface of SWNTs. The intra-molecular structures are clearly visible for each fullerene. Some of them suffer a considerable deformation and deviate from the spherical shape ((d) and (g) for example).

3.5 *In Situ* Observation of Nano-Carbon Growth

One of the central topics in the nano-carbon research field has been the growth mechanism of the carbon nanostructures. Most important question is how the extra carbon atoms can be incorporated into the carbon networks during the growth. Do they need an open edge with the dangling bonds to accommodate the carbon atoms? How can the catalytic particles help the carbon atoms to be incorporated into the carbon network? We have started a systematic study by using an *in situ* HR-TEM to answer these fundamental questions.

Jin *et al.* reported a non-catalytic growth of carbon nanotube.[15] In this report, a growing carbon nanotube with the "closed cap" has been directly observed for the first time at high temperatures inside HR-TEM. An asymmetric cap of the growing nanotube (attributed to the accumulated pentagons) has been identified as the growing sites, where the carbon dimers from the vapor can be incorporated into the carbon networks. It has been therefore proved that any open edge with the carbon dangling bonds is not necessary for the nanotubes to grow.

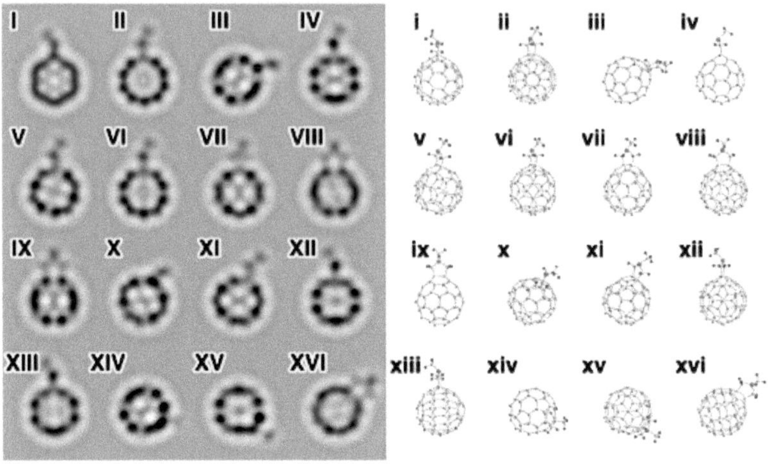

Figure 3.8. Image simulations of C_{60} fullerene derivatives for various orientations (left) to be compared with the Fig. 3.7. Corresponding atomic models are also shown (right). The pyrrolidine type functional groups are attached arbitrarily in the image simulations. The simulated image for the molecule oriented to the six-fold axis (I) fits quite well with the observed images of Fig. 3.7a for example.

In the case of fullerene growth, we have introduced the tungsten (W) particles as a catalyst to promote its enlargement.[16] Figure 3.8 shows a series of HR-TEM images for the W catalyzed enlargement of fullerenes. A W cluster (marked as white arrowhead) suddenly jumped onto a large fullerene ($\sim$0.9 nm in diameter, roughly $C_{84\pm4}$) as shown in Fig. 3.9a. Upon the adsorption of this W cluster, the fullerene immediately started to grow as shown in Fig. 3.9b–e. The W cluster was found to migrate continuously on the fullerene cage and induced some local distortions on the cage. The fullerene cage grew radially (inflation in its diameter), instead of being elongated, confirming that the fullerene energetically prefers to keep a round shape. Formation and annihilation of sharp edges on the fullerene cage were also frequently observed during the growth. After the W cluster detached, the fullerene did not grow any more as shown in Fig. 3.9f. The final diameter of the fullerene reached $\sim$1.1 nm, which can be roughly assigned as $C_{136\pm8}$, corresponding to an average growth speed of about 0.5 atom/s.

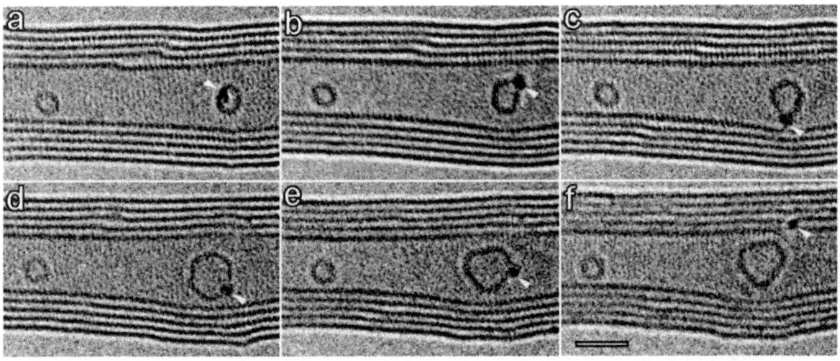

Figure 3.9. *In situ* HR-TEM images of the fullerene growth at high temperatures. The W clusters act as catalyst (indicated by arrows) [16]. Scale bar = 2 nm.

From the experiments shown above we can reasonably derive that a major growth mechanism of fullerene or nanotube should be the carbon atoms incorporation into the adjacent pentagon sites and the re-arrangement of carbon networks afterwards possibly due to the Stone-Wales transformations. However, the open edges with the carbon dangling bonds have also been identified for the nanotube and graphene layers during or after a high-temperature heat treatments.[17,18]

3.6 Summary

Here we have shown how a C_S-corrected HR-TEM at a moderate accelerating voltage (120 kV) can be applied to visualize the carbon nanostructures. Visualization of carbon network is indispensable to correlate directly the atomic structure and the physical properties of carbon nanostructures. The chiral index assignment of individual carbon nanotubes after separation is of great consequence to corroborate the optical property measurements of specific carbon nanotubes.[3] We also emphasize here the importance of atomic defects in carbon nanotubes. They do affect the physical and chemical properties of carbon nanotube and need to be fully investigated before its practical applications.

Acknowledgments

The research presented here has been supported by JST-CREST, NEDO, JST-ERATO, and Grant-in-aid from MEXT.

References

1. J. M. Zuo, I. Vartanyants, M. Gao, R. Zhang, and L. A. Nagahara, *Science* **300**, 1419–1421 (2003).
2. R. R. Meyer *et al., J. Microsc.* **212**, 152–157 (2003).
3. Y. Sato, K. Yanagi, Y. Miyata, K. Suenaga, H. Kataura, and S. Iijima, *Nano Lett.* **8**, 3151–3154 (2008).
4. A. Hashimoto, K. Suenaga, A. Gloter, K. Urita, and S. Iijima, *Nature* **430**, 870–873 (2004).
5. K. Suenaga, H. Wakabayashi, M. Koshino, Y. Sato, K. Urita, and S. Iijima, *Nat. Nanotechnol.* **2**, 358–360 (2007).
6. C. Jin, K. Suenaga, and S. Iijima, *Nat. Nanotechnol.* **3**, 17–21 (2008).
7. C. Jin, K. Suenaga, and S. Iijima, *Nano Lett.* **8**, 1127–1130 (2008).
8. K. Urita, K. Suenaga, T. Sugai, H. Shinohara, and S. Iijima., *Phys. Rev. Lett.* **94**, 155502 (2005).
9. K. Suenaga *et al., Science* **290**, 2280–2282 (2000).
10. Z. Liu, M. Koshino, K. Suenaga, A. Mrzel, H. Kataura, and S. Iijima, *Phys. Rev. Lett.* **96**, 088304 (2006).
11. M. Koshino, T. Tanaka, N. Solin, K. Suenaga, H. Isobe, and E. Nakamura, *Science* **316**, 853 (2007).
12. Z. Liu, K. Yanagi, K. Suenaga, H. Kataura, and S. Iijima, *Nat. Nanotechnol.* **2**, 422–425 (2007).
13. Y. Sato, K. Suenaga, S. Okubo, T. Okazaki, and S. Iijima, *Nano Lett.* **7**, 3704–3708 (2007).
14. Z. Liu, K. Suenaga, and S. Iijima, *J. Am. Chem. Soc.* **129**, 6666–6667 (2007).
15. C. Jin, K. Suenaga, and S. Iijima, *ACS Nano* **2**, 1275–1279 (2008).
16. C. Jin, H. Lan, K. Suenaga, L.-M. Peng, and S. Iijima, *Phys. Rev. Lett.* **101**, 176102 (2008).
17. C. Jin, K. Suenaga, and S. Iijima, *Nano Res* **1**, 434–439 (2008).
18. Z. Liu, K. Suenaga, P. J. F. Harris, and S. Iijima, *Phys. Rev. Lett.* **102**, 015501 (2009).

Chapter 4

Electronic and Optical Properties of Carbon Nanotubes

Christian Kramberger and Thomas Pichler
University of Vienna, Faculty of Physics,
Strudlhofgasse 4, A-1090, Vienna, Austria
christian.kramberger-kaplan@univie.ac.at
thomas.pichler@univie.ac.at; epm.univie.ac.at

4.1 The Electronic Ground State

The electronic properties of matter are of fundamental relevance for the function and behavior of the physical world as we know it. Normal matter is built up of atoms, where literally all mass is focused in the nucleus, but the majority of its interaction with the local environment is mediated by the engulfing cloud of electrons. The mutual electric interaction between the individual atoms, or more precisely between bound electrons, is the key ingredient for accessing the intrinsic physical properties of matter. On a very fundamental level there are throughout physics two different descriptions of bound electrons. Very interestingly, we have to combine, but not mix, these two opposing concepts for the complete description of electrons in carbon nanotubes. The first case is bound

Advances in Carbon Nanomaterials: Science and Applications
Edited by Nikos Tagmatarchis

ISBN 978-981-426-78-78 (Hardcover), 978-981-426-78-85 (eBook)
www.panstanford.com

and localized electrons, as for instance in an isolated atom. There the electron can exist only in discrete states with well-defined quantum numbers. The discrete electronic transitions between this states give rise to the emission and absorption spectra of glowing gases. The second case is bound but delocalized electronic states. These exist within condensed matter, where the electrons can behave as quasi-free particles. Here they are allowed to propagate and possess momenta that correspond to continuous energies [Kittel (1963)]. Still, the connection between energy and momentum is no longer a parabola as in homogeneous free space, but it gets strongly modified by the discrete crystal structure inside matter. An electron in free space has a constant rest mass, which just adds to its kinetic energy, but in a solid there are additional energetic contributions stemming from the interaction of the electron with the lattice. The actual momentum of an electron determines the wavelength of the corresponding electronic wavefunction and thus also the spatial overlap of the electron with the surrounding crystal. The electronic dispersion relation, viz. the electrons energy as a function of their momentum, is called the electronic band structure. Typically it consists of several branches that originate from the different symmetries of the allowed electronic wavefunctions. The material specific shape of the band structure determines the electronic density of states (DOS). The DOS just tells how many electronic states can be there per unit volume with a certain energy, regardless of their actual momentum or their spin state. In an isolated atom the DOS is a discrete set of infinitesimally sharp peaks (δ functions), but in solids the DOS, which is readily derived by taking the inverse slopes of the dispersion relation, is a smooth function. In a two-dimensional sheet or in a one-dimensional wire the DOS is a staircase function or a series of sharp van Hove singularities (VHS). The latter one-dimensional VHS are a fingerprint of truly one-dimensional electronic systems. The general shape of the energy-dependent DOS in one, two, and three dimensions is recapitulated in Fig. 4.1.

The band structure is occupied with electrons up to the Fermi level, which separates the occupied valence band and the empty conduction band. The knowledge of the detailed band structure and the Fermi level in a material means nothing less than knowing the

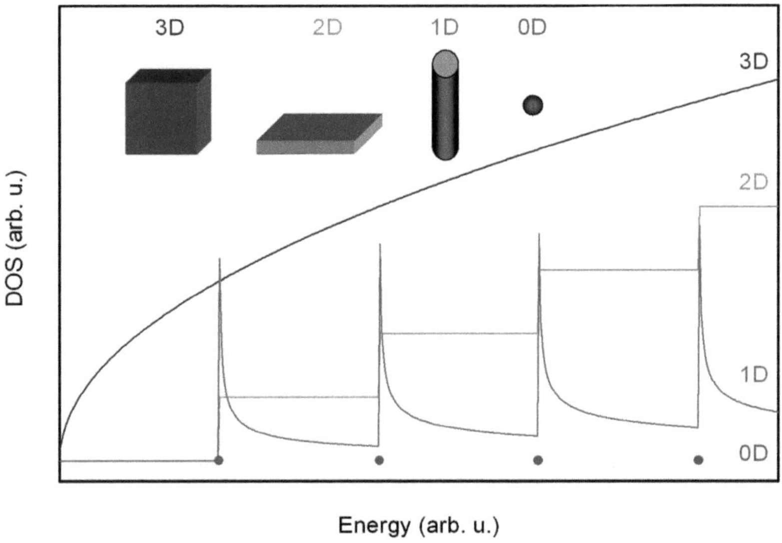

Figure 4.1. Characteristic shape of the electronic density of states (DOS) of a free electron gas in one, two, and three dimensions. In a 0D quantum dot there is only a discrete spectrum. See also Color Insert.

electronic ground state among all possible electronic configurations. The allowed electronic transitions between these configurations comprise the response of the electronic system to any impinging probe from the outer world.

Elucidating the electronic structure of carbon nanotubes will be a solid basis to describe their electronic and optical properties and how these may be experimentally accessed by spectroscopic techniques. Here we will just briefly discuss the physical process behind the spectroscopic techniques and look into their application on SWNTs. A more comprehensive introduction to spectroscopy on solids may be found elsewhere [Kuzmany (1998)].

In the following sections, we will take a look at the electronic structure of an isolated sheet of graphite or graphene in the intuitive tight binding scheme and then, in section 4.1.1, perform the notional roll-up of a graphene ribbon into a cylinder surface. The detailed outcome of this roll-up will crucially depend on the actual geometry of the tube, which fully determines all of its physical properties and is also the basis for the classification of (single-walled) carbon nanotubes, as elucidated in section 4.1.2. Lastly, in section 4.1.3, we will address the quantitative comparison of the tight binding approach with more elaborate *ab initio* calculations.

4.1.1 *From Graphene to Carbon Nanotubes*

Graphene is a single layer of graphite, where the individual carbon atoms are arranged in a flat hexagonal honeycomb structure. The electronic band structure of graphene was for the first time calculated in the tight binding scheme more than six decades ago [Wallace (1947)]. In this allotrope of carbon the atoms are in the flat sp^2 configuration. Each atom has two core electrons that fill the first atomic C1s shell with the spin-up and the spin- down state. The core electrons are well localized at the individual atoms. The remaining four electrons per carbon form three horizontal σ orbitals and a perpendicular π orbital. The σ electrons form the very strong in-plane carbon-carbon bonds that determine the hexagonal geometry. The π electrons form comparably weaker π bonds to the three nearest neighbors. The σ bands cover the energy range from $\sim$7 to $\sim$14 eV, but the π electrons cover the whole low energy range from the Fermi level up to $\sim$9 eV. The π electrons are therefore the only ones in the relevant energy range (several meV) for electrical and thermal transport, and also interaction with infrared to ultraviolet light. The two- dimensional π band structure of graphene is presented in Fig. 4.2.

The peculiarity of the π bands is that the Fermi surface, where the populated valence band and the empty conduction band are just touching, consists only of a pair of inequivalent points at the corners **K** and **K'** of the hexagonal Brillouin zone. Locally, the low-energy band structure around the **K** points are linear cones. The beautifully simple linear dispersion relation is well known from relativistic physics as the dispersion of light. For photons the energy scales as the constant speed of light times the momentum. In analogy the energy of π electrons near the **K** point scales with the constant slope of the dispersion, which is the Fermi velocity. The low-energy electrons in graphene behave like massless Dirac particles, and the linear part of the electronic band structure is commonly referred to as the 'Dirac cone'.

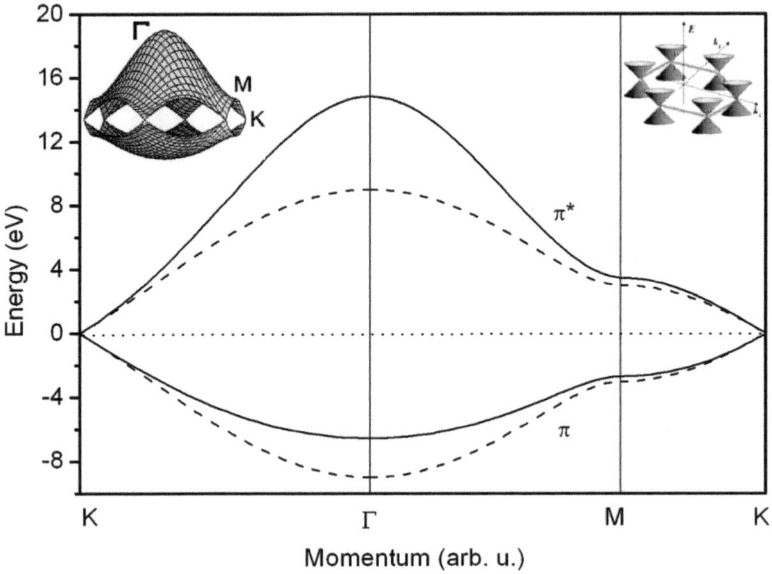

Figure 4.2. The main panel shows a cut along **K-M-Γ-K** of the two-dimensional π band structure of a graphene layer, according to Painter and Ellis [Painter and Ellis (1970)]. The solid line is with an asymmetry parameter s of 0.13, whereas $s = 0$ for the symmetric dashed line. The insets to the left and to the right illustrate the full two-dimensional π band structure and the linear cones around **K** and **K'**. See also Color Insert.

Apparently there is an intimate relation between graphene and carbon nanotubes, since the latter are simply made of the former. The conceptual idea is that the periodicity of the circumference of a nanotube can be mimicked by simply imposing the periodic boundary conditions of a rolled-up graphene layer [Hamada *et al.* (1992)]. This approach considers rolled-up nanotubes as locally flat and omits curvature effects. The electronic wavefunction can no longer have arbitrary wavelengths, but only integer fractions of the circumference are possible. Electrons on a nanotube may still have a continuous on-axis momentum like quasi-free electrons in a solid, but their angular momentum is quantized as in a quantum dot. The periodic boundary conditions slice the two-dimensional Brillouin zone of graphene into a finite set of parallel lines, with a spacing of just the inverse nanotube radius. The roll-up of a stripe

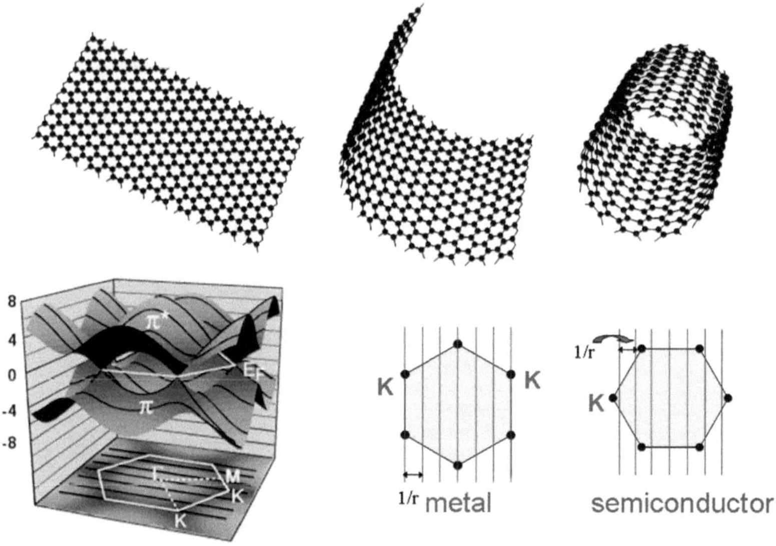

Figure 4.3. A rolled-up stripe of graphene imposes periodicity along the circumference of a resulting SWNT (top row). The 2D hexagonal Brillouin zone is sliced into parallel lines that either hit a **K** point in the metallic SWNT or miss it in the semiconducting SWNT (bottom row).

of graphene as well as the cutting lines in the Brillouin zone are illustrated in Fig. 4.3. The π band structure of a carbon nanotube consists of the sub-bands with different angular momenta. The orientation of the cutting lines is along the axis of the nanotube, which is just perpendicular to the roll-up vector, e.g., the lattice vector of graphene, that goes along the circumference of the nanotube.

There are as many choices of the roll-up into a cylinder as there are inequivalent lattice vectors in the graphene sheet. Each lattice vector may be uniquely expressed by a pair of two integer numbers (n, m). The resultant roll-up vector is defined via $n \cdot \vec{a}_1 + m \cdot \vec{a}_2$. The hexagonal lattice and the two basis vectors $\vec{a}_1$ and $\vec{a}_2$ that span the diatomic unit cell of the graphene sheet are shown in Fig. 4.4. Taking into account the C–C bondlength $a_0 = 0.142$ nm the diameter of a (n, m) SWNT is readily obtained via $d = \frac{a_0}{\pi} \cdot \sqrt{3(n^2 + n \cdot m + m^2)}$. There are two special directions of high symmetry in the graphene

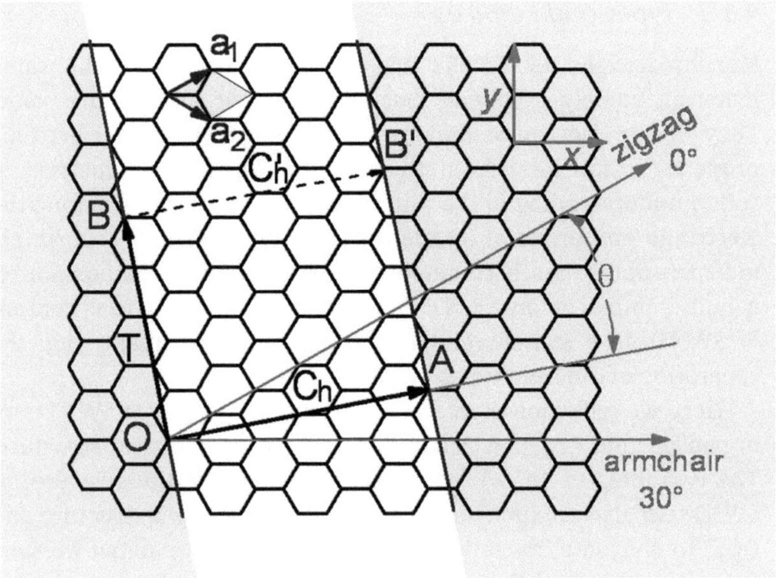

Figure 4.4. The hexagonal graphene sheet is built from its diatomic unitcell (blue). The highlighted stripe is an unrolled (4,2) SWNT. The chiral angle Θ of a $(n, m < n)$ SWNT lies in between $0°$ in zig-zag direction and $30°$ in armchair Direction.

sheet along $(n, 0)$ and $(n, m = n)$. As shown in Fig. 4.4, these two directions are labeled *zig-zag* and *armchair* for apparent reasons. Any other lattice vector that lies in between these two delimiting directions has the form $(n, 0 < m < n)$. Each of these lattice vectors results in a so-called chiral SWNT, with a chiral angle $0° < \Theta < 30°$. Chiral SWNTs appear as mirror pairs of right-handed and left-handed SWNTs with positive and negative chiral angles, respectively. So every chiral SWNT is either right or left handed, the only mirror symmetric ones are the achiral armchair and zigzag SWNT. Unless stated otherwise we will always disregard the mirror degeneracy of chiral SWNTs, since they will (in the absence of extreme magnetic fields) always share exactly the same electronic and optical properties. From a spectroscopic point of view there is no need to discriminate between left-handed and right-handed SWNTs.

4.1.2 Types and Families

Morphologically, all SWNTs are rolled up stripes of the same material, namely graphene, but they are not just all the same. They are in fact quite different. They have versatile electronic properties that stem from their helical (screw) symmetries. A sound understanding of the differences and similarities among the electronic properties of all the different types of SWNT is simply indispensable when it comes to (i) identifying the composition of a bulk sample of an SWNT, (ii) determining the actual content of SWNTs in a sample, and (iii) confirming and quantifying the separation of different types of SWNT.

Here we will show how the electronic properties of SWNTs are organized and grouped according to their geometrical structure. The two integers (n, m) that uniquely determine the structure of SWNTs do also uniquely define their electronic band structure and DOS. To elucidate this intimate relationship in more detail we start from the hexagonal Brillouin zone of graphene and the symmetry of its peculiar band structure. The two-dimensional band structure of graphene from Fig. 4.2 is plotted as equi-energy contours in Fig. 4.5. The high symmetry points Γ, **M**, and **K** are at the center, the edge, and the corners of the hexagons. Note that the hexagonal Brillouin zone is mirrored with respect to the hexagonal lattice in real space. The reciprocal basis vectors $\vec{b}_{1,2}$ have to be orthogonal to the original basis $\vec{a}_{1,2}$ and scale as the inverse length. This is readily satisfied by $\vec{a}_i \cdot \vec{b}_j = 2\pi\delta_{i,j}$. As depicted in Fig. 4.4 the vector $(n, m = -n)$ points along the (vertical) zigzag direction in real space. The corresponding reciprocal vector runs just along the (horizontal) armchair direction in the Brillouin zone. The reciprocal $(n, m = -n)$ vector points along the straight dashed line connecting to equivalent Γ points via **K**, **M**, **K** in Fig. 4.5. As a consequence of the hexagonal symmetry the **M** point lies at half the distance and there are two mirrored **K** points at one- and two-third of the whole distance. As every arbitrary (n, m) vector may be decomposed into a (m, m) and an orthogonal $(n - m, 0)$ vector, the number $n - m$ defines how many parallel cutting lines of a (n, m) have to cross these dashed lines at equidistant spacings. Here one can distinguish three different situations that can be readily identified

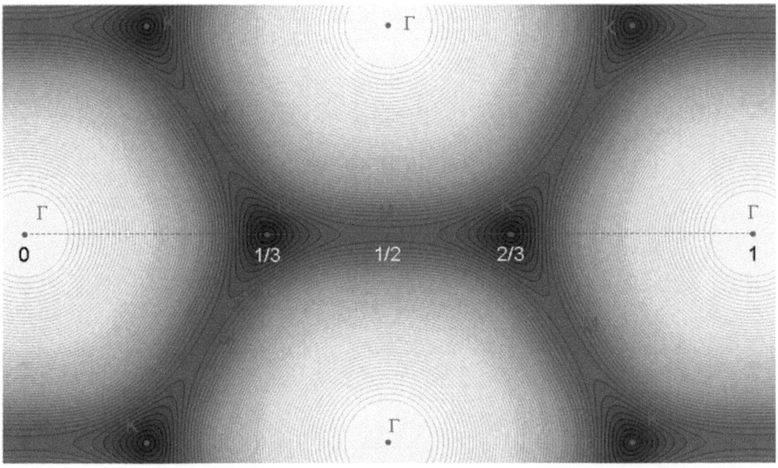

Figure 4.5. Equi-energy contour plot of the conduction band of graphene. The high symmetry points at the center (Γ), edges (**M**), and corners (**K**) of the two-dimensional Brillouin zones are labeled. The dashed line connects two equivalent Γ points. The relative distances of the **M** and **K** points are derived from the honeycomb structure. See also Color Insert.

by the remainder of the integer division of mod $(n - m, 3)$. In case the remainder evaluates to zero and $n - m$ is an exact integer multiple of 3, the cutting lines hit the **K** points. All (n, m) SWNTs with mod $(n - m, 3) = 0$ are metallic. If the remainder is, however, 1 or 2, the two **K** points will lie just at one-third in between the cutting lines. All these SWNTs are semiconductors. There is a further distinction of type I and type II semiconductors. The difference between them is whether the nearest cutting line to **K** crosses the dashed line from Fig. 4.5 at the flatter **K M** or the steeper **K** Γ side and the second closest-cutting line, and vice versa. Going on the dashed line in Fig. 4.5 from one **K** point to the other, the closest and second-closest cutting lines will always switch from left to right, but so does the entire band structure around **K**. In rolled-up SWNTs the two mirrored **K** points are degenerate.

The linear cone around the **K** point is not exactly circular but it has a trigonal shape, which arises from the symmetry of the next three neighboring bonds in sp^2 carbon. The equi-energy contours reveal this threefold rotation symmetry around the **K** point.

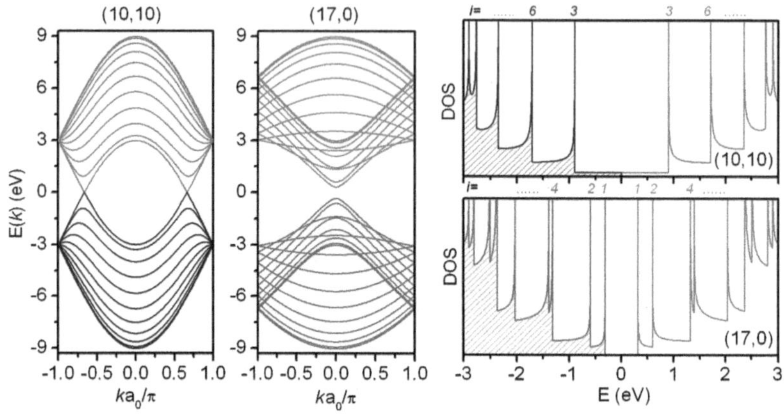

Figure 4.6. *Left panel*: Examples of the electronic band structure of a metallic (10,10) armchair SWNT and a semiconducting (17,0) zigzag SWNT. *Right panel*: Electronic DOS with the characteristic VHS of the (10,10) and the (17,0) SWNT, respectively.

As the 2D Brillouin zone of graphene is reduced to the parallel cutting lines of a (n, m) SWNT each of the 1D subbands will run through local maxima and minima. At zero slope in the dispersion relation the related DOS has a discontinuity and a van Hove singularity (VHS) arises. In any 1D electronic system the VHS are very sharp, well-defined spikes. These fingerprints of one-dimensionality in the electronic DOS are visualized in Fig. 4.1. Figure 4.6 shows the band structure and DOS of a representative pair of a metallic (10, 10) and a semiconducting type II (17, 0) SWNT. The regular pattern of metallic, semiconducting I, and semiconducting II roll-up vectors is presented in Fig. 4.7.

The constant DOS around the Fermi level in metallic SWNTs is a direct consequence of the linear cone (constant slope) in the band structure around the **K** point. Because of this linearity the spacing of the parallel cutting lines relates directly to the spacing of the VHS, and as the spacing of the cutting lines scales as the inverse diameter, so do the energies of the VHS. The whole sequence of semiconducting and metallic VHS S_1, S_2, M_1, S_3, and so on scales thus as the inverse SWNT diameter. The quite different sequence of the semiconducting and metallic VHS in Fig. 4.6 (for two SWNTs

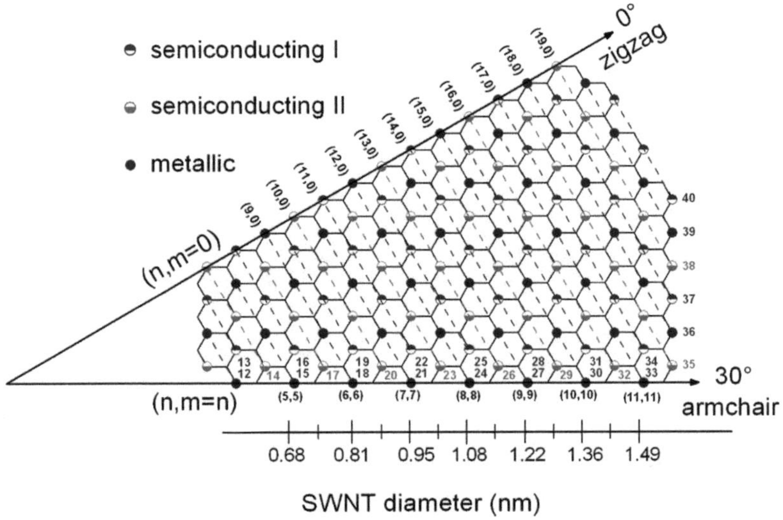

Figure 4.7. The Hamada map displays the 30° wide angle between the zigzag and the armchair direction in the graphene sheet and covers (disregarding left- and right-handed chiralities) all inequivalent roll-up vectors (n, m) of any SWNT. The remainder of $(n - m)$ mod 3 defines the pattern of metallic, semiconducting I, and semiconducting II SWNT. The straight dashed lines connect the families of constant $2 \cdot n + m$.

with 1.4 nm diameter) arises from an almost identical spacing of the cutting lines in the two SWNTs. The difference is that in the metallic (10, 10) SWNT the central cutting line hits the **K** point and the cutting lines to the left (+) and the right (-) are in units of $1/r$ at a distance of $\pm 1, \pm 2, \pm 3$, and so on to **K**. In the semiconducting SWNT the **K** point lies at $1/(3r)$ between the closest, e.g., left (+) and the second closest right (-) cutting line. So in the same units the sequence of minimum distances to the **K** point works out to be just $+1/3, -2/3, +4/3, -5/3$, and so on.

The aforementioned trigonal warping around the **K** point lifts in chiral metallic SWNTs the degeneracy of the closest left and right cutting lines. It also leads to a detailed modulation on top of the overall $1/r$ scaling of the energies of the VHS. The slope of the dispersion varies with the chiral angle. Since left and right closest and second-closest cutting lines are just flipped between semiconducting I and II SWNTs the sign of the modulation is also just

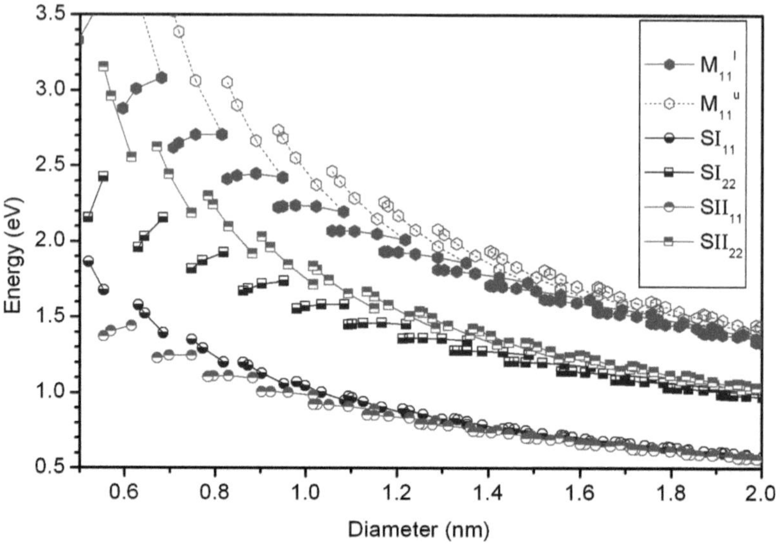

Figure 4.8. The Kataura plot shows the direct transition energies S_{11}, S_{22}, M_{11}, ... between the one-dimensional van Hove singularities in SWNTs as a function of their diameter. Families of SWNTs with constant $2 \cdot n + m$ are connected by straight lines. For non-armchair metallic SWNTs, where the cutting lines do not run parallel to **K M** the M_{11} is split due to the trigonal warping around **K**. See also Color Insert.

flipped. The behavior is best viewed in Fig. 4.8, which plots the even optical transitions between the mirror-like VHS in the conduction and the valence band (e.g., $S_{11} = S_1^{\star}\text{-}S_1$) in all different SWNTs as a function of their diameter. The overall shape of the blurred hyperbolic bands in Fig. 4.8 resembles the $1/r$ scaling, which stems from the linear cones around **K**. Within the overall blurred behavior, the type I and II semiconductors are grouped in short branches curling away from the overall trend. The actual transition energies that are plotted in Fig. 4.8 were determined by fitting an extended chirality-dependent, tight-binding scheme to a comprehensive set of experimental transition energies from resonant Raman spectrocopy [Araujo *et al.* (2007)].

Each of these strikes is connected with a line and represents one family [Bachilo *et al.* (2002); Telg *et al.* (2004)] of SWNTs with a constant $2 \cdot n + m$. The families just correspond to the dashed

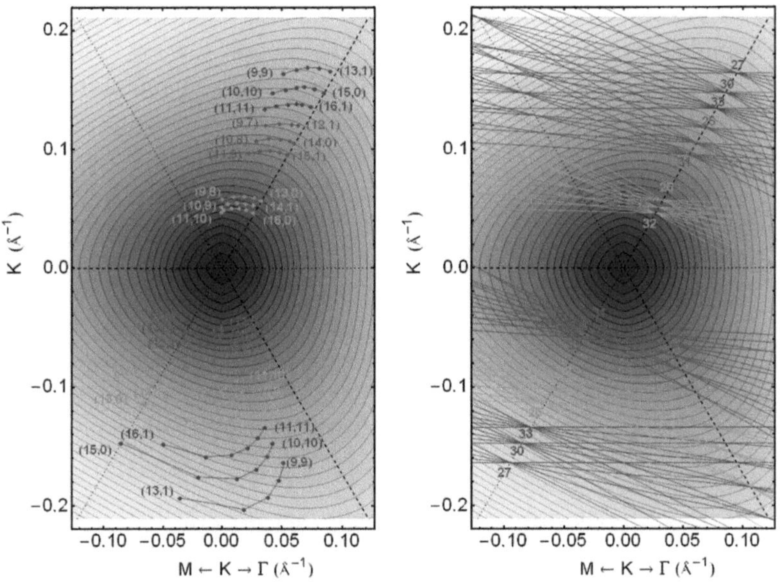

Figure 4.9. Contour plot of the vertical joint DOS of graphene around **K**. Equi-energy contours are separated by 0.1 eV. Dashed and dotted lines run along **KM** and **KΓ**. *left panel*: Position of the first and second VHS in semiconducting I (red) and II (blue) and metallic (green) SWNTs. The families of constant $2 \cdot n + m$ are connected. *right panel*: The hosting cutting lines of the VHS in the left panel. All even/odd cutting lines of a family meet at one point along the **KM** and **KΓ** direction. See also Color Insert.

lines in Fig. 4.7. The families are linear subsets of SWNTs, where the smallest diameter has also the smallest chiral angle. The number of the families members is given by the length of the dashed lines in Fig. 4.7. The different electronic characters of the three types of SWNT and the familiarities in the electronic structure within a family are vividly displayed in Fig. 4.9.

The equi-energy contours around the **K** point display the trigonal warping in the direct JDOS. In the top half of the viewgraph the curvature of the equi-energy contours and the chiral rotation are greatly compensating for one another, which gives rise to the rather flat branches in the family behavior in the Kataura plot Fig. 4.8, whereas in the lower half of Fig. 4.9 the curls are out of phase, which gives rise to the steepened branches in the Kataura plot.

Another remarkable symmetry of the families is that all homologous cutting lines from all members meet in one single point on the high symmetry line Γ**KM**. At these common invariant points of every family the geometrical effects of increasing chiral angle and diameter, viz. the angle of the cutting lines with respect to the Γ**KM** line and their spacing, are just in balance. Since the dashed lines in Fig. 4.7 are orthogonal to the zigzag direction, all members of a family collapse into a single point if the roll-up vectors are projected onto the zigzag direction. This real space direction is in reciprocal space by the orthogonal mirror inversion mapped just onto the armchair direction along Γ**KM**. The intrinsic family pattern of SWNTs is a unique fingerprint that is preserved in bulk spectroscopy on chirality mixed SWNTs.

4.1.3 *Tight Binding versus First Principles*

There are two very different approaches toward the calculation of the electronic band structure of carbon nanotubes. One the one hand there are huge efforts in parameter-free *ab initio* calculations and on the other hand there is a well-established framework of parameterized tight binding models. The key difference here is the ratio of feasibility versus complexity. Tight binding assumes that each bond in the hexagonal carbon lattice gives one separate contribution to an electronic Bloch state (plane waves). The simplification of the elegant step is to neglect all mutual interactions between the equivalent carbon-carbon bonds. The tight-binding scheme generates the overall shape of the band structure according to the symmetry of the lattice, but there are always energy parameters that are *a priori* unknown. Mathematically, they appear as overlap (often also transfer) integrals γ between the neighboring atoms. The amount of the parameters γ_{ij} depends on how many of the inequivalent pair interactions in the lattice are included in the model. So, for instance, in graphene there are three equivalent overlap integrals γ_{01} with the nearest in-plane neighbors and a sixfold degenerate overlap integral γ_{02} with the second-nearest neighbors. The obvious feasibility of tight binding is that the entire shape of the band structure may be readily derived. The yet unknown overlap integrals γ_{ij} may be derived from experiments. A thorough example of this procedure may be found elsewhere [Gruneis *et al.* (2008)]. The fitting values from therein were also used to calculate the band structure of graphene and the VHS of carbon nanotubes in Fig. 4.9.

In the first-principle approaches to the electronic band structure there is no room for unknown parameters, but there is the need to find a solution of the many-body Schrödinger equation. While this is in principle the most accurate state-of-the art formalism to describe (nonrelativistic) condensed matter, it imposes a truly inconceivable level of complexity. Still great efforts have been put into developing working assumptions for the calculations of the electronic band structure in solids. One or maybe the major breakthrough on this way is density functional theory (DFT) [Kohn and Sham (1965); Onida *et al.* (2002); Charlier *et al.* (2007, 2008)]. The multi-electron wave function, which is the solution to the Schrödinger equation, describes the correlated microscopic physical state of myriads of electrons in a piece of matter. However, the bulk properties of this piece of matter do not actually depend directly on the very detailed underlying microscopic electronic state. The situation is very much like describing and eventually even accurately predicting atmospheric conditions, without any need to know the location and momentum of each and every molecule in the atmosphere. We will not go into the heavy mathematical formalism behind this concept, but the important note here is that DFT only takes into account the electronic density (and often also its gradient). For predicting material properties it is relevant to know how many electrons there are on average at a specific site. In the framework of any DFT there is no phase information that would be needed to describe quantum interference.

There is since many years no general recipe on how to sacrifice the more complex phase information while preserving the spatial density distribution. This does by no means say that *ab inito* methods are futile from the beginning, but they have to always include some *approximations*, and Mother Nature is, to say the least, not a reliable friend. That means there is no way of really knowing beforehand if a certain set of approximations and/or numerical techniques will finally yield the correct answer. The experimental verification will always be needed to judge the calculations. But anyway, neither experimentalists nor theoreticians should ever deny the scientific method of cross-checking one another. First-principle methods are doubtlessly far more elaborate than simple tight-binding schemes and the calculations done by experienced theoreticians stand to their tests on experiments quite frequently. Still tight-binding models are widely used because of their feasibility.

A common strategy is to combine *ab initio* calculations and a tight-binding model. This procedure has been exemplified in great depth for carbon nanotubes [Spataru *et al.* (2008)] only recently. If the tight-binding model is fit to the calculated band structure this provides a facile model with *known* parameters. An overview of the electronic band structure in an SWNT as calculated from *ab initio* methods as well as the nearest and third-nearest neighbor tight binding is presented in Fig. 4.10. All three ways yield hardly discernible band structures with even more akin to VHS that dominate and dictate the electronic and optical properties of SWNTs. Owing to the manyfold efforts in theory and experiment over more than a decade the detailed band structure of SWNTs is today probably as well established as the band structure of silicon.

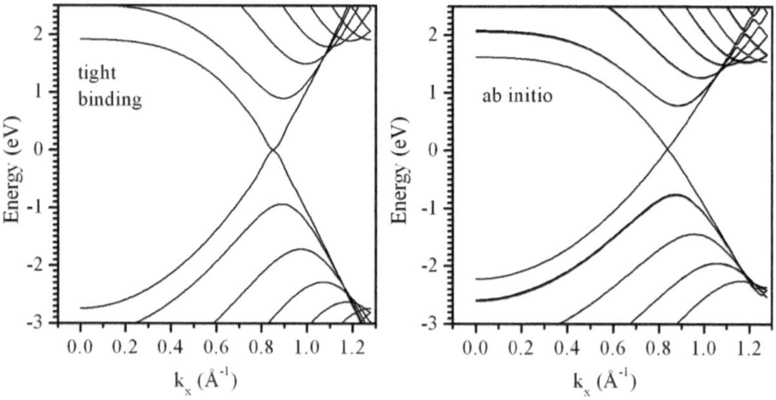

Figure 4.10. Band structure of a metallic SWNT as computed by *ab inito* methods, first-neighbor and third-neighbor tight binding. The horizontal dashed lines mark the energies of the VHS in the local minima of the band structure.

4.2 Electronic Excitations

Any interaction of a piece of matter with a probing electromagnetic wave or an impinging electron may be described in terms of elementary excitations in the solid. The most important quantized excitations for the electronic and optical properties of carbon nanotubes are electronic excitations and lattice dynamics. The different electronic excitations cover a wide energy range. A very schematic overview over the various possible electronic transitions that may be involved in different spectroscopic techniques is presented in Fig. 4.11. In a solid the densely packed individual atomic potentials are joined together and form an engulfing potential well with local dimples at the individual atoms. The DOS within this potential is filled up to the Fermi level, which lies below the free vacuum state.

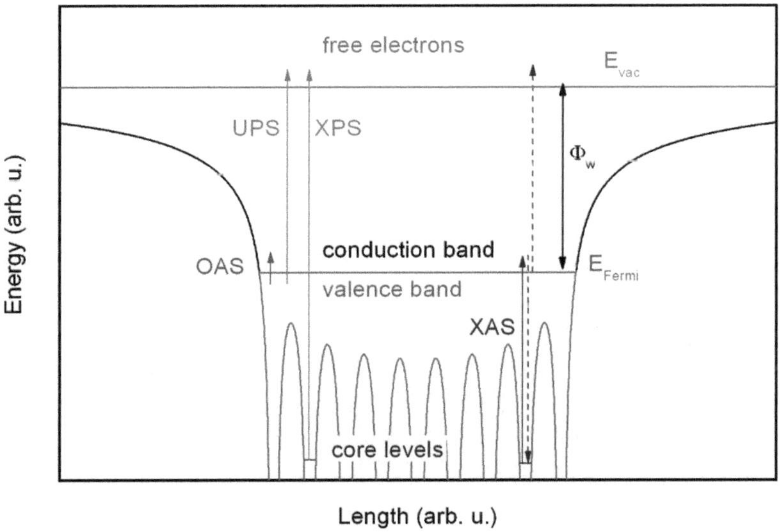

Figure 4.11. In a small cluster or even a bulk solid the atomic Coulomb potentials add up to a collective well with individual dimples. The collective well is occupied up to the Fermi level, which marks the borderline between the conduction and valence band, respectively. The atomic core levels are localized within the individual dimples. The electronic transitions associated with OAS, UPS, XPS and XAS are represented by vertical arrows.

The latter difference is the so-called workfunction Φ_W of a material. Each elementary electronic transition has to go from an occupied into an unoccupied state. The product of the involved occupied DOS and the accessible DOS is the joint DOS (JDOS) at a certain transition energy. The *direct* JDOS is relevant for scattering events with a broad range of electromagnetic waves. It refers to direct transitions where there is no (noticeable) additional momentum transfer involved as the electron is lifted from the valence to the conduction band. The momentum of visible light is typically $\sim$ 1000 times smaller than the size of a solid Brillouin zone. This ratio may also be deduced from the typical atom distances, which are between 1 and 3 Å, and the wavelength of the light, which is the order of a few hundred nanometers. The momentum of visible light is thus comparable to the pixel size in the contour plot of the hexagonal Brillouin zone of graphite in Fig. 4.5.

In case of impinging probes with considerable momenta (e.g. high voltage electrons or hard x-rays) dispersive electronic transitions that do involve a momentum transfer have to be considered. They are the constituents of the *dispersive* JDOS, that is a function of the momentum transfer q. Inter-band transitions from the valence to the conduction band give rise to absorbance in the infra-red to ultra-violet range. Optical absorption spectroscopy (OAS) maps out the inter-band JDOS. If the photon energy suffices to extract a valence electron from the solid and put it into a free vacuum state a photoemission process can occur. Ultra-violet photoemission spectroscopy (UPS) maps out the valence band. X-ray photoemission spectroscopy (XPS) probes the atomic core levels. And X-ray absorption spectroscopy (XAS) maps out the JDOS of the atomic core levels and the conduction band. All of the electronic transitions involved in these methods are illustrated by labeled arrows in Fig. 4.11. The dashed arrows for XAS indicate a secondary Auger process.

4.2.1 *Excitonic Inter-Band Excitations*

Till this point we did consider electrons in a solid as independent particles with a specific band structure. This concept assigns energies and momentum to all possible electronic states, and any electronic transition that can satisfy the conservation of energy and momentum may occur due to an impinging probe. In principle this concept is only applicable to the static ground state in equilibrium. The band structure approach incorporates the mutual Coulomb interaction of all charges into a description of effectively independent *quasi*-particles with an effective mass that mimics the drag in the charged medium. In this picture an excited electronic state consists of two distinct particles. The first obvious one is the electron in the otherwise empty conduction band, and the other one is the *hole* in the otherwise occupied valence band. Both of these co-exist in the surrounding electronic system, which mediates their mutual interaction. If a *real* electron is scattered and changes its energy and momentum, there is neither a mysterious hole nor, even stranger, any reason why the excited electron should interact with that hole. The important distinction here is that both the electron and the hole are understood as quasi-particles, which in turn are elementary excitations with respect to the groundstate of the solid. The groundstate of the solid is nothing else but the vacuumstate of elementary excitations that may be created there, and that are ultimately the measurable quantities.

The *electron-hole* concept is fully consistent *within* this quasi-particle picture of the solid. The excited electronic state is in fact described as the electronic ground state with an extra pair of an electron in the conduction band and a hole in the valence band. The fate of these two quasi-particles depends very much on the actual experimental circumstances. For instance, an electric transport measurement of a photocurrent will pull apart the oppositely charged electron and hole and they will exist as independent free particles. In semiconductor physics this precondition for a photocurrent is commonly termed *charge separation*. If the excitation occurs due to optical absorption and there is no bias current or chemical potential ripping the excitation apart, the electron and the hole can form a bound, hydrogen-like, state. The binding energies of such excitonic states are greatly influenced by their environment. The two most important environmental factors are the dielectric screening, viz. the density of available electrons, and a possible spatial confinement, as for instance on a carbon nanotube. The density of screening electrons just defines how much one single electronic excitation matters. The more electrons there are, the less are the exciton binding energies. In metals the excitonic corrections to the band structure description are essentially negligible. In conventional bulk semiconductors like Si the excitonic corrections are in the order of a few meV. So one might think that excitonic effects will not be too crucial in semiconducting and metallic carbon nanotubes. But just the opposite is true [Wang *et al.* (2005)]. Due to the narrow confinement on the nanometer scale, the excitons in SWNTs exhibit binding energies in the order of several 100 meV [Spataru *et al.* (2008)]. This magnitude is rather typical for molecular systems and unprecedented for solids. The exciton binding energy in carbon nanotubes is another molecular reminiscence that is present in the one-dimensional solid. The effect is so strong because one elementary excitation on a nanotube causes already a significant disturbance of the electronic groundstate. Such high exciton binding energies are typical for molecules, as for instance in the C_{60} fullerene [Lof *et al.* (1992)]. The situation of bound excitonic states within the band gap of a semiconducting SWNT is illustrated in Fig. 4.12. The excitonic level lies below the bare transition energy and causes a red-shift of the optical transition. The confined excitons in carbon nanotubes do strongly depend on the diameter as they scale with $1/r$, and they are also very sensitive to the environment of the SWNT. The observable optical transition energies in dispersed SWNT material depends crucially on the dielectric constant of the solvent [Ohno *et al.* (2006); Lefebvre *et al.* (2008)].

In analogy to the hydrogen atom, excitons exist not only in the spherical symmetric 1s groundstate but also in higher energy states, with more allowed angular momenta. These energy levels lie between the bound excitonic groundstate and the free electron-hole state in the bare band structure. In addition, the electron and the hole are both fermions with a spin of $\pm 1/2$. The absorption of a photon, which is a boson with an integer spin ± 1, does always require a change of the systems angular momentum by ± 1. The angular momentum is composed of the orbital contribution and the particles' individual spins. For the lowest excitonic 1s states the triplet state with parallel electron and hole spin cannot fulfill the optical selection rules. The triplet state is a *dark* exciton, while the singlet state with anti-parallel spins is a *bright* exciton. Dark excitons

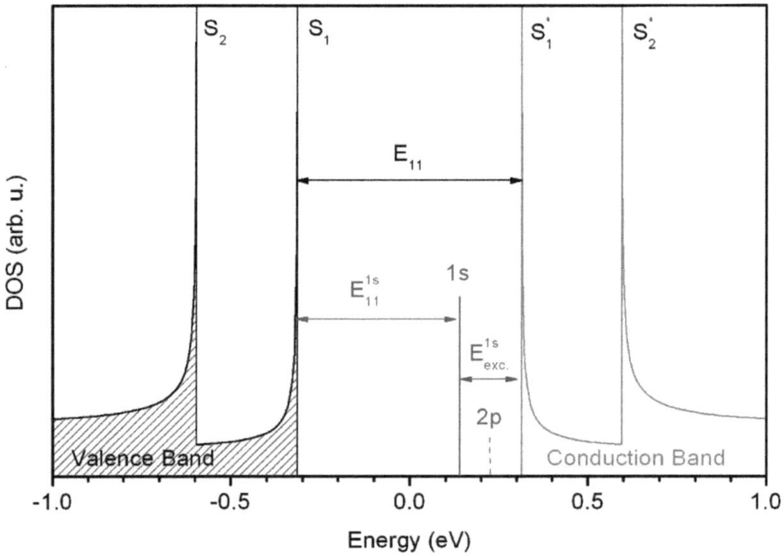

Figure 4.12. Schematic representation of the valence and conduction DOS and the excitonic states. The latter bound electron-hole states lie by the exciton binding energy $E_{\text{exc.}}$ below the bare optical transitions E_{ii} and cause a red-shift of the optical transitions.

are forbidden for individual elementary absorption events. Higher excitonic orbitals and even dark excitons can be experimentally accessed by two photon processes [Maultzsch *et al.* (2005)], which allows to map all the bound excitonic energy levels.

4.2.2 *Valence and Core Holes*

If a photon with the quantum energy $\hbar\omega$ successfully extracts an electron from the solid in a photoemission process the electron will be in a vacuum state with a kinetic energy according to energy conservation $E_{kin} = \hbar\omega - E_B - \Phi_W$. The material-specific workfunction can be determined by measuring the Fermi edge, which is just at the binding energy of the Fermi level. The actually measured solid is missing one electron either in the valence band or an atomic core level. This missing electron in the solid is commonly referred to as the $N - 1$ final state. For interband excitations this energy-dependent correction is often treated with a semi-empirical parameterized functional called self-energy or directly tackled on the *ab initio* level in the long-established GW approximation [Hedin (1965); Spataru *et al.* (2008)]. The general effects of the $N - 1$ final state are a renormalization of quasiparticle energies as well as a concomitant lifetime broadening. Regarding core level excitations quantitative *ab inito* methods have been developed more recently [Wessely *et al.* (2005)]. Core holes are relevant not only in XPS but also in XAS and core level EELS as well as IXS, since all of the latter do involve the excitation of a core electron.

4.2.3 *Collective Plasma Excitations*

Plasmons are next to excitonic single-particle excitations another class of electronic excitations. The electrons in a material form also a medium, a gas of charged particles, or a plasma. Plasmons are the discrete energy levels of the collective density waves in the plasma. They may be envisaged as the electronic analogue to the discrete lattice vibrations (phonons) in a solid. Collective phenomena as density waves emerge in a many-body system and are naturally beyond the realm of a microscopic description of independent quasi-particles. In many classical bulk metallic systems these excitations are reasonably well described within a continuum model that was pioneered at the beginning of the twentieth century by P. Drude [Drude (1900)]. A key characteristic of plasmons is that they are, as a collective phenomenon, upshifted by the bulk charge density. More charges mean more Coulomb interaction and a stiffening of the medium, which in turn raises the resonator frequencies. Another characteristic of density waves, such as sound waves or plasmons, is that they are longitudinal. As such they cannot directly couple to electromagnetic waves that are transversal. However, the selection rules are easily engineered for surface plasmons via adequately shaped and sized structures. A popular example of every-day technological relevance are radio waves that couple very well to electric density waves in radio antennas.

The intrinsic material specific bulk plasmons cannot directly couple to electromagnetic radiation. There are no direct absorption or emission events. Plasmons may only be observed in a material's *loss-function*, which stems from inelastic scattering events. The loss-function is defined as the relative fraction of probing projectiles that undergo specific inelastic scattering events with a certain energy loss and momentum transfer. With contemporary technologies the probing projectiles, which can actually probe a solid's Brillouin zone, may either be X-rays, neutrons or fast electrons with acceleration voltages of the order of 100 keV.

In an inelastic scattering event, any electronic excitation that can fulfill the conversation of energy and momentum is accessible and thus enters in the momentum-dependent loss-function [Lindhard (1954)]. A significant difference to absorption events is that the plasmons in the loss-function have, like the electrons in the solid, a dispersion with the momentum transfer q. The momentum of a plasmon corresponds to the propagation of a density wave. In a regular three-dimensional free electron gas the plasmon dispersion is isotropic and quadratic [Lindhard (1954)]. Which was, for instance, experimentally verified in elemental Al [Fink (1989)]. Plasmon excitations are not only bound to free charge carriers but may also occur for bound charges. These may be described phenomenologically by the inclusion of discrete Lorentz oscillators. More elaborate ways to describe the loss-functions and plasmon excitations of solids lie beyond this brief overview and may be found elsewhere [Onida *et al.* (2002)].

The dimensionality of an electronic system greatly affects the Coulomb interaction driving the charge density dynamics [DasSarma and Hwang (1996)]. For instance in a three-dimensional free electron gas, the plasmon resonance will have a finite value in the *optical limit*. The optical limit of a plasmon is its energy for diminutive momentum transfers or diverging wavelengths, respectively. The finite resonator energy in the optical limit follows directly from considering two infinitely extended charged sheets that will always interact with an exactly constant force, regardless of their distance. The electric field in our ideal capacitor is always a constant. The charge densities in these sheets can oscillate harmonically at the bulk plasma frequency of free electrons $\omega_P = \sqrt{ne^2/\varepsilon_0 m^\star}$. For charged stripes in a plane or even charged disks on a wire, the analogous capacitor models yield a very different result. In the latter geometries the Coulomb interaction

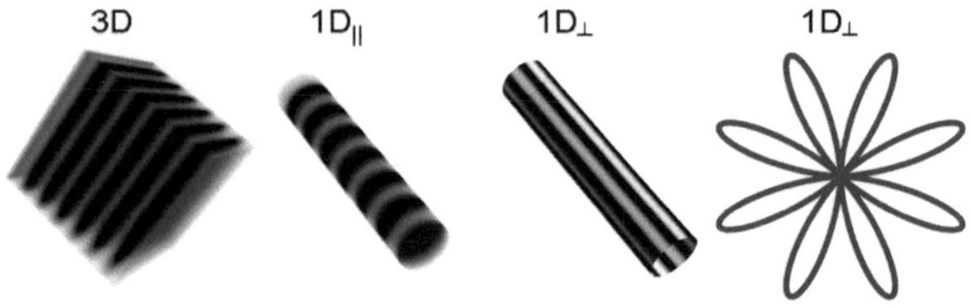

Figure 4.13. Charge density patterns of plasmons. From left to right: plane waves in the bulk, plane wave on a wire, circumferential localized plasmon mode, and polar plot of the charge distribution in a localized plasmon mode. See also Color Insert.

actually fades out in the optical limit of diverging separation. The fully separated charges do no longer interact, and can thus no longer resonate. In low-dimensional systems the free charge carrier resonance vanishes in the optical limit. In low-dimensional nanostructured materials, as for instance carbon nanotubes, there are significantly altered plasmon dispersions. In particular there is a splitting of every plasmon into one localized circumferential and another one-dimensional plasmon mode [Kramberger *et al.* (2008)]. Localized modes cannot propagate and have hence no defined momentum state. The conceptual distinction of localized and dispersive plasmons on a wire is visualized in Fig. 4.13. The splitting of plasmons into longitudinal density waves running along the axis and static modes with angular momenta is a direct consequence of the tubular symmetry in an SWNT.

4.3 Spectroscopic Methods

Every spectroscopic technique is based on a scattering experiment. In an actual experiment the (nanotube) sample is exposed to an incident beam, and the experimentalist observes a secondary beam coming from the sample. The secondary beam may consist either of scattered or transmitted particles of the primary beam or of newly formed secondary particles. In the vast majority of methods the primary and secondary particles each may either be electromagnetic waves or free electrons. In this scheme spectroscopic methods can be further divided into *first-order* and *second-order* scattering events. In a first-order experiment the probe particle is directly absorbed and creates an elementary excitation in the sample. These are the electronic transitions from the JDOS as they are described in section 4.2.1. The signature of the primary absorption event is accessible either via the attenuation of the primary beam or by detecting subsequently created secondary particles. In a second-order scattering event the primary particle is not absorbed, but it is inelastically scattered as it creates another type of elementary excitation in the sample. The latter collective excitations are from the sample's loss function, which is described in section 4.2.3. The signature of a second-order scattering event is imprinted in the inelastically scattered primary beam.

In the following sections we will briefly introduce the general reader to the common physical scattering events behind a variety of different spectroscopic methods. The methods that are collected in this section are naturally only a limited choice of the numerous methods suitable for investigations on carbon nanotubes. A wider and more detailed overview of spectroscopy on condensed matter may be found elsewhere [Kuzmany (1998)]. We will start with optical absorption spectroscopy (OAS) in section 4.3.1 and angle-resolved electron energy loss spectroscopy (AR EELS) in section 4.3.2. Next are photoluminescence spectroscopy (PLS) in section 4.3.3 and Raman spectroscopy (RS) in section 4.3.4. Finally, we will introduce photoemission spectroscopy (PES) in section 4.3.5 and X-ray absorption spectroscopy (XAS) in section 4.3.6.

4.3.1 *Optical Absorption Spectroscopy*

Optical absorption spectroscopy (OAS) measures the frequency-dependent optical absorption of a sample. In SWNTs the absorbing electronic transitions in the near-visible infrared (NIR) to ultraviolet (UV) spectral range are excitonic interband transitions between VHS. The relevant electronic transitions are the direct JDOS. The (spectral) weight of transitions between the diverging VHS outmatches that of all other nonresonant transitions. The optical absorption in SWNT [Kataura *et al.* (1999)] is very well described by considering just transitions between the VHS on top of the smooth response from graphite [Taft and Philipp (1965)]. The latter is reminiscent of all sp^2 carbon. In rolled-up SWNTs the optical transition may either occur vertically from a VHS in the valence band to the VHS in the conduction band on the same cutting line or occur between two neighboring cutting lines. Light that is polarized along the axis of an SWNT can only be absorbed by direct transitions that preserve the angular quantum number of the electron. If the polarization is crossed (i.e., perpendicular to the SWNT axis), the circumferential angular momentum may be changed by ± 1. Then the electron and the hole are at two adjacent cutting lines [Gruneis *et al.* (2004)]. In this case an odd transition (e.g., $E_{12} = S_1^\star - S_2$), for instance from the second occupied VHS to the first unoccupied VHS, may occur.

Nowadays OAS is a very well established tool for the bulk characterization of SWNTs. It retrieves information on the diameter distribution, sample purity, and content of metallic and semiconducting SWNTs.

4.3.2 *Electron Energy Loss Spectroscopy*

Electron energy loss spectroscopy (EELS) is a powerful experimental technique with a very wide dynamic range. The scattering process in the experimental setup is a highly energetic electron beam (typically $\sim$170 keV) going through about a 100 nm thick sample. Under typical conditions most of the electrons just go straight through the sample. Eventually, some of them will be scattered once by the creation of a plasmon (see section 4.2.3). The fast electron emits a plasmon into the surrounding medium. The Drude plasmon of quasi-free electrons in metals just depends on the charge carrier density. If plasmons are a resonance of bound electrons, for instance π electrons in SWNTs [Pichler *et al.* (1998)], the plasmon energy is upshifted with respect to the underlying excitonic electronic transition. The upshift is a general behavior that results from the overall stiffening of the electronic plasma at finite densities.

EELS can probe the atomic core excitations at a few keV as well as the entire energy range down to $\sim$ 0.5 eV. It can access electronic intra- and interband excitations. The lowest accessible excitation is typically the free charge carrier or Drude plasmon of metals. The accessible excitations in the energy range of several eV come from the interband loss-function. The excitations at energy losses of hundreds of eV are atom-specific core-level excitations. These are well-localized excitations, where the loss-function is no longer distinguishable from absorbtion spectra. These are the same electronic transitions that give rise to the element-characteristic X-rays. Their resonances are, for instance, probed in XAS. EELS is a commonly applied method to determine site-specific elemental compositions on the nanometer scale in analytical transmission electron microscopy (TEM).

If EELS is performed in a purpose-built setup, it may also be done with a collimated unfocused electron beam. This does not facilitate any spatial resolution, but it allows to accurately measure the angle of deflection along with the energy loss of the scattered electrons. Currently there is one unique purpose-built angle resolved AR-EELS spectrometer in operation, which was thoroughly described earlier [Fink (1989)]. Measuring the deflection angle of a scattered electron in AR-EELS corresponds to measuring simultaneously the energy *and* momentum of the plasmon that was created in the scattering event. AR-EELS can directly map out the full electronic momentum dependent loss-function in solids or, in the present context, in SWNTs [Kramberger *et al.* (2008)].

4.3.3 *Luminescence Spectroscopy*

Photoluminescence spectroscopy (PLS) relies on the luminescence process. An absorbed photon first creates an excitonic electron hole pair in a solid. The excitonic state may cool down rapidly by emitting manyfold low-energy excitations like phonons. On the edge of the bandgap, there is no further way to continuously dissipate energy. The only possible decay channel left to the excitonic electron-hole pair is their radiative recombination. The sample luminesces with an energy corresponding to the bandgap. Since PL requires a gap in the direct JDOS, it can never occur in metals. Typically SWNTs form thick bundles, where the SWNTs are hexagonally packed. If a metallic SWNT is next to a semiconducting SWNT, it will simply short-circuit the excitation gap in the semiconducting SWNT and quench the luminescence. The electron and the hole will recombine via the metallic channel in a nonradiative, and nonluminescent, way. Thus bulk SWNT material is not suited for PL studies. The key prerequisite for PLS on SWNTs is the preparation of stable dispersions of isolated SWNTs. This may be achieved by wrapping the nanotubes with surfactants and ultracentrifugation. In that way macroscopic amounts of dispersions of luminescing semiconducting SWNTs can be provided [Bachilo *et al.* (2002)]. PLS offers the unique possibility to attain correlated pairs of lowest and second-lowest electronic transitions between VHS on the very same SWNT. Different (semiconducting) chiralities can be individually fingerprinted in a macroscopic suspension of SWNTs. The field of PL on SWNTs has expanded and was thoroughly reviewed only recently [Lefebvre *et al.* (2008)].

4.3.4 *Raman Spectroscopy*

Raman scattering is the coherent inelastic (or superelastic) scattering of visible or near-visible light. A photon is scattered on the electronic system while another quasi-particle is created (or annihilated). If a quasi particle is created upon the recoil of the electromagnetic wave, the light is red-shifted. If the scattered photon takes up an excitation from the solid the scattered light is blue-shifted. The first case is called Stokes scattering, and the latter anti-Stokes scattering. The optical Raman spectrum of a material is by definition its loss function for monochromatic illumination with visible light. Visible light can only provide diminutive momentum transfers in solids. Raman active excitations must be from the center of the Brillouin zone, and they must be able to couple to light. These requirements are met by Raman active phonons. Raman activity of a phonon implies that the atomic displacement pattern of the phonon changes the dielectric polarizability. If the atomic displacement can affect the effective electric field, then the oscillating electromagnetic field couples to that phonon. Inelastic X-ray scattering (IXS) is at its heart exactly the same scattering process and commonly termed X-ray Raman. Hard X-rays with several keV provide sufficient momenta to access the full phonon dispersion across the Brillouin zone. Regarding plasmons, the technique is capable of covering the same range of energy losses and momentum transfers as AR-EELS.

Raman spectroscopy in the near-visible and visible spectral range is widespread and commonly applied. It is a versatile, nondestructive characterization tool for SWNTs [Kuzmany *et al.* (2001)]. Raman spectroscopy was used literally right from the beginning to characterize multi-walled and single- walled nanotubes [Hiura *et al.* (1993); Eklund *et al.* (1995)].

4.3.5 *Photoemission Spectroscopy*

The photoemission process is the experimental foundation that originally inspired the physical concept of electromagnetic quanta, viz. photons [Einstein (1905)]. The process may occur only if the energy of a photon suffices to extract an electron from the solid. The solid is left behind with an unpaired hole (see section 4.2.2). In order to be of spectroscopic value the photoelectron has to escape the solid, which is only possible from the first few atomic layers. On its way out, photoelectrons may pick up additional signatures from the loss function. Even in a metal, the electrons from the Fermi level still need to overcome the work function Φ_W. The latter is the potential step at the surface of the material. A very rough sketch of the situation is given in Fig. 4.11. The electronic conduction and valence band exist within the realm of the joint macroscopic potential well, where quasi-free electrons may be delocalized. Valence band photoemission (UPS) is conducted at ultraviolet photon energies of several electron-volts. With X-ray energies of several 100 eV to keV, X-ray photoemission spectroscopy (XPS) is suited to probing the localized atomic core levels. This analytic technique not only identifies elements by their characteristic X-ray fingerprints but can even reveal bonding-specific chemical shifts in the atomic core levels.

4.3.6 *X-Ray Absorption Spectroscopy*

X-ray absorbtion spectroscopy can only be conducted with a tuneable X-ray source. If the quantum energy of the X-ray beam is tuned to an electronic transition from an atomic core level into the conduction band (see Fig. 4.11), resonant X-ray absorption may occur. The initially excited electronic state is not directly visible to the experimentalist, but it is only short-lived and will quickly decay. The freed energy in the decay of the highly excited electronic state can cause the emission of a secondary electron. Only the secondary emission process is readily observable in an experimental setup. Since the measurement of XAS relies on secondary emitted electrons it is ultimately as surface sensitive as PES. The stronger the resonant absorption, the more X-rays are absorbed by the first atomic layers, and the stronger is the collected signal. X-ray photons that make their way beyond the surface layers will simply heat up the sample, but they will not contribute to the signal.

The resonance profile that is obtained by tuning the quantum energy $\hbar\omega$ across the transitions from the core level into the conduction band is a polarized atomic site-selective local projection. The effect of the core hole is generally a shift toward smaller transition energies with a concomitant compression of the bandwidth. The quantitative treatment of core holes in the XAS response of sp^2 carbon has been described elsewhere [Wessely *et al.* (2005)]. XAS yields information on the unoccupied conduction band and is therefore a complement to UPS that probes the valence band.

4.4 Spectroscopy on Nanotubes

Spectroscopy on carbon nanotubes may be roughly divided into two regimes: fundamental studies on the spectroscopic response of carbon nanotubes and the utilization of feasible spectroscopic tools to characterize samples. Naturally, there is an ongoing transition from one of these two domains to the other. The border line is continuously shifting since today's spectroscopic characterization tools have been yesterday's fundamental studies, and today's fundamental studies might turn to be tomorrow's established standards. This section describes how the methods introduced in section 4.3 may be employed to experimentally explore the elementary excitations from section 4.2 in carbon nanotubes. These offer an unprecedented experimental access to fundamental studies on one-dimensional physics of van Hove singularties in section 4.4.1 and the charge carrier response in one-dimensional electronic liquids in section 4.4.2. The thoroughly checked knowledge about this excitations is the scientific foundation of characterization techniques for the alignment of carbon nanotubes in section 4.4.4, their relative bulk fractions of semiconducting and metallic SWNTs in section 4.4.5, the SWNT diameter distribution in section 4.4.6, their crystallinity insection 4.4.7, as well as their purity in section 4.4.8.

4.4.1 *Van Hove Singularities*

A key fingerprint of carbon nanotubes are the VHS in the one-dimensional electronic system. In a bulk sample the position and width of the macroscopic VHS are determined by the average SWNT diameter and the spread of SWNT diameters. The macroscopic VHS are the sum of all individual VHS weighted with the diameter distribution. VHS may be independently observed in electronic inter-band transitions or separately in either the valence or conduction band.

The absorption spectrum of bulk SWNTs comprises [Kataura *et al.* (1999)] a broad absorption peak centered at $\sim$ 4.6 eV as well as a sequence of peaks that stem from first and second semiconducting as well as the first metallic VHS in SWNTs. In samples with different mean diameters the VHS shift in energies. The characteristic absorption peaks due to VHS are shown in the right panel of Fig. 4.14. The comparison of OAS with EELS in the left panel of Fig. 4.14 reveals a relative upshift of the peaks. The higher peak positions in EELS are due to the free charge carrier density in the bulk SWNT samples. The comparative analysis of sample purity as well as of the diameter distribution from OAS and EELS is described in great depth by Liu *et al.* [Liu *et al.* (2002)].

The luminescence process is another way to access VHS in the JDOS of SWNTs [Bachilo *et al.* (2002)]. PLS yields a two-dimensional map of the luminescence yield as a function of the incident and the luminesced wavelength. A peak in this map corresponds to the absorption in the second optical transition E_{22} and an emission from the first optical transition E_{11} on the same semiconducting nanotube. The distinct pattern of all the PL peaks did allow for the first time to structurally assign whole families of carbon nanotubes [Bachilo *et al.* (2002)]. This work spectroscopically confirmed the

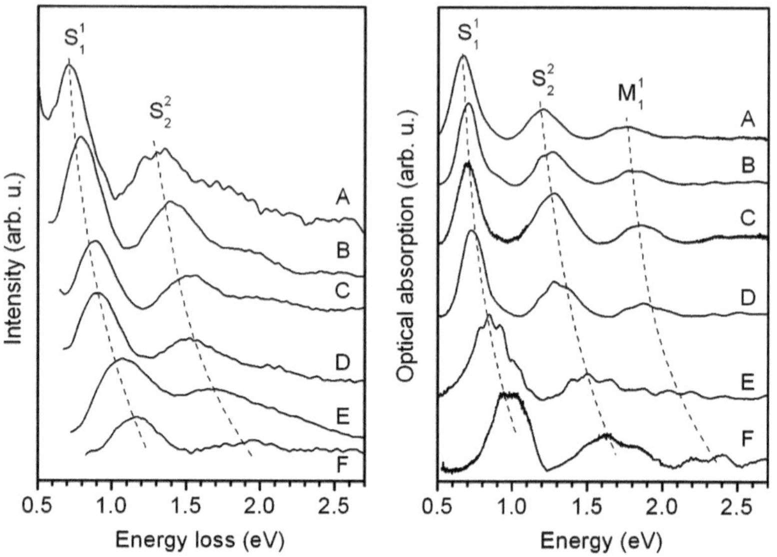

Figure 4.14. The peaks in the left panel (OAS) and the right panel (EELS) shift with the mean diameter of the SWNT material. A, B, C, D, E, and F label different samples with mean diameters of 1.46, 1.37, 1.34, 1.30, 1.09, and 0.91 nm, respectively. The image is reproduced from [Liu *et al.* (2002)].

entire concept of type I and type II semiconducting SWNTs and the families of constant $2m + n$, which were introduced in section 4.1.2.

Electronic transitions can only go from an occupied to an unoccupied state. The latter are separated by the Fermi level E_F of SWNTs. If either electron acceptors or donators are inserted into the interstitial channels of a hexagonally packed bundle of SWNTs E_F will either be lowered into the valence band or raised into the conduction band [Ugawa *et al.* (1999); Itkis *et al.* (2002)]. The shift of E_F causes the opening of an additional excitation gap. Then there is a minimum threshold energy required for the smallest possible inter-band transition in metallic SWNTs. In progressively $FeCl_3$ doped SWNTs the E_F shifts into the valence band. A series of OAS spectra of doped SWNTs is shown in Fig. 4.15. As the Fermi level drops below the VHS in the valence band the corresponding peaks in the direct JDOS are successively depleted. The opening of the excitation gap is

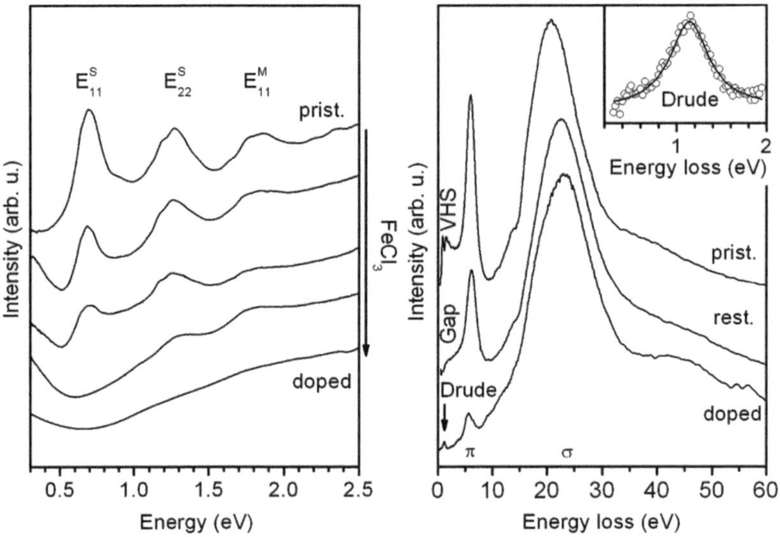

Figure 4.15. The OAS spectra in the left panel of pristine (top) and progressively $FeCl_3$ intercalated SWNTs (underneath) reveal the depletion of the electronic states in the VHS in the valence band. The loss functions in the right panel uncover the additional Drude plasmon in doped SWNTs and a remaining excitation gap at intermediate doping levels after partial de-intercalation. This figure is reproduced from [Liu *et al.* (2004)].

visible in the onset of the loss function in the right panel of Fig. 4.15, which will be discussed in more detail in section 4.4.2.

The VHS in the valence band may be directly accessed by the photoemission process. The two valence band UPS spectra in Fig. 4.16 are of SWNTs and a clean Au surface. The overall shape of the spectra is again reminiscent of sp^2 carbon. SWNTs exhibit three additional peaks due to the first and second semiconducting $S_{1,2}$ and the first metallic M_1 VHS [Ishii *et al.* (2003); Rauf *et al.* (2004)]. In semiconducting nanotubes there is no accessible Fermi level. There are only the top of the valance band and the bottom of the conduction band and an arbitrary energy in between the charge neutrality level. In pristine carbon nanotubes the latter is found to be above the metallic Fermi level by 0.1 eV [Kramberger *et al.* (2009)]. The resultant difference in the work functions is more readily accessed by core-level XPS in Fig. 4.17. In case of doping

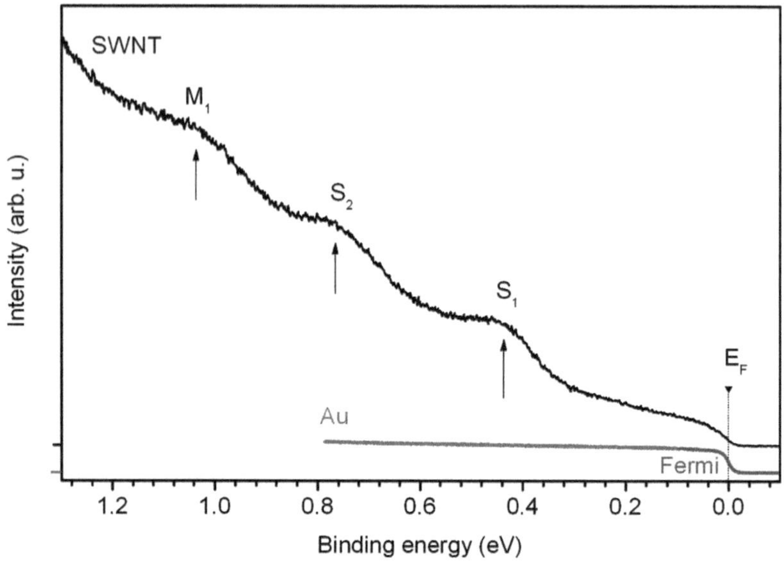

Figure 4.16. Valence band UPS of an SWNT (black) and Au (brown). A monochromatized He Kα lamp with $\hbar\omega = 21.22$ eV was used for illumination. The arrows mark the macroscopic semiconducting ($S_{1,2}$) and metallic (M_1) VHS in the bulk SWNT material. The binding energies are calibrated to the fitted Fermi function (red).

with an electron donor E_F is raised and the VHS are shifted to lower binding energies. Of course, the latter lowering of the binding energies of the VHS is merely a consequence of the fact that in UPS the Fermi edge always marks zero binding energy. UPS on alkaline intercalated SWNTs reveals, in the left panel of Fig. 4.17, a shift of the VHS away from the Fermi level. The concomitant changes [Kramberger *et al.* (2009)] at the Fermi edge as well as the core-level XPS response in the right panel will be discussed in section 4.4.2.

The VHS in the conduction band may be probed via an X-ray absorption process, where a transition from the C1s core level into the conduction band occurs. The XAS response of bulk isotropic SWNT material is presented in Fig. 4.18. The C1s edge is composed of individual resonances from the π and the σ conduction band. These features are typical for any sp^2 carbon and are well known, for instance, from bulk graphite [Batson (1993)].

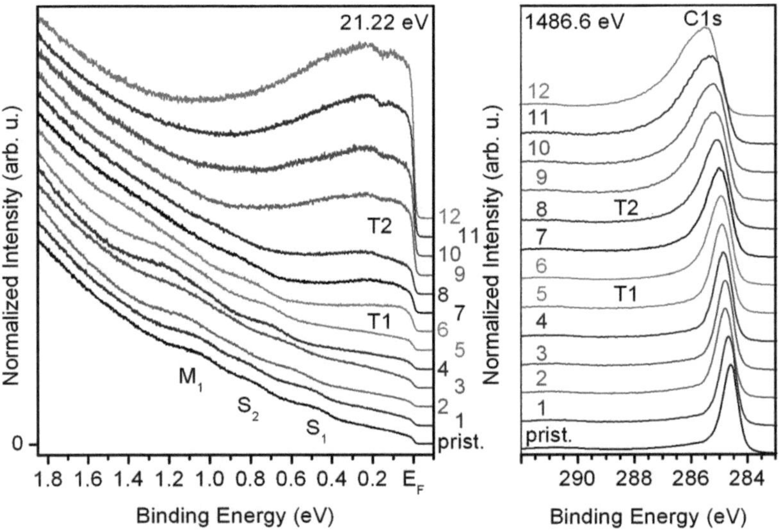

Figure 4.17. UPS and XPS of pristine and successively K^+ doped (1, ..., 12) SWNTs. At the first transition (T1) between steps 5 and 6 the Tomanaga Luttinger liquid (TLL) becomes a one-dimensional Fermi liquid. At T2 a bulk three-dimensional Fermi liquid emerges. The figure is adapted from [Kramberger *et al.* (2009)].

High-resolution measurements resolve fine structures in the π band that are exclusively observed in SWNT material with narrow-diameter distributions. The comparison to the π absorption edge in the right panel of Fig. 4.18 shows that the sequence of peaks is simply missing in highly ordered pyrolytic graphite (HOPG), which was measured at the very same resolution [Kramberger *et al.* (2007a)]. The fine structures originate from resonant transitions from the C1s core level into the VHS in the conduction band of SWNTs. The left panel of Fig. 4.18 shows the evolution of the fine structures on the C1s$\rightarrow$ $\pi^\star$ absorption edge [Kramberger *et al.* (2009)]. The VHS are successively depleted but not shift upon potassium intercalation. The X-ray resonances stay at fixed energies. Note that the evolution of VHS in the conduction band (stationary and depleting) is just complementary to the evolution of the VHS in the valence band (shifting without depletion) with increasing potassium intercalation. Naturally, the roles of the conduction and

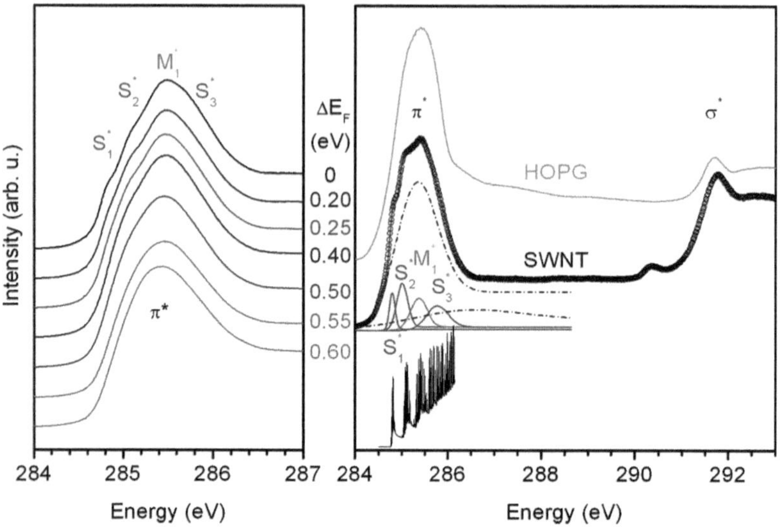

Figure 4.18. *Left panel*: Upon successive K^+ intercalation E_F is raised into the CB and the fine structures are depleted progressively. *Right panel*: The C1s absorption edge of SWNTs and HOPG shows two resonances due to π as well as σ states in the conduction band. The π band contains four fine structures due to the semiconducting $S_{1,2,3}$ and the metallic M_1 VHS. Data points are from experiment and solid lines from line shape analysis. The spiky diameter cumulative DOS from parameterized tight bind is displayed underneath for a qualitative comparison to the fine structures in the π^* resonance.

valence band are just flipped as one goes from n-type to p-type charge transfer. This antisymmetrical interchangeability is not present in symmetrical inter-band transitions from the valence to the conduction band. The JDOS is never shifted but always depleted as either the valence band is emptied or the conduction band is filled.

4.4.2 *Electronic Response*

In carbon nanotubes there is a remarkable connection between their geometrical structure and their electronic character. Slight changes in the chiral twist decide whether they are a semiconductor or a metal. Moreover, neither a one-dimensional semiconductor nor a one-dimensional metal is simply the one-dimensional projection of the bulk. The metallic phase in carbon nanotubes is not even a regular Fermi liquid but rather a Tomonaga–Luttinger liquid (TLL) [Tomonaga (1950); Luttinger (1963)]. Besides the direct study of the metallic states in carbon nanotubes there are also numerous cases where metallicity tunes the spectroscopic response significantly. Tuning the balance between semiconducting and metallic abundances via intercalation allows to control these effects in bulk SWNT material.

The strength of the metallicity (e.g., the free charge carrier density) as well as all other electronic transitions scales in a straightforward manner with the macroscopic density. The direct comparison of the loss function in sp^2 carbon at different densities is presented in Fig. 4.19. All sp^2 shows the collective plasmons of the electronic π and the σ system. The comparison of the loss function of bulk graphite, consolidated bundled SWNTs, and woolly isolated SWNTs explicitly demonstrates the scaling of the π as well as of the σ plasmon with the materials density.

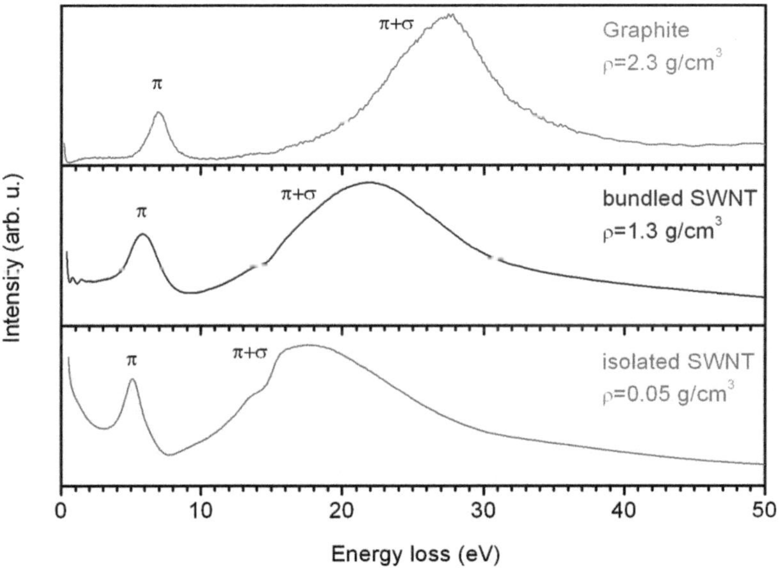

Figure 4.19. The archetypical loss function of any sp^2 carbon comprises the collective π and σ plasmons. Their positions scale down with the lowering density in graphite, consolidated bundled SWNTs, and woolly isolated SWNTs.

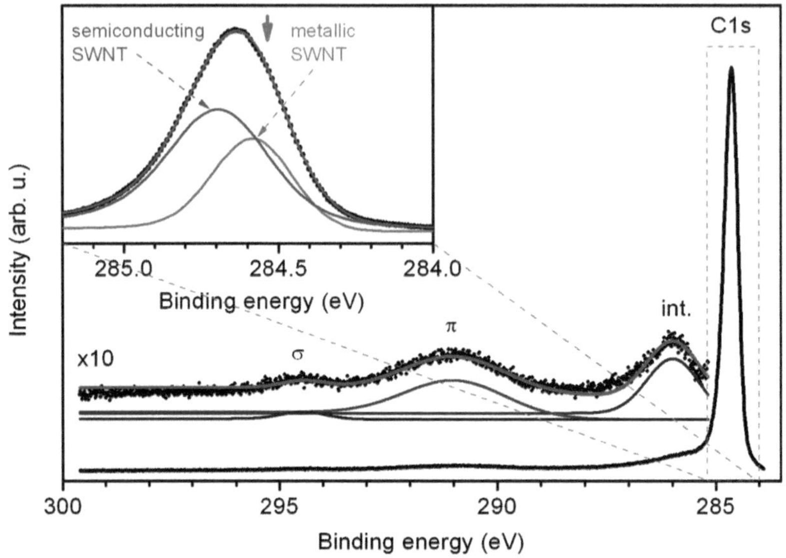

Figure 4.20. High-resolution PES of the C1s region as measured (data points) on a bulk SWNT with an excitation energy of $\hbar\omega = 400$ eV. Solid lines are fits to the individual peak and their sum. The figure has been adapted from [Kramberger *et al.* (2007a)].

The C1s core level XPS response of carbon nanotubes is presented in Fig. 4.20. There are another three comparably weaker structures next to the main C1s peak. The latter are so-called shake-ups. The photoelectron that escapes the sample creates an electronic excitation and undergoes an additional energy loss. The three shake-ups originate from low-energy inter-band scattering as well as from the π and σ plasmons, respectively. The C1s XPS line of bulk SWNTs is noticeably split by about $\sim$0.1 eV which originates from the section 4.4.1 mentioned different work functions in the semiconducting and the metallic SWNTs [Kramberger *et al.* (2007a)]. The intrinsic line shape of a photoemission peak is a symmetric Lorentzian. But in a metallic system a continuum of low-energy electronic excitations is accessible, which causes a dissipative asymmetry in the line shape. The asymmetry α of the Doniach–Sunjic line profile [Doniach and Sunjic (1970)] is a measure of the metallic DOS at the Fermi level. The increasing metallicity in

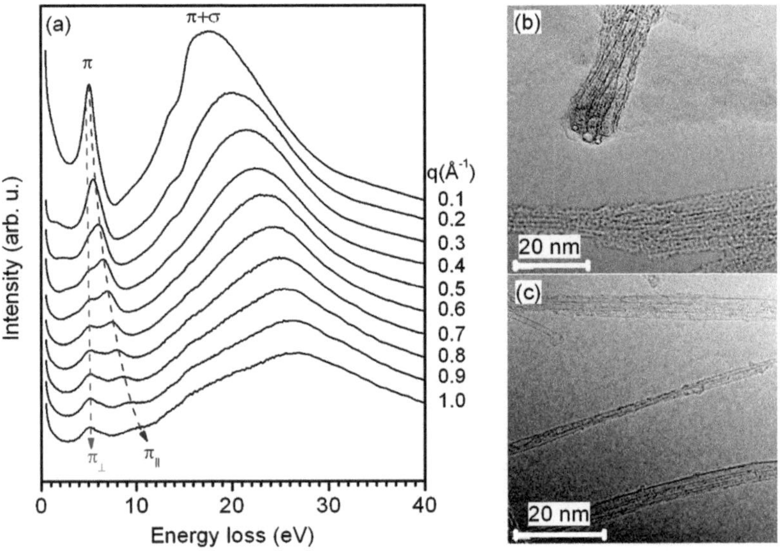

Figure 4.21. The left panel (a) shows the dispersion of the Loss function of isolated SWNTs. The arrows mark the localized $\pi_\perp$ and dispersive $\pi_\|$ plasmons. The TEM micrographs on the right show the cross section (b) and the side view (c) of the freestanding SWNTs.

gradually K^+ intercalated SWNTs is traced in the evolution of C1s XPS line shape in Fig. 4.17.

Owing to their low density, isolated free-standing SWNTs can be envisaged as an archetypical case of isolated nanowires. TEM micrographs of these nanowires are displayed on the right part of Fig. 4.21. In this material the individual SWNTs form thin wires of only a few (<10) SWNTs [Einarsson *et al.* (2007)]. In this network of aligned nanowires the plasmon excitations are confined to very narrow one-dimensional channels. The splitting of plasmons into dispersive modes propagating along the axis and localized surface was experimentally found in AR-EELS [Kramberger *et al.* (2008)].

The measured loss functions of isolated SWNTs are presented in the left part of Fig. 4.21. The dispersive loss function reveals two distinct π plasmons that are identified as the localized $\pi_\perp$ plasmon and the on-axis propagating $\pi_\|$ plasmon.

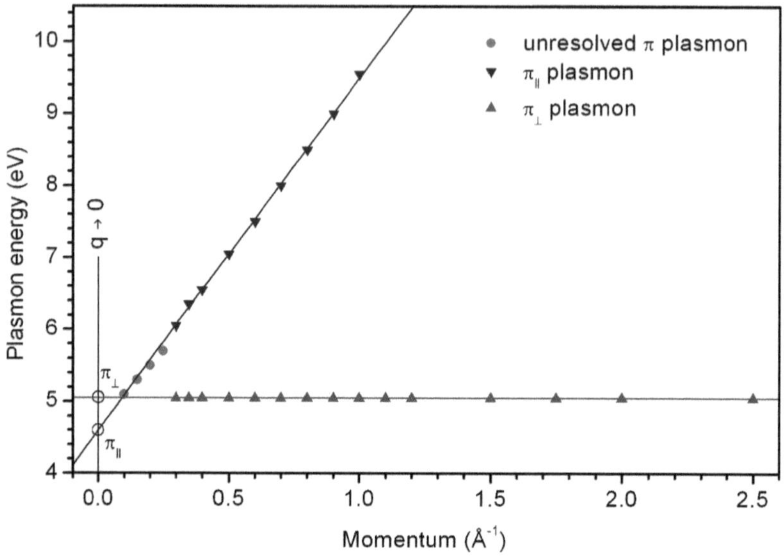

Figure 4.22. The linear dispersion of the one-dimensional on-axis $\pi_{\parallel}$ and the constant dispersion of the localized $\pi_{\perp}$ plasmon cross each other at a finite momentum. At low q the two π plasmons are no longer resolved.

The dispersions of the two distinct $\pi_{\perp}$ and the $\pi_{\parallel}$ plasmon are presented in Fig. 4.22. The localized $\pi_{\perp}$ consistently shows a perfectly flat dispersion. The $\pi_{\parallel}$ plasmon dispersion is linear. The constant slope is a fingerprint of plasmons in low-dimensional electron systems. The dispersions may be extrapolated back into the optical limit. This limit can be compared to absorption spectroscopy. In the bulk the optical limit of a plasmon will always have higher energies than the corresponding excitonic inter-band excitation. The difference between absorption peaks [Kataura *et al.* (1999)] and loss features [Pichler *et al.* (1998)] has been shown for the VHS in bulk bundled SWNTs as well as the ultraviolet absorption [Taft and Philipp (1965)] and the loss function [Marinopoulos *et al.* (2002)] of graphite. A direct comparison [Liu *et al.* (2002)] of absorption and loss spectra on various SWNT samples is presented in Fig. 4.14. Indeed, low-density hierarchical media with a diminutive electron density can be spectroscopically fingerprinted by a match between absorption spectra and their loss function. The quantitative match of the loss peaks [Kramberger *et al.* (2008)] to the UV absorption peaks [Murakami *et al.* (2005b)] is an immediate consequence of the diminutive macroscopic density of the well-separated nanotubes. The nanotubes are, with bulk spectroscopic confirmation, truly an archetypical case of isolated nanowires.

As already pointed out earlier, metallic SWNTs are unlike bulk metals — not an electronic Fermi liquid but a one-dimensional Tomonaga–Luttinger liquid (TLL) [Tomonaga (1950); Luttinger (1963)]. Its characteristic spectroscopic signature is a power law renormalization at the Fermi edge. This was experimentally found and confirmed in SWNTs [Ishii *et al.* (2003); Rauf *et al.* (2004); Dora *et al.* (2008)]. The onsets of the two valence band PES spectra in Fig. 4.16 compare the bulk Fermi liquid from gold to that of metallic SWNTs. The latter is a power law scaling with an exponent of about 0.4, which is readily distinguished from the symmetrically smeared-out step function in gold. The TTL behavior in metallic carbon nanotubes only differs from a normal Fermi liquid for energies close to E_F. Therefore it does not affect the overall band structure or the specific VHS in carbon nanotubes. Their energy range of a few eV is clearly far beyond the realm of the TLL.

In alkaline doped SWNTs, PES reveals a shift of the VHS in the valence band (see Fig. 4.17) and a concomitant increase in the asymmetry of the C1s line profile in XPS. At small shifts of E_F the DOS of metallic SWNTs stays constant and the gap of the semiconducting SWNT remains open. The power law scaling in the onset of the UPS spectra is preserved and there is also not yet any significant change in the electronic loss function [Liu *et al.* (2004)]. The first-phase transition occurs at T1 in Fig. 4.17. The shift of the charge neutrality level sets E_F to the bottom of the conduction band of the semiconducting SWNT. At this stage the electronic phase of the bulk sample is an intriguing composition of a metallic Fermi liquid in doped semiconducting SWNTs and a TLL in metallic SWNTs. The characteristic power law renormalization drops to zero as the metallic SWNTs are no longer separated by the no longer gapped semiconducting SWNTs. At T2 the charge transfer also reaches the M1 VHS in metallic SWNTs. Only then an unpercolated three-dimensional Fermi liquid is established across the bulk SWNT material. This is evidenced by the sudden emergence of a Drude plasmon at T2 in the electronic loss function of intercalated carbon nanotubes [Liu *et al.* (2004)]. A recent comprehensive review of photoemission, X-ray absorption, and also EELS on doped SWNTs may be found elsewhere [Kramberger *et al.* (2009)].

4.4.3 *Opto-Mechanical Response*

The opto-mechanical response of carbon nanotubes lies in their very special conditions for the Raman process. Raman active phonons are optically excited lattice vibrations. The phonons themselves posses structural information on the tubes' diameter or degree of crystallinity. The interaction of the phonons with the incoming as well as the outgoing photons is only mediated via electronic transitions. The cross section of the Raman process is thus resonantly enhanced as either the incoming photon energy or the scattered photon energy coincides with the VHS in the electronic JDOS.

Figure 4.23 shows a typical Raman spectrum of SWNTs. The Raman spectrum of SWNTs is composed of several peaks due to Raman active phonons. The most prominent Raman active features are the radial breathing mode (RBM), the D and the G line, as well as their overtones. The RBM is a unique mode in the tubular structure. Its frequency scales with the inverse SWNT diameter; higher RBM frequencies belong to narrower SWNTs. Thus the convoluted line shape of the RBM of a macroscopic sample offers a self-sustained way to probe its diameter distribution. The evaluation of diameter distributions from Raman spectroscopy is thoroughly compared to X-ray diffraction elsewhere [Kuzmany *et al.* (2001)]. The strong G line is well known from graphite, where it is also the strongest Raman line. As depicted in Fig. 4.23 the G line belongs to an in-plane stretching mode, while there is only one G peak in graphite. The G line in SWNTs is split into an on-axis G^+ and a circumferential G^- peak; the first always locally resembles the ideal flat situation of graphite, but the latter is softened with smaller nanotube diameters. The D line is also known from other sp^2 carbon. It originates from a phonon at the **K** point. The direct creation of such a phonon may never be allowed due to momentum conversation. However, this rule may be overcome in the presence of a structural defect, which can

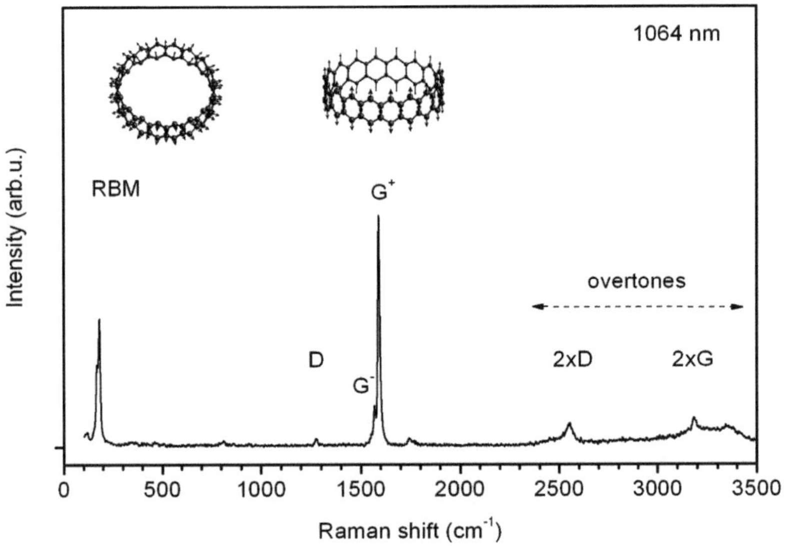

Figure 4.23. Raman spectrum of SWNTs. The excitation wavelength is 1064 nm. The insets show the displacement patterns of the RBM and the G mode, respectively.

provide an anchor point for the required backscattering to cancel the total momentum transfer. The D/G ratio is readily obtainable from Raman spectra and commonly used as an indicator for the crystallinity of nanotubes. It should be noted that the overtone of the D line does not require a defective anchor point since the two phonons can always be antiparallel and cancel one another in the total momentum.

While the Raman shift in the scattered photons is an imprint of the phonon spectrum of SWNTs, the cross section, viz. the intensity of the peaks, is an additional probe for the electron–phonon coupling and the strength of the optical transitions, respectively. The latter are strongly enhanced if the transition energy matches a VHS in the JDOS. The Raman cross section in SWNTs depends strongly on the wavelength of the excitation laser line. If either the incoming or the outgoing photon energy matches closely to an optical transition between VHS, resonant Raman scattering will occur. If the resonance condition is closely fulfilled for specific chiralities, their Raman

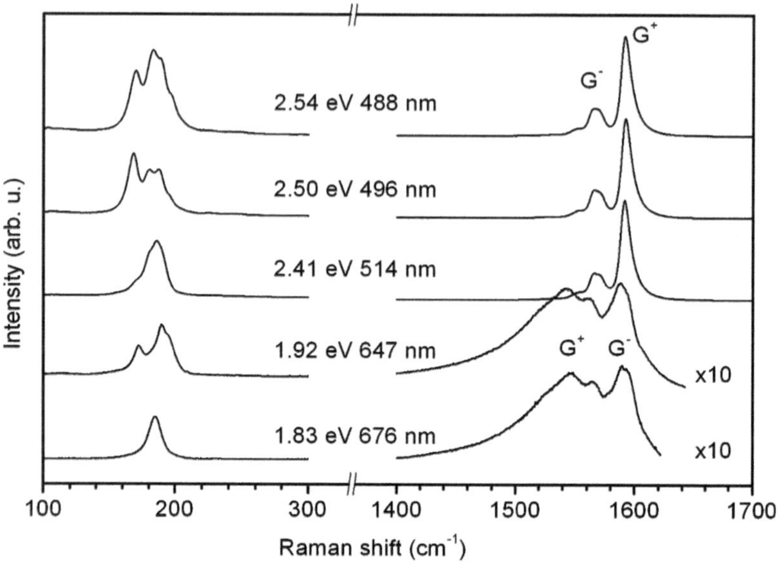

Figure 4.24. Multi-frequency resonance Raman spectra of a bulk SWNT sample. At each excitation wavelength, only the resonant SWNTs show up in the low-frequency RBM or the high-frequency G line.

cross section will dominate over all other nonresonant chiralities. A Raman experiment on a bulk sample of mixed SWNTs is always a very selective experiment, where only the resonant fraction of SWNTs shows up. Every time the same nanotube material is measured with another laser energy, another fraction of chiralities will be visible in the Raman spectrum. The spectral shape of a sample can undergo significant changes. A vivid example for the convoluted RBM of a bulk sample of SWNTs is shown in Fig. 4.24. The structured line shape of the convoluted RBM of the bulk material undergoes huge changes, in the visible spectral window.

The commonly observed intrinsic line shape of phonons is a Lorentzian profile. The Lorentzian is the natural line shape, provided the mechanical oscillations of a solid are only gradually damped. The damping directly enters the width of the Lorentzian line shape. If a discrete oscillator level (e.g., a phonon) can intermix with a continuous DOS (e.g., low-energy metallic states in metallic SWNTs), then the line shape changes to an asymmetric Fano profile [Fano

(1961)]. The G line in metallic SWNTs shows a broad asymmetric Fano line shape, which extends to lower Raman shifts compared with the symmetric Lorentzian line shape of semiconducting SWNTs. The symmetries of G^+ and G^- are exchanged with one another due to the metallic electron–phonon interaction [Piscanec *et al.* (2007)]. If a bulk sample of SWNT with a narrow diameter distribution is probed by different laser energies, the resonance window will shift between semiconducting and metallic SWNTs. This is shown in Fig. 4.24. In the blue and the green spectral range semiconducting SWNTs with $d \sim 1.4$ nm are in resonance and the G^- and G^+ peaks are Lorentzians. In the red spectral range the metallic SWNTs come in resonance. G^- and G^+ are exchanged, and the latter becomes a broadened asymmetric Fano line.

Raman spectroscopy can access either effect in doped SWNTs. Firstly, the VHS in the direct JDOS will be eventually depleted and the Raman cross section will fade out. Secondly, a shift or a change in a phonon's line shape can trace its hardening, viz. softening or changes, in the electron–phonon coupling. Raman spectroscopy on n- and p-type doped SWNTs was thoroughly investigated by Rao *et al.* [Rao *et al.* (1997)] Another example on DWNTs is given in Fig. 4.25. Here bundles of DWNTs have been intercalated with K^+ cations. The ions are located next to the surface of the outer shell of the DWNT. In the RBM region the response of the inner and the outer tube shell can be distinguished in two distinct regions. With increasing interstitial K intercalation the direct JDOS of the outer shell is rapidly quenched. The two components of the G line undergo a softening, viz. stiffening. A detailed discussion of the Raman spectra of doped DWNTs is given elsewhere [Rauf *et al.* (2006)].

4.4.4 *Alignment*

The orientation of aligned nanotubes may be experimentally accessed either by diffraction experiments or by optical methods. Diffraction experiments can be done either with electrons or X-rays and give a direct image of the polar distribution in the sample. If the sample are perfectly aligned nanotubes their diffraction pattern will just show sharp point-like Bragg reflexes. In case

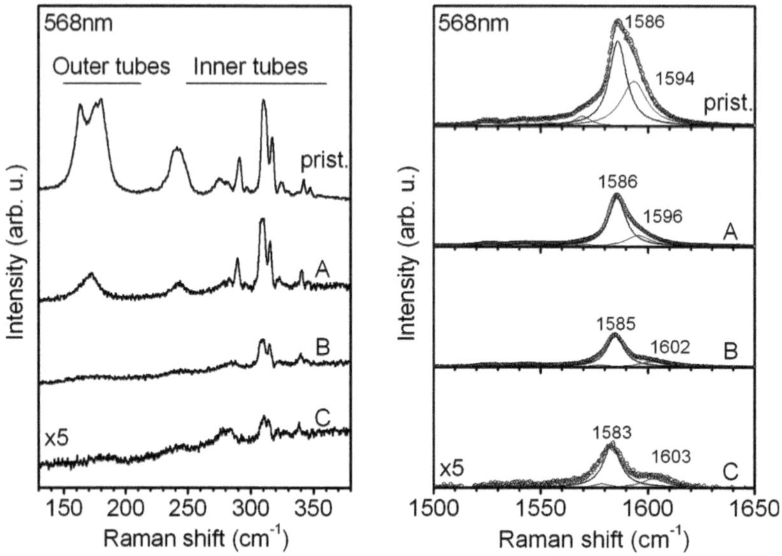

Figure 4.25. Raman spectra of pristine and successively (A, B, C) K^+ intercalated doped double-walled nanotubes (DWNTs). The left panel shows the RBM of the inner and the outer nanotubes, respectively. The right panel shows the G line of n-doped DWNT. The figure is reproduced from [Rauf *et al.* (2006)].

of dipole transitions due to electromagnetic absorption events, the image of the alignment will no longer be a direct cut. The dipole transitions in a perfectly aligned sample will scale with the square of the cosine. An important consequence is that dipole transitions can very well quantify the width or sharpness of an alignment, but they can never look into its detailed shape. The evaluation of direct projections of the angular distribution in an SWNT sample from diffraction experiments is straightforward and not directly related to their unique electronic and optical properties. The angular dependence of dipole transitions is, however, directly linked to the electronic properties. Despite their principal inferiority to diffraction methods, spectroscopic investigations of the alignment of SWNTs very often have their place. Diffraction experiments may only be conducted if free-standing samples of the right thickness can be prepared. In many cases (for instance, on any substrate) these requirements cannot be met. Then dipole transitions from absorption spectroscopies as OAS [Walters *et al.* (2001); Murakami *et al.* (2005b)] and XAS [Li *et al.* (2007); Kramberger *et al.* (2007b)] or inelastic scattering like AR-EELS [Liu *et al.* (2001)] or Raman [Murakami *et al.* (2005a)] have to be employed.

An individual SWNT is, due to its enormous aspect ratio and nanometer size, predestined for anisotropic optical properties. On the individual tube scale the absorption or scattering of light is very much pronounced for polarizations (viz. electric field vectors of the electromagnetic wave) along the nanotube axis and attenuated for crossed polarizations. The so-called antenna effect is well observed on aligned SWNTs that may be either obtained by post-synthesis filtration in strong magnetic fields [Walters *et al.* (2001)] or the direct growth of mats of vertically aligned forests of SWNTs by chemical vapor deposition [Murakami *et al.* (2005b)].

The luminescence event in an isolated SWNT is always a site-selective response to a previous absorption event on the *same* SWNTs. If a specific isolated semiconducting nanotube in the dispersion absorbs a photon, then the very same nanotube will emit another photon from the electron–hole recombination at the bandgap. The local correlation allows to distinguish luminescence events where the incoming and outgoing photons have the same polarization from those where they have crossed polarization. In such experiments parallel and cross-polarized absorption events on carbon nanotubes can be identified. For instance, the $E_{22} \rightarrow E_{11}$ process is visible if the incoming and the outgoing photons are polarized in parallel. If their polarizations are crossed then the $E_{12} \rightarrow E_{11}$ PL peak is observed at a longer excitation wavelength [Miyauchi *et al.* (2006)]. Polarized PLS not only confirms the existence of odd optical transition in SWNT [Gruneis *et al.* (2004)], but it simultaneously shows that they are, locally on the individual nanotube level, polarized perpendicularly to direct optical transitions.

In anisotropic matter the X-ray absorption edge can vary strongly with the polarization, depending on how well the individual dipole transitions into different orbitals actually point along the photons polarization. For instance, in graphite an in-plane polarized X-ray photon can only be absorbed by transitions into the empty σ states. If the electric field vector of the X-ray is pointing out of plane, then only transitions into unoccupied π states are possible. In carbon nanotubes the in-plane and the out-of-plane components are only fully separated for polarizations along the tube axis. Then just the in-plane component is visible, whereas cross-polarized X-rays see a geometric average of both components [Li *et al.* (2007); Kramberger *et al.* (2007b)]. The isotropic mixture of π and σ states in the XAS response of SWNTs is displayed in Fig. 4.18. The polarization dependence in the XAS resonance in vertically aligned SWNTs shows a consistent behavior of the entire π as well as the σ electronic system. It is thus a full extension to polarized OAS experiments, which are limited to just the π states.

4.4.5 *Metallic and Semiconducting Abundances*

As elucidated in the section 4.4.2 the metallicity of carbon nanotubes causes numerous spectroscopic fingerprints in various methods. However, the principal drawback of all these metallic spectroscopic modulations is that they cannot be quantitatively compared to the (nonexistent) semiconducting modulations. The indirect spectrocopies are well suited to detect either a semiconducting or metallic enrichment in a sample, but they can only show trends. The quantitative ratio of metallic to semiconducting SWNTs in a sample may only be deduced from an equally shared, yet distinct, spectroscopic signature. The macroscopic VHS in the bulk just fulfill these two critical requirements. They are equally and simultaneously accessible for either type of SWNT and yet they may be distinguished because they differ in energy. The separated peak areas directly reflect the relative abundances in metallicity selected SWNTs. Up to date the macroscopic VHS may be quantitatively accessed with OAS, EELS, UPS, and XAS. In terms of feasibility OAS is the clear favorite for characterization purposes on separated nanotubes [Miyata *et al.* (2008)]. Sorting SWNTs into metallic and semiconducting fractions is a crucial prerequisite for scalable integration into electronic devices. So far, there are three different and reproducible strategies to separate metallic and semiconducting SWNTs. These are dielectrophoresis [Krupke *et al.* (2003)], trapping in agarose gel [Tanaka *et al.* (2009)], and density-gradient ultracentrifugation [Arnold *et al.* (2006)]. The latter can not only separate semiconducting and metallic SWNTs, but it can sort the SWNTs very narrowly by their diameter.

4.4.6 *Diameter Distribution*

The diameter of an SWNT may be spectroscopically probed via its VHS and its RBM frequency. The macroscopic VHS in the interband transitions can appear in only samples with reasonably narrow diameter distributions. They are typically well resolved for mean diameters of less than 2 nm and a distribution width of less than ± 0.2 nm. The macroscopic VHS are best and most facilely observed in OAS [Liu *et al.* (2002)]. An alternate facile way to access the diameter distribution in a bulk sample of SWNTs is multi-frequency Raman spectroscopy [Kuzmany *et al.* (2001)]. Here several excitation wavelengths have to be combined to a comprehensive sampling of the actual diameter distribution. After their careful calibration to diffraction experiments, OAS and Raman spectroscopy are nowadays well-established methods for a reliable determination of the diameter distribution in bulk SWNTs.

If a diameter distribution is too broad or the mean diameter is too large, then there is no bulk spectroscopic way to access it in detail. In such cases, counting statistics from numerous TEM micrographs are typically applied.

4.4.7 *Crystallinity*

The crystallinity of carbon nanotubes may be feasibly accessed by the D/G ratio from Raman spectroscopy. As this is a unique stand-alone bulk characterization method, it may only be used for the relative comparison of otherwise similar samples. Lattice distortions such as pentagon/heptagon pairs, add-atoms, or vacancies are known from aberration-corrected TEM imaging of nanotubes, but their concentration cannot be quantitatively deduced from the D/G ratio of bulk samples. Microscopic studies on individual nanotubes show that the D/G ratio has local variations that are connected to tube ends and possibly local defects [Hartschuh (2008)]. Local optical probes unlock valuable insight from fundamental studies, but in terms of bulk characterization they simply lack representability.

Although up until now there is simply no quantitative spectroscopic method available to access the crysallinity of SWNTs, the D/G ratio is commonly used for process monitoring. It gives valuable feedback in the synthesis and purification of SWNT material.

4.4.8 *Purity*

Sample purity is always a crucial issue if carbon nanotubes are either subject to spectroscopic studies or being used in completely other studies. There are different types of impurities that have to be distinguished. Amorphous carbon and carbon-coated metal particles can be common, but not necessarily definite, by-products of the synthesis. For fundamental studies on the intrinsic spectroscopic features of SWNTs, either of them have to be removed in a purification procedure [Ishii *et al.* (2003)]. Traces from catalyst metals can, with enormous sensitivity, be evidenced by XPS or X-ray luminescence. Even boron and nitrogen heteroatoms incorporated into the walls of nanotubes [Ayala *et al.* (2007, 2008)] can be identified and quantified from their site-dependent chemical shifts. Quantitatively distinguishing amorphous carbon from carbon nanotubes is a hard challenge since both share with an *a priori* unknown portion the general reminiscence of any sp^2 carbon. A good visual impression of the ratio of carbon nanotubes to amorphous species is given by TEM overview micrographs. Individual aspects of a sample's purity can be accurately measured, but there is always the need to combine several methods for a comprehensive picture of a given sample. The complete picture of the purity of an SWNT sample with its morphology and elemental composition and elemental distribution can never fit into a single percentage. Percentages provided by commercial suppliers typically refer to the result from one chosen method under defined conditions.

From a spectroscopists point of view, more pure samples that have a higher content of nanotubes will always show more and stronger pronounced signatures of nanotubes. These are macroscopic VHS, TLL behavior, RBM, and a reasonable G/D ratio. There will also be less unstructured response from impurities. Besides clean TEM overview micrographs without too many dark spots from metal particles or heavily carbon-coated nanotubes, the most compelling and commonly available spectroscopic signature of good SWNT materials are VHS in the OAS comparable to Fig. 4.14 and a clean Raman spectrum as in Fig. 4.23.

4.5 Summary

Owing to the fundamental one-dimensional physics they hold in them, carbon nanotubes are a truly exceptional material. Due to their enormous aspect ratio, they reside at the borderline of point-like molecules or clusters with well-localized properties and bulk, extended solids with itinerant properties. The structural roll-up of a stripe of graphene introduces the distinction of longitudinal and circumferential modes. The parent compound graphene is a semimetal, but nanotubes may either be metals or semiconductors depending on the detailed symmetry of their chiral twist. Their roll-up splits phonons and plasmons (mechanical and electrical waves) and also the fundamental electronic wavefunctions in nanotubes into continuous on-axis and quantized circumferential modes. The discrete circumferential electronic states give rise to diverging van Hove singularities in the nanotubes' optical transitions. They lead to a very strongly enhanced coupling to light if their color matches narrowly to the electronic transition energy. The resonant coupling to electromagnetic waves leads to a broad variety of spectroscopic signatures in absorbing spectroscopies such as photoemission, optical absorption, luminescence, resonance Raman, and X-ray absorption. Inelastic scattering methods such as electron energy loss and Raman spectroscopy can access the unique phonon and plasmon modes in carbon nanotubes. The wealth of spectroscopic evidence presented here firmly corroborates the intriguing concept of nanotubes possessing the duality of molecular localized and bulk itinerant properties.

Acknowledgments

Great thanks for many fruitful and stimulating discussions go to our colleagues here at the University of Vienna: Rudolf Pfeiffer, Paola

Ayala, Ferenc Simon and Wolfgang Plank. They were always willing to share their thoughts and ideas with us, which helped a lot on our way from the concept of the wide spread topic into a compact review on the *Electronic and optical properties of carbon nanotubes.*

References

Araujo, P. T., Doorn, S. K., Kilina, S., Tretiak, S., Einarsson, E., Maruyama, S., Chacham, H., Pimenta, M. A. and Jorio, A. (2007). Third and fourth optical transitions in semiconducting carbon nanotubes, *Phys. Rev. Lett.* **98**, 6, p. 067401.

Arnold, M. S., Green, A. A., Hulvat, J. F., Stupp, S. I. and Hersam, M. C. (2006). Sorting carbon nanotubes by electronic structure using density differentiation, *Nat. Nanotec.* **1**, 1, pp. 60–65.

Ayala, P., Grueneis, A., Gemming, T., Grimm, D., Kramberger, C., Ruemmeli, M. H., Freire, F. L., Kuzmany, H., Pfeiffer, R., Barreiro, A., Buechner, B. and Pichler, T. (2007). Tailoring n-doped single and double wall carbon nanotubes from a nondiluted carbon/nitrogen feedstock, *J. Phys. Chem. C* **111**, 7, pp. 2879–2884.

Ayala, P., Plank, W., Gruneis, A., Kauppinen, E. I., Rummeli, M. H., Kuzmany, H. and Pichler, T. (2008). A one step approach to b-doped single-walled carbon nanotubes, *J. Mater. Chem.* **18**, 46, pp. 5676–5681.

Bachilo, S. M., Strano, M. S., Kittrell, C., Hauge, R. H., Smalley, R. E. and Weisman, R. B. (2002). Structure-assigned optical spectra of single-walled carbon nanotubes, *Science* **298**, 5602, pp. 2361–2366.

Batson, P. E. (1993). Carbon-1s near-edge-absorption fine-structure in graphite, *Phys. Rev. B* **48**, 4, pp. 2608–2610.

Charlier, J.-C., Blase, X. and Roche, S. (2007). Electronic and transport properties of nanotubes, *Rev. of Mod. Phys.* **79**, 2, 677, and references therein.

Charlier, J. C., Eklund, P. C., Zhu, J. and Ferrari, A. C. (2008). *Electron and phonon properties of graphene: Their relationship with carbon nanotubes, Topics in applied physics,* Vol. 111 (Springer Verlag, Berlin Heidelberg).

DasSarma, S. and Hwang, E. H. (1996). Dynamical response of a one-dimensional quantum-wire electron system, *Phys. Rev. B* **54**, 3, pp. 1936–1946.

Doniach, S. and Sunjic, M. (1970). Many-electron singularity in x-ray photoemission and x-ray line spectra from metals, *Jour. of Phys. C* **3**, 2, p. 285.

Dora, B., Gulacsi, M., Koltai, J., Zolyomi, V., Kurti, J. and Simon, F. (2008). Electron spin resonance signal of luttinger liquids and single-wall carbon nanotubes, *Phys. Rev. Lett.* **101**, 10, p. 106408.

Drude, P. (1900). Zur elektronentheorie der metalle, *Ann. d. Phys.* **306 (3)**, pp. 566–613.

Einarsson, E., Shiozawa, H., Kramberger, C., Rummeli, M. H., Gruneis, A., Pichler, T. and Maruyama, S. (2007). Revealing the small-bundle internal structure of vertically aligned single-walled carbon nanotube films, *J. Phys. Chem. C* **111**, 48, pp. 17861–17864.

Einstein, A. (1905). Über einen die Erzeugung und Verwandlung des Lichtes betreffenden heuristischen Gesichtspunkt, *Ann. d. Phys.* **17**, p. 132.

Eklund, P. C., Holden, J. M. and Jishi, R. A. (1995). Vibrational-modes of carbon nanotubes - spectroscopy and theory, *Carbon* **33**, 7, pp. 959–972.

Fano, U. (1961). Effects of configuration interaction on intensities and phase shifts, *Phys. Rev.* **124**, 6, p. 1866.

Fink, J. (1989). Recent developments in energy-loss spectroscopy, *Adv. Elec. and Elec. Phys.* **75**, pp. 121–232.

Gruneis, A., Attaccalite, C., Wirtz, L., Shiozawa, H., Saito, R., Pichler, T. and Rubio, A. (2008). Tight-binding description of the quasiparticle dispersion of graphite and few-layer graphene, *Phys. Rev. B* **78**, 20, p. 205425.

Gruneis, A., Saito, R., Jiang, J., Samsonidze, G. G., Pimenta, M. A., Jorio, A., Souza, A. G., Dresselhaus, G. and Dresselhaus, M. S. (2004). Resonant raman spectra of carbon nanotube bundles observed by perpendicularly polarized light, *Chem. Phys. Lett.* **387**, 4-6, pp. 301–306.

Hamada, N., Sawada, S.-i. and Oshiyama, A. (1992). New one-dimensional conductors: Graphitic microtubules, *Phys. Rev. Lett.* **68**, 10, pp. 1579–1581, doi:10.1103/PhysRevLett.68.1579.

Hartschuh, A. (2008). *New techniques for carbon-nanotube study and characterization*, *Topics in applied physics*, Vol. 111 (Springer Verlag, Berlin Heidelberg).

Hedin, L. (1965). New method for calculating 1-particle greens function with application to electron-gas problem, *Phys. Rev.* **139**, 3A, p. A796.

Hiura, H., Ebbesen, T. W., Tanigaki, K. and Takahashi, H. (1993). Raman studies of carbon nanotubes, *Chem. Phys. Lett.* **202**, 6, pp. 509–512.

Ishii, H., Kataura, H., Shiozawa, H., Yoshioka, H., Otsubo, H., Takayama, Y., Miyahara, T., Suzuki, S., Achiba, Y., Nakatake, M., Narimura, T., Higashiguchi, M., Shimada, K., Namatame, H. and Taniguchi, M. (2003). Direct observation of Tomonaga-Luttinger-liquid state in carbon nanotubes at low temperatures, *Nature* **426**, pp. 540–544.

Itkis, M. E., Niyogi, S., Meng, M. E., Hamon, M. A., Hu, H. and Haddon, R. C. (2002). Spectroscopic study of the fermi level electronic structure of single-walled carbon nanotubes, *Nano Letters* **2**, 2, pp. 155–159.

Kataura, H., Kumazawa, Y., Maniwa, Y., Umezu, I., Suzuki, S., Ohtsuka, Y. and Achiba, Y. (1999). Optical properties of single-wall carbon nanotubes, *Synth. Met.* **103**, 1-3, pp. 2555–2558.

Kittel, C. (1963). *Quantum Theory of Solids* (John Wiley & Sons, Inc.).

Kohn, W. and Sham, L. J. (1965). Self-consistent equations including exchange and correlation effects, *Phys. Rev.* **140**, 4A, p. 1133.

Kramberger, C., Hambach, R., Giorgetti, C., Rummeli, M. H., Knupfer, M., Fink, J., Buchner, B., Reining, L., Einarsson, E., Maruyama, S., Sottile, F., Hannewald, K., Olevano, V., Marinopoulos, A. G. and Pichler, T. (2008). Linear plasmon dispersion in single-wall carbon nanotubes and the collective excitation spectrum of graphene, *Phys. Rev. Lett.* **100**, 19, p. 196803.

Kramberger, C., Rauf, H., Knupfer, M., Shiozawa, H., Batchelor, D., Rubio, A., Kataura, H. and Pichler, T. (2009). Potassium-intercalated single-wall carbon nanotube bundles: Archetypes for semiconductor/metal hybrid systems, *Phys. Rev. B* **79**, 19, p. 195442.

Kramberger, C., Rauf, H., Shiozawa, H., Knupfer, M. and Buchner, B. (2007a). Unraveling van hove singularities in x-ray absorption response of single-wall carbon nanotubes, *Phys. Rev. B.* **75**, p. 235437.

Kramberger, C., Shiozawa, H., Rauf, H., Gruneis, A., Rummeli, M. H., Pichler, T., Buchner, B., Batchelor, D., Einarsson, E. and Maruyama, S. (2007b). Anisotropy in the x-ray absorption of vertically aligned single wall carbon nanotubes, *Phys. Stat. Sol. B* **244**, 11, pp. 3978–3981.

Krupke, R., Hennrich, F., von Lohneysen, H. and Kappes, M. M. (2003). Separation of metallic from semiconducting single-walled carbon nanotubes, *Science* **301**, 5631, pp. 344–347.

Kuzmany, H. (1998). *Solid-state spectroscopy* (Springer Verlag, Berlin Heidelberg New York).

Kuzmany, H., Plank, W., Hulman, M., Kramberger, C., Gruneis, A., Pichler, T., Peterlik, H., Kataura, H. and Achiba, Y. (2001). Determination of swcnt diameters from the raman response of the radial breathing mode, *Eur. Phys. J. B* **22**, 3, pp. 307–320.

Lefebvre, J., Maruyama, S. and Finnie, P. (2008). *Photoluminescence: Science and applications, Topics in applied physics*, Vol. 111 (Springer Verlag, Berlin Heidelberg).

Li, Z. R., Zhang, L., Resasco, D. E., Mun, B. S. and Requejo, F. G. (2007). Angle-resolved x-ray absorption near edge structure study of vertically aligned single-walled carbon nanotubes, *Appl. Phys. Lett.* **90**, 10, p. 103115.

Lindhard, J. (1954). On the properties of a gas of charged particles, *Kgl. Dans. Videsk. Selsk. Mat. Fys. Medd.* **28**, 8, pp. 1–57.

Liu, X., Pichler, T., Knupfer, M., Fink, J. and Kataura, H. (2004). Electronic properties of fecl3-intercalated single-wall carbon nanotubes, *Phys. Rev. B* **70**, 20, p. 205405.

Liu, X., Pichler, T., Knupfer, M., Golden, M. S., Fink, J., Kataura, H. and Achiba, Y. (2002). Detailed analysis of the mean diameter and diameter distribution of single-wall carbon nanotubes from their optical response, *Phys. Rev. B* **66**, 4, p. 045411.

Liu, X., Pichler, T., Knupfer, M., Golden, M. S., Fink, J., Walters, D. A., Casavant, M. J., Schmidt, J. and Smalley, R. E. (2001). An electron energy-loss study of the structural and electronic properties of magnetically aligned single wall carbon nanotubes, *Synth. Met.* **121**, 1-3, pp. 1183–1186.

Lof, R. W., Vanveenendaal, M. A., Koopmans, B., Jonkman, H. T. and Sawatzky, G. A. (1992). Band gap, excitons, and coulomb interaction in solid c-60, *Phys. Rev. Lett.* **68**, 26, pp. 3924–3927.

Luttinger, J. M. (1963). An Exactly Soluble Model of a Many-Fermion System, *Jour. of Math. Phys.* **4**, 9, p. 1154.

Marinopoulos, A. G., Reining, L., Olevano, V., Rubio, A., Pichler, T., Liu, X., Knupfer, M. and Fink, J. (2002). Anisotropy and interplane interactions in the dielectric response of graphite, *Phys. Rev. Lett.* **89**, 7, p. 076402.

Maultzsch, J., Pomraenke, R., Reich, S., Chang, E., Prezzi, D., Ruini, A., Molinari, E., Strano, M. S., Thomsen, C. and Lienau, C. (2005). Exciton binding energies in carbon nanotubes from two-photon photoluminescence, *Phys. Rev. B* **72**, 24, p. 241402.

Miyata, Y., Yanagi, K., Maniwa, Y. and Kataura, H. (2008). Optical evaluation of the metal-to-semiconductor ratio of single-wall carbon nanotubes, *J. Phys. Chem. C* **112**, 34, pp. 13187–13191.

Miyauchi, Y., Oba, M. and Maruyama, S. (2006). Cross-polarized optical absorption of single-walled nanotubes by polarized photoluminescence excitation spectroscopy, *Phys. Rev. B* **74**, 20, p. 205440.

Murakami, Y., Chiashi, S., Einarsson, E. and Maruyama, S. (2005a). Polarization dependence of resonant raman scattering from vertically aligned single-walled carbon nanotube films, *Phys. Rev. B* **71**, 8, p. 085403.

Murakami, Y., Einarsson, E., Edamura, T. and Maruyama, S. (2005b). Polarization dependence of the optical absorption of single-walled carbon nanotubes, *Phys. Rev. Lett.* **94**, 8, p. 087402.

Ohno, Y., Iwasaki, S., Murakami, Y., Kishimoto, S., Maruyama, S. and Mizutani, T. (2006). Chirality-dependent environmental effects in photoluminescence of single-walled carbon nanotubes, *Phys. Rev. B* **73**, 23, p. 235427.

Onida, G., Reining, L. and Rubio, A. (2002). Electronic excitations: density-functional versus many-body Green's-function approaches, *Rev. Mod. Phys.* **74**, p. 601.

Painter, G. S. and Ellis, D. E. (1970). Electronic band structure and optical properties of graphite from a variational approach, *Phys. Rev. B* **1**, 12, p. 4747.

Pichler, T., Knupfer, M., Golden, M. S., Fink, J., Rinzler, A. and Smalley, R. E. (1998). Localized and delocalized electronic states in single-wall carbon nanotubes, *Phys. Rev. Lett.* **80**, 21, pp. 4729–4732.

Piscanec, S., Lazzeri, M., Robertson, J., Ferrari, A. C. and Mauri, F. (2007). Optical phonons in carbon nanotubes: Kohn anomalies, peierls distortions, and dynamic effects, *Phys. Rev. B* **75**, 3, p. 035427.

Rao, A. M., Eklund, P. C., Bandow, S., Thess, A. and Smalley, R. E. (1997). Evidence for charge transfer in doped carbon nanotube bundles from Raman scattering, *Nature* **388**, 6639, pp. 257–259.

Rauf, H., Pichler, T., Knupfer, M., Fink, J. and Kataura, H. (2004). Transition from a tomonaga-luttinger liquid to a fermi liquid in potassium-intercalated bundles of single-wall carbon nanotubes, *Phys. Rev. Lett.* **93**, 9, p. 096805.

Rauf, H., Pichler, T., Pfeiffer, R., Simon, F., Kuzmany, H. and Popov, V. N. (2006). Detailed analysis of the raman response of n-doped double-wall carbon nanotubes, *Phys. Rev. B* **74**, 23, p. 235419.

Spataru, C. D., Ismail-Beigi, S., Capaz, R. B. and Louie, S. G. (2008). *Quasiparticle and excitonic effects in the optical response of nanotubes and nanoribbons*, *Topics in applied physics*, Vol. 111 (Springer Verlag, Berlin Heidelberg).

Taft, E. A. and Philipp, H. R. (1965). Optical properties of graphite, *Phys. Rev.* **138**, 1A, p. A197.

Tanaka, T., Jin, H., Miyata, Y., Fujii, S., Suga, H., Naitoh, Y., Minari, T., Miyadera, T., Tsukagoshi, K. and Kataura, H. (2009). Simple and scalable gel-based separation of metallic and semiconducting carbon nanotubes, *Nano Lett.* **9**, 4, pp. 1497–1500.

Telg, H., Maultzsch, J., Reich, S., Hennrich, F. and Thomsen, C. (2004). Chirality distribution and transition energies of carbon nanotubes, *Phys. Rev. Lett.* **93**, 17, p. 177401.

Tomonaga, S. (1950). Remarks on blochs method of sound waves applied to many-fermion problems, *Prog. of Theo. Phys.* **5**, 4, pp. 544–569.

Ugawa, A., Rinzler, A. G. and Tanner, D. B. (1999). Far-infrared gaps in single-wall carbon nanotubes, *Phys. Rev. B* **60**, 16, pp. R11305–R11308, doi: 10.1103/PhysRevB.60.R11305.

Wallace, P. R. (1947). The band theory of graphite, *Phys. Rev.* **71**, 9, pp. 622–634.

Walters, D. A., Casavant, M. J., Qin, X. C., Huffman, C. B., Boul, P. J., Ericson, L. M., Haroz, E. H., O'Connell, M. J., Smith, K., Colbert, D. T. and Smalley, R. E. (2001). In-plane-aligned membranes of carbon nanotubes, *Chem. Phys. Lett.* **338**, 1, pp. 14–20.

Wang, F., Dukovic, G., Brus, L. E. and Heinz, T. F. (2005). The optical resonances in carbon nanotubes arise from excitons, *Science* **308**, 5723, pp. 838–841.

Wessely, O., Katsnelson, M. I. and Eriksson, O. (2005). Ab initio theory of dynamical core-hole screening in graphite from x-ray absorption spectra, *Phys. Rev. Lett.* **94**, 16, p. 167401.

Chapter 5

Fullerene-Based Electronics

James M. Ball, Paul H. Wöbkenberg, and Thomas D. Anthopoulos
Department of Physics, Imperial College London, Prince Consort Road, London, SW7 2BZ, UK
thomas.anthopoulos@imperial.ac.uk

The family of hollow ellipsoid fullerenes is an important and widely studied class of small molecule semiconducting materials used in various electronic devices. In this chapter, we will outline the electronic properties of fullerenes, their preparation in thin-films, and the physics of devices in which they are used. Furthermore, we will highlight important advances in the field of fullerene-based electronics and offer an outlook on future directions and challenges.

5.1 Introduction

Although electronic conduction was demonstrated in molecular solids early in the 20th century,[1] these materials received little interest until the discovery of electroluminescence in anthracene crystals in the 1960s.[2] Following important studies on the conductivity of doped polyacetylene,[3] several demonstrations of organic

Advances in Carbon Nanomaterials: Science and Applications
Edited by Nikos Tagmatarchis

ISBN 978-981-426-78-78 (Hardcover), 978-981-426-78-85 (eBook)
www.panstanford.com

semiconductor devices based on small molecules and polymers were presented in the 1980s.[4–9] Since then, organic semiconductors have remained a highly popular subject of interest for both fundamental and applied research.

The appeal of organic semiconductor materials, in comparison to their conventional inorganic counterparts, lies in their ease of processing for device fabrication near room-temperature on low-cost substrates such as glass or plastics.[10] With the use of solution-based semiconductor deposition procedures such as spin-coating, spray-coating, and ink-jet printing, the manufacturing cost of devices can potentially be significantly reduced. Solvent-free vapor phase deposition and stamping techniques also offer lower-cost alternatives for depositing organic materials in comparison to conventional photolithography.

Organic solids are relatively soft materials characterized by weak intermolecular van der Waals interactions compared to stronger intramolecular covalent bonds found in hard, brittle inorganic materials. This disparity in electronic coupling between charge transport sites means that organic materials will never reach the electrical performance of highly crystalline inorganic semiconductors. Development of organic semiconducting thin-films[11–14] has enabled room-temperature charge carrier mobilities to reach $\sim$1 cm^2/Vs, significantly lower than typical mobilities of $\sim 10^3$–10^4 cm^2/Vs observed in crystalline inorganic films.[15] However, the ability to deposit these materials over large areas at low-cost with additional properties such as flexibility[16] and transparency[17] means that organic semiconductors will be suitable for applications currently inaccessible to conventional inorganic semiconductors. With the plethora of materials that can be chemically synthesized and tailored for purpose, organic semiconductors are expected to command attention in future electronic devices.

A variety of electronic and optoelectronic applications have been explored with these semiconductors including organic photovoltaics (OPV),[18–20] organic light emitting devices (OLEDs),[8,21] photodetectors,[22] memory devices,[23] and organic field-effect transistors (OFETs).[24–26] At the time of writing, the first generation of applications implementing OLEDs, OPV, and OFETs has reached or is on the verge of reaching commercialization with many more examples fast approaching.

OFETs are anticipated to meet the performance requirements for commercial implementation into integrated circuits,[27] back-planes for optical displays,[28,29] and large volume microelectronic applications such as radio-frequency identification tags.[30,31] Several studies of charge transport in OFETs have suggested that many materials display conduction of only a single polarity of charge carrier,[24,32–34] typically holes. As a result, a number of hole transporting (p-channel) organic semiconductors have been developed and studied extensively and are readily employed in organic unipolar logic circuits.[31] However, it has been shown that complementary inverters, a fundamental building block for logic circuits, comprising both electron (n-channel) and hole transporting OFETs, provide lower power consumption and wider noise margins compared to their unipolar counterparts.[27] This presents the requirement for the development of high-performance n-channel organic semiconductors.[35]

A significant technological challenge associated with organic complementary circuits is the deposition and patterning of the p- and n-channel semiconductors with high resolution. Solution processing techniques, e.g., ink-jet printing,[36] could potentially provide cheap and simple methods of patterned deposition. Many research groups have therefore focused on the synthesis of soluble organic semiconductors. This has resulted in a number of examples of soluble p-channel semiconductors with field-effect mobilities comparable to amorphous silicon.[11,12] In contrast, progress on development of soluble n-channel semiconductors has yielded relatively few examples with high-performance.[37–40] Ambient stability of n-channel materials is also a significant problem.[35] The study and development of soluble n-channel organic materials therefore carries importance towards the advancement of high-performance organic integrated circuits.

OPV devices present the prospect of supplying low-cost energy that can help alleviate the global dependence on non-renewable sources. Current state-of-the-art cells have a power conversion efficiency of >5%[41,42] with upper-limit estimates in the range 10–15%[43,44] for solution processed systems. Bulk-heterojunction (BHJ) OPV cells are the most widely studied class of organic devices for extracting useful energy from the conversion of light to electrical current.[18,19] BHJ cells comprise an interpenetrating network of both n-channel (electron acceptor, A) and p-channel (electron donor, D) semiconductors to provide a charge separation surface area within the blend larger than can be obtained with a bilayer. Excitons are photoinduced in the donor and diffuse to the D–A interface where the electrons and holes are separated and transported to the cathode and anode, respectively, producing a current. An efficient exciton dissociation step is crucial for generating useful current and depends significantly on many properties associated with the acceptor material. These includes both its electronic structure with respect to the donor and anode[18] as well as its thermodynamic properties within the blend.[45] As the electron acceptor, it is favorable that the material has a high electron affinity but it also needs to be easily processable, preferably from solution, to allow control over the thin-film nanostructure and optimization of the blend morphology. Relatively few materials fulfill these requirements.

The implementation of fullerenes into OFETs and OPV cells as electron transporters has proven fruitful in several examples[18,37,39,45] towards overcoming the aforementioned difficulties. It is the aim of the present chapter to discuss the properties of fullerenes and why they are suitable for OFET and OPV applications. We will review the important discoveries and studies that have been achieved with this family of molecules and present our perspective on future directions and challenges.

5.2 Properties of Fullerenes

In 1985, Kroto *et al.* reported the discovery of the third allotrope of carbon.[46] Known as Buckminsterfullerene, C_{60} is a hollow truncated icosahedron comprised exclusively of carbon atoms at each of its 60 vertices as shown in Fig. 5.1a. This molecule was named after the architect Richard Buckminster Fuller, who popularized the use of its shape in geodesic domes prior to the scientist's discovery.

The family of closed ellipsoidal fullerenes, analogues of the originally discovered molecule, are composed of 12 pentagons completely surrounded by *n* hexagons (isolated pentagon rule, IPR) as required by Euler's theorem.[47] C_{60}, with 20 hexagons, is the smallest and most abundant stable fullerene for which this rule is

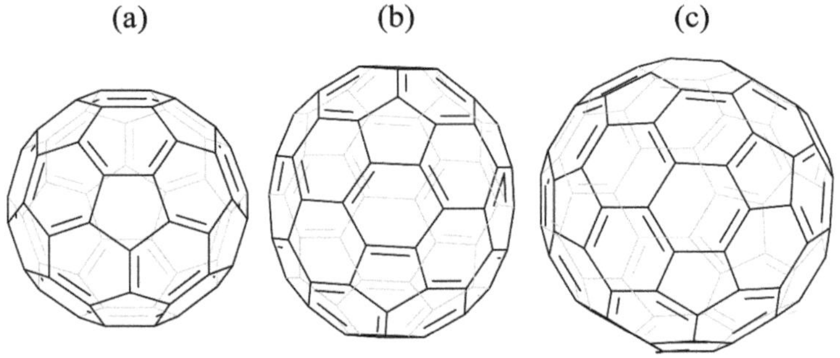

Figure 5.1. Molecular structures of the three most abundant fullerenes. (a) C_{60}, (b) C_{70}, and (c) C_{84} isomer with D_{2d} symmetry.

obeyed. Several smaller fullerenes exist with fewer carbon atoms for which connected pentagons are required to close the cage. Higher fullerenes that satisfy the IPR with more carbon atoms will be discussed with particular emphasis on the second and third most abundant fullerenes, C_{70} and C_{84} (shown in Fig. 5.1b and Fig. 5.1c respectively).

The unique electronic properties of fullerenes that give rise to their favorable implementation into devices are outlined in this section. Chemical modification of the basic cage is also described as a route towards tailoring fullerenes for purpose and several prominent examples are explained.

5.2.1 *Electronic Properties*

The electronic properties of a fullerene carbon cage arise from the confinement of the constituent electrons, resulting in a structure that is electronically zero-dimensional. For an individual fullerene, this gives rise to an electronic structure that is composed of discrete energy levels. In C_{60}, each carbon atom is bound to three others at the intersection between two hexagons and one pentagon. The pentagons allow sufficient curvature for the cage to close introducing pyramidalization of the σ-bonds of each vertex. This pyramidalization modifies the sp^2 orbital hybridization that would be expected from a planar conjugated system. The diminished p character of the remaining electron orbital at each vertex contributes to the modified delocalized molecular π-orbital that is extended further beyond the outer surface of the cage than within the interior.[48] The energetically low lying 2s orbital of each carbon mixes with the 2p orbital leading to a lowest unoccupied π-orbital with a higher electron affinity than that which results from purely 2p orbitals in planar systems. This effect is diminished as the size of the fullerene cage increases because the extent of pyramidalization at each vertex is reduced.[49]

The highest occupied molecular orbital (HOMO) of C_{60} is completely filled (closed shell). The lowest unoccupied molecular orbital (LUMO) is triply degenerate and therefore capable of accepting up to six electrons. Electrochemical measurements in solution have detected all six reductions reversibly[50] in qualitative agreement with predictions of the electronic structure calculated by Hückel molecular orbital theory.[51] The extent of LUMO level degeneracy in extractable fullerenes is closely related to the fact that 12 pentagons with a dimerized arrangement are required to close the cage.[52] Common to the most abundant fullerenes is energetic bunching of unoccupied molecular orbitals in groups of three, spatially distributed around dimerized pentagons.[52] The result is that all fullerenes that fulfill the IPR have six low lying unoccupied energy levels even in larger, less symmetric structures than C_{60}.

The C_{70} molecule can be envisioned by adding a ring of five hexagons along the equatorial plane of C_{60}, reducing its relative symmetry.[47] For fullerenes higher than C_{76}, addition of further carbon atoms results in an increase of the number of structural isomers for that fullerene. In these cases the symmetry of the isomer determines its electronic properties. Synthesis of C_{84}, which has 24 isomers,[53] is predicted to produce two stable isoenergetic structures with D_2 and D_{2d} symmetry[54] consistent with NMR spectra that suggest a 2:1 weight ratio of the respective isomers.[55] This means that C_{84} is typically processed in devices as an isomeric mixture. It should be noted that only 3–4% by weight[56] of fullerenes produced by the graphite arc process[57] are fullerenes other than C_{60} and C_{70}. This has resulted in relatively few reports of devices based on alternative fullerenes.

The quasi-spherical surface of the carbon cage adds strain energy to the bonding between carbon atoms that is not encountered in

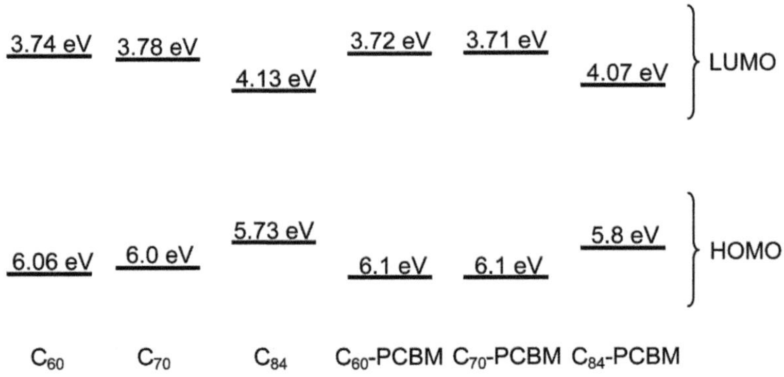

Figure 5.2. HOMO and LUMO levels for C_{60}, C_{70}, C_{84},[58] and their PCBM derivatives[39,59,60] extracted from cyclic voltammetry data. Energies are given with respect to the vacuum level.

planar systems.[48] Relief from this strain is the main driving force for exohedral chemical reactions of fullerenes. A reactivity comparison of fullerenes to other aromatic molecules such as benzene cannot be made because the absence of hydrogen prevents the possibility of substitution reactions. This means that all chemical changes to fullerenes result in a change of structure and therefore a change in the energy levels of molecular orbitals as the pyramidalization of the vertices is modified. The HOMO and LUMO energies for C_{60}, C_{70}, C_{84}, and some important soluble derivatives (see Fig. 5.3 for molecular structures) are shown in Fig. 5.2.

The electronic properties of fullerenes can also be modified by both endohedral encapsulation[61] and doping.[49] The incorporation of metals and metal compounds into the C_{60} lattice can give rise to metallic and even superconductive behavior. Indeed, reasonably high critical temperatures for the onset of superconductivity of $\sim$40 K for cesium doped C_{60} have been observed.[62] However, applications based on these properties are not the topic of the current chapter.

5.2.2 *Thin-Film Processing*

There are three broad categories of deposition procedures for fullerenes: epitaxy, vapor phase, and solution processing. Early

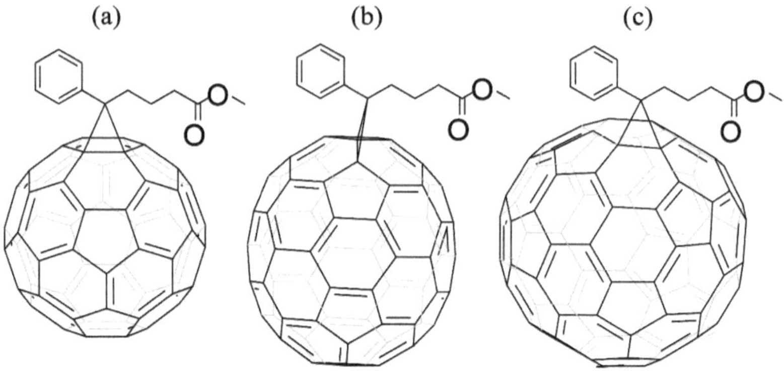

Figure 5.3. Molecular structures of the most common soluble fullerene derivatives based on analogues of [6,6]-phenyl-C_n-butyric acid methyl ester (PCBM). (a) C_{60}-PCBM, (b) C_{70}-PCBM, and (c) expected C_{84}-PCBM isomer based on a D_{2d} carbon cage.

studies on fullerene thin-films were investigated from epitaxial growth due to interest in the potential properties of C_{60} as a quasi-element or super-atom due to its high symmetry.[63] These studies require ultra-high vacuum and high substrate temperatures to form highly ordered films. Observed crystal structures included the most common face-centered cubic (fcc) in addition to a hexagonal close packed (hcp) phase.[63] Lattice matching has also been shown possible on appropriate substrates.[63] This technique, however, is rarely used in device fabrication because of its impracticality.

Vapor phase deposition procedures are widely used to fabricate highly ordered polycrystalline films. Common techniques include physical vapor deposition (PVD), chemical vapor deposition, pulsed laser deposition, and ion sputtering. PVD films are formed by heating the source material into the vapor phase under vacuum after which molecules are transported to the substrate where they are deposited to form a solid film. PVD films of C_{60} that have exhibited high electron mobility in transistors[14] were suggested to exhibit a similar microstructure to films formed by molecular-beam deposition[64] and hot wall epitaxy.[65] That is a polycrystalline structure with average grain sizes between 25 and 125 nm. The grains were established to be composed of several crystallites with a dimension of ~10 nm with an fcc lattice.

Solution processing of fullerenes (or fullerene:polymer/small molecule blends) is the simplest method of thin-film formation. Pristine fullerene cages, however, are almost insoluble in many common organic solvents.[66] It is therefore most appropriate to employ fullerene derivatives that can be processed in this manner. Addition of particular side-chains to the cage generally leads to solubility in weakly polar organic solvents,[67] e.g., chloroform, toluene, chlorobenzene. Specific deposition procedures include spin-coating, drop casting, spray-coating, ink-jet printing, gravure-printing, and stamping. These quick deposition procedures lead to relatively disordered films in comparison to those grown epitaxially or from the vapor phase. Despite this, the order within the film can vary dramatically depending on the specific processing conditions and the choice of side-chain and solvent. For example, [6,6]-phenyl-C_{61}-butyric acid methyl ester (C_{60}-PCBM, Fig. 5.2a) deposited by spin-coating from a chloroform solution can be amorphous or composed of randomly orientated nanocrystallites.[68] In these cases the films are optically isotropic and show no features on their X-ray diffraction (XRD) pattern or Atomic force microscopy (AFM) images.[68] Conversely, fullerenes with a long fluorinated side-chain can yield polycrystalline films from spin-coating from a chloroform solution with clear crystal domains observable with polarized optical microscopy.[37] Reports on similar molecules spin-cast from a chlorobenzene solution showed no scattering intensity during XRD measurements.[69]

5.2.3 *Why These Properties are Desirable for Electronics and Optoelectronics*

In OFETs, there are several reasons why these properties are favorable for n-channel transport. The relatively deep LUMO energy suggests that injection of electrons is possible with a minimum barrier from atmospherically stable metal contacts. The near spherical symmetry of fullerene molecules can enable isotropic charge transfer not typically displayed in well-ordered high-mobility semiconductors. This simplifies deposition of the semiconductor because controlled molecular orientation is, in principle, unnecessary for obtaining the maximum mobility. In addition, the versatility of chemical control over fullerene derivatives allows simple processing of fullerene films from solution.

The properties of fullerenes also make them almost ideal acceptor materials for blending with a donor polymer in BHJ OPV cells. From a processing perspective, the choice of side-chain can enable dissolution of the acceptor in the same solvent as the polymer donor allowing simultaneous deposition of both blend components. From an energetic perspective, fullerenes have a deep LUMO energy (high electron affinity) relative to the majority of potential donor materials. This favors efficient exciton dissociation and charge transfer from the donor. This charge transfer has been shown to be ultrafast in several polymer:fullerene blends with radiative and non-radiative decay channels of the excited state several orders of magnitude slower.[70] Additionally, the LUMO is triply degenerate and can exhibit reversible reduction of six electrons demonstrating its ability to stabilize negative charge. Finally fullerene films can exhibit a crystalline structure with high electron mobility which is important for maximizing the photocurrent and hence power output of the device.

5.3 Thin-Film Transistors, Integrated Circuits, and OPV

This section provides an overview of the device physics and operating characteristics of organic transistors, circuits, and photovoltaic cells. A brief summary of charge transport models for organic semiconductors is also presented. The aim is to provide the requisite background for subsequent sections that describe the results obtained from fullerene devices.

5.3.1 *Thin-Film Transistors*

The following description of transistor device operation is based on refs. 25, 35, 71, and 72. A field-effect transistor is a three-electrode structure where the third electrode, the gate, modulates the current between the other two. The transistor consists of a gate electrode; a semiconducting layer; a gate insulation layer (dielectric) separating the gate from the semiconductor; and two contact electrodes that

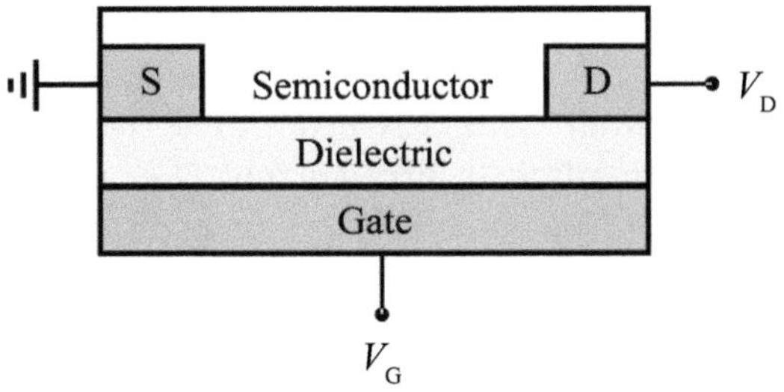

Figure 5.4. Schematic of bottom gate, bottom contact OFET device architecture.

inject (source electrode) and collect (drain electrode) a given species of charge carrier. The width of the source and drain electrodes defines the channel width (W) and their separation defines the channel length (L). The channel is the region in which charge carriers are transported between the source and drain electrodes. This structure can be built upon a glass or flexible plastic substrate although it is also common for a highly doped silicon wafer to act as both gate electrode and substrate. A schematic is shown in Fig. 5.4.

Voltage is applied to both the gate (V_G) and drain (V_D) electrodes whereas the source (V_S) is typically grounded. The potential difference between the drain and source electrodes is referred to as the drain–source voltage (V_{DS}). When $V_{DS} = 0$ V and a gate voltage is applied, charge carriers are accumulated at the semiconductor-insulator interface with uniform charge density along the channel.

For positive V_G electrons are accumulated and for negative V_G holes are accumulated because the source and drain electrodes normally have a more negative or positive potential than the gate, respectively. However, not all charges are mobile and free to contribute to the drain–source current. Any traps at the semiconductor-insulator interface will need to be filled if additional charges are to be mobile.[73] Therefore, the effective gate voltage inducing mobile charges above the threshold (V_T) is given by $V_G - V_T$, where V_T is the gate potential at which all traps are filled. It has also been observed that interface dipoles or impurities etc. can generate free charges in the channel[74] even at $V_G = 0$ V. These devices are referred to as normally on and require the opposite polarity potential to that of the expected accumulation potential to fully turn the device off.

When a drain–source voltage is applied such that $V_{DS} \ll V_G - V_T$, a linear gradient in charge density exists across the channel from the source to the drain. This is referred to as the linear regime of operation where the drain–source current (I_{DS}) increases linearly with increasing V_{DS}. The potential $V(x)$ along the channel increases linearly from $V(x) = 0$ V at $x = 0$ to V_{DS} at $x = L$.

Increasing V_{DS} further results in the formation of a depletion region at the drain electrode when $V_{DS} = V_G - V_T$. This occurs because the potential $V(x)$ at this point becomes lower than the threshold. A space-charge-limited current flows and the device is operating in the saturation regime. Since the potential $V(x)$ at the pinch off point remains approximately constant, the potential between that point and the source electrode remains constant, saturating the drain current. Any further increase in V_{DS} leaves the potential at the pinch off point unaltered. The operating regimes of an OFET are depicted in Fig. 5.5.

The above operation is only achieved when the gradual channel approximation is satisfied. This condition requires that the electric field parallel to the drain–source current due to V_{DS} is much smaller than the perpendicular field generated by the gate electrode. This

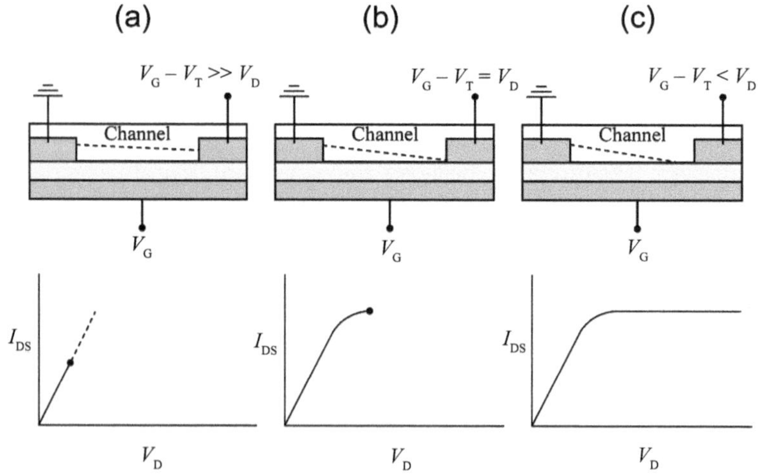

Figure 5.5. Channel profile and corresponding current output for (a) the linear regime, (b) at pinch-off, and (c) the saturation regime.

ensures that the charge density in the channel is controlled by the gate and is achieved typically for $L > 10d_{\text{insulator}}$.

The density of mobile charge (Q_{mob}) induced by a gate voltage above the threshold is proportional to the geometrical capacitance (C_i) of the insulator. However, the effective voltage also depends on the potential at a given point along the channel. Thus, the density induced in the channel is

$$Q_{\text{mob}} = C_i \left(V_{\text{G}} - V_{\text{T}} - V(x)\right). \tag{5.1}$$

The drain–source current attained on application of an electric field along the channel is therefore given by

$$I_{\text{DS}} = WQ_{\text{mob}}\frac{dV}{dx} = WC_i \left(V_{\text{G}} - V_{\text{T}} - V(x)\right)\frac{dV}{dx}. \tag{5.2}$$

Integrating both sides of Eq. 5.2 along the channel in the x direction from $x = 0$ to $x = L$ and thus $V(x) = 0$ to $V(x) = V_{\text{DS}}$ provides the general equation for the drain current in the transistor channel,

$$I_{\text{DS}} = \frac{W\mu C_i}{L}\left[\left(V_{\text{G}} - V_{\text{T}}\right)V_{\text{DS}} - \frac{V_{\text{DS}}^2}{2}\right]. \tag{5.3}$$

In the linear regime where $V_{\text{DS}} \ll V_{\text{G}} - V_{\text{T}}$, Eq. 5.3 can be simplified to

$$I_{\text{DS,lin}} = \frac{W\mu_{\text{lin}}C_i}{L}\left(V_{\text{G}} - V_{\text{T}}\right)V_{\text{DS}}. \tag{5.4}$$

This equation has linear dependence on the gate voltage so the linear charge carrier mobility (μ_{lin}) and V_{T} are extracted from the gradient and x-axis intercept of the straight line that fits $I_{\text{DS,lin}}$ as a function of V_{G}.

At the pinch-off point, $V_{\text{DS}} = V_{\text{G}} - V_{\text{T}}$, the channel current cannot increase significantly and saturates. In the saturation regime the drain–source current is given by

$$I_{\text{DS,sat}} = \frac{W\mu_{\text{sat}}C_i}{L}\left(V_{\text{G}} - V_{\text{T}}\right)^2. \tag{5.5}$$

The saturation charge carrier mobility (μ_{sat}) and V_{T} are extracted from the gradient and x-axis intercept of the straight line that fits the square-root of $I_{\text{DS,sat}}$ as a function of V_{G}.

It is common for the charge carrier mobility in OFETs to exhibit gate voltage dependence leading to deviations from the aforementioned linear fitting. In such cases it is appropriate to express the mobility as an effective gate-dependent value that can be obtained by simple re-arrangement of Eqs. 5.4 and 5.5.[75,76] In this case the threshold voltage should be substituted for the switch-on voltage (V_{ON}), i.e., the gate voltage at which I_{DS} begins to increase.

5.3.2 *Integrated Circuits*

The ultimate aim of transistor development is their implementation within integrated circuits. It is therefore worthwhile to assess their performance within circuit elements. The standard element for this assessment is the inverter which is an important building block for logic gates. It is a two-transistor device and can be used itself as a NOT-gate. The truth table for the inverter is given in Table 5.1.

Three families of organic logic are considered here: unipolar, where both transistors are made from the same material that transports either holes or electrons; complementary, where one transistor is n-channel and the other is p-channel; and complementary-like, where both transistors are made from the same ambipolar material that can transport both holes and electrons. There are strengths and weaknesses to these approaches.

Unipolar inverters are easy to fabricate because the same semiconductor material can be deposited everywhere on the substrate. However, their performance is hindered, as illustrated in Fig. 5.6a, by low noise margins, low gain ($= dV_{OUT}/dV_{IN}$) and high static power consumption ($P = V_{DD}I_{DD}$) because both transistors are functioning in the high output state.[77] Complementary inverters, although more difficult to fabricate because they require patterning of two different materials, have improved performance in all areas because only one transistor is operating in each output state resulting in power dissipation only when the inverter is switching,[78] as shown in Fig. 5.6b.

Table 5.1. Truth table for an inverter (NOT gate)

Input	Output
1	0
0	1

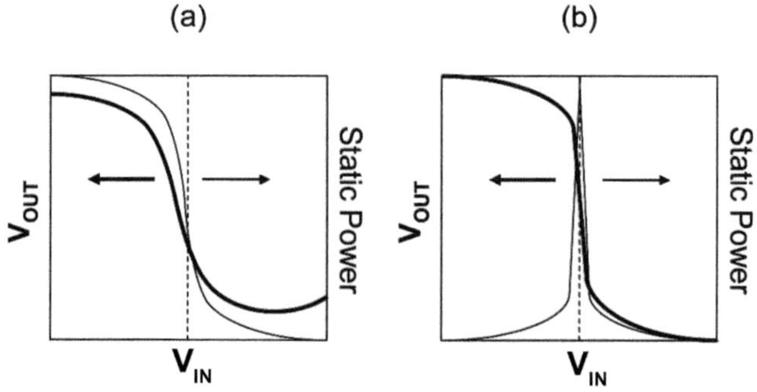

Figure 5.6. Output voltage and power consumption of (a) a unipolar and (b) a complementary inverter as a function of input voltage.

Complementary-like inverters combine the advantages of both the simple fabrication of unipolar devices and the intrinsic improved performance of complementary logic. However, a lack of suitable high-performance ambipolar materials means examples are scarce.

The OFET connected to the load voltage (V_{DD}) is the load transistor and the OFET connected to ground is the driving transistor as shown in Fig. 5.7. In static operation the inverter circuit can be considered as a potential divider. For a low input signal, $V_{IN} = 0$ V, the driving transistor is switched off and thus behaves as

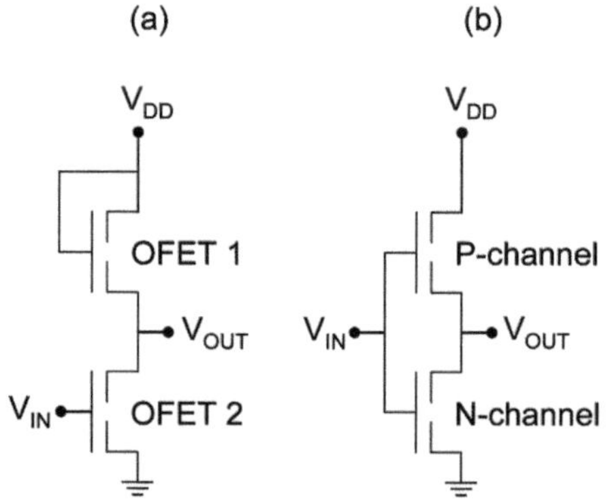

Figure 5.7. Example circuit diagrams of (a) unipolar and (b) complementary inverters.

a resistor with $R = \infty$. The load transistor has a finite resistance, R_1, and the output (V_{OUT}) is the potential drop across it, i.e., the high state. For a high V_{IN}, the load transistor is switched off and has $R = \infty$. V_{OUT} is therefore the voltage drop across the switched on driving transistor (R_2) which produces the low output state.

In the unipolar case, the load transistor cannot fully switch off for high V_{IN} but its resistance can be made higher than the driving transistor by scaling its channel width. Although the load transistor does not have infinite resistance in this state, most of the voltage drop will still be across the driving transistor. This situation gives rise to the aforementioned problems with unipolar logic performance. Because both transistors are switched on in this state, there is a constant current flowing from the load to ground, which means that the circuit is consuming power. It also means that when switching between states as V_{IN} is increased, the change in V_{OUT} is slow, giving rise to low gain and low noise margins.

In the complementary and complementary-like cases this problem does not arise. In either V_{IN} state, one of the transistors is switched off fully so power is only consumed when switching between states. Because one of the transistors switches off while the other one switches on when V_{IN} changes state, the change in V_{OUT} is more abrupt providing a higher gain and higher noise margins. These parameters are of course also strongly dependent on the charge carrier mobility and geometry of the transistors.

The noise margin is an important inverter metric that represents the range of voltages that will be recognized as the high and low states by the elements of a circuit. It therefore determines the reliability of the circuit and its tolerance to signal fluctuations and noise. Although the trip point (V_{IN} at maximum gain) can be controlled by geometric scaling of the transistor channels, the noise margin is ultimately limited by the inverter gain.

Another key feature of an organic circuit element is the speed at which it can operate. This will critically influence the dynamic response of the digital circuit in which it is used. To test this, the inverters can be combined in series to produce ring oscillators as shown in Fig. 5.8. A ring oscillator consists of an odd number of inverters where the output of each stage is connected to the input of the following stage. If the output of the last stage is

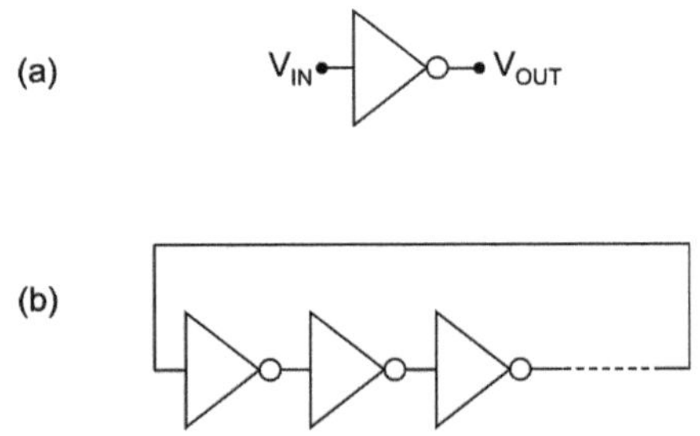

Figure 5.8. (a) Circuit element symbol for an inverter and (b) schematic of ring oscillator circuit containing an odd number of inverters in series.

connected back to the input of the first, the output at each stage will spontaneously oscillate between the high and low states. The stage delay (τ_d), and thus the operating frequency, is limited by the charging and discharging of the capacitive load of the output node of each inverter. In addition to the channel conductivity (dependent on carrier mobility), this is also determined by the parasitic capacitance and series resistance (associated with contact resistance) as well as the driving voltage and channel lengths.[79] It is therefore necessary to optimize OFET design as well as maximize mobilities to enable faster charging of the subsequent inverter input and hence reduce τ_d.

5.3.3 *Organic Photovoltaics*

The basic structure of a BHJ OPV cell is shown in Fig. 5.9. The device is typically built on transparent indium tin oxide (ITO)-coated glass or plastic substrates. The ITO is usually coated with the transparent conducting polymer poly(3,4-ethylenedioxythiophene)-polystyrene sulfonate (PEDOT-PSS). The PEDOT-PSS-coated ITO

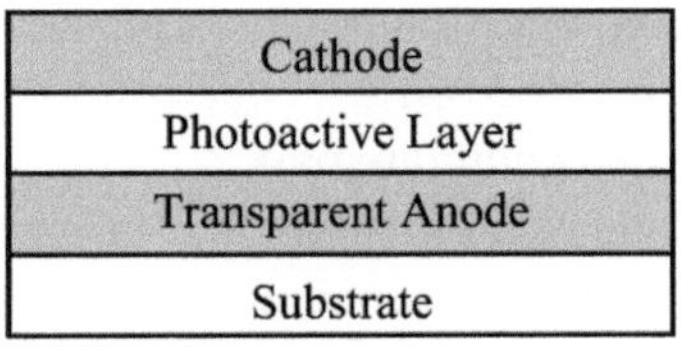

Figure 5.9. Schematic profile of basic solar cell device structure.

acts as the anode for extraction of holes. The next layer is the photon-absorbing, charge transport layer and is composed of an interpenetrating network of the acceptor and donor materials. The device is completed with a cathode.

When a photon is absorbed in the donor material it creates an electron-hole pair bound state known as an exciton. The exciton has an electron promoted to the LUMO level leaving a hole in the HOMO level of the donor. Because the excitonic bound state is associated with an electrostatic distortion, the energy levels occupied by the electron and hole lie within the LUMO-HOMO gap. The exciton can diffuse to the donor–acceptor interface where, if energetically favorable, the electron occupying the donor LUMO will fall into the LUMO of the acceptor material. The electron is then transported through the acceptor to the cathode and the hole is transported through the donor to the anode to produce current. This process is shown in Fig. 5.10.

The dissociation of the exciton is a critical step for extracting useful current from BHJ OPV cells. When the exciton has reached the D–A interface a downhill energetic driving force must exist to favor transfer of the electron from the donor to the acceptor LUMO levels. In general it is considered that there must be a favorable change in the free energy of the system by transferring from the two neutral states to the separated charged states.[80]

The energetic difference must also be large enough to overcome the Coulombic binding energy of the exciton, typically 0.4–0.5 eV.[81] Following the transfer of the electron from D to A (forming a geminate pair), a Coulombic attraction between the donor cation and the acceptor anion must also be overcome to separate the free charges. This is driven both thermally and by the device intrinsic electric field. However, alternative dissociation mechanisms have also been observed in certain systems such as Förster resonance energy transfer from the donor. This generates an exciton in the acceptor followed by electron transfer from the donor to acceptor HOMO levels to form a geminate pair.[82]

The power conversion efficiency (η) of an OPV cell is given by the ratio of the maximum output power density (P_{OUT}) to that of the input power density (P_{IN}). P_{OUT} is given as the product of the short circuit current density (J_{sc}), the open circuit voltage (V_{OC}),

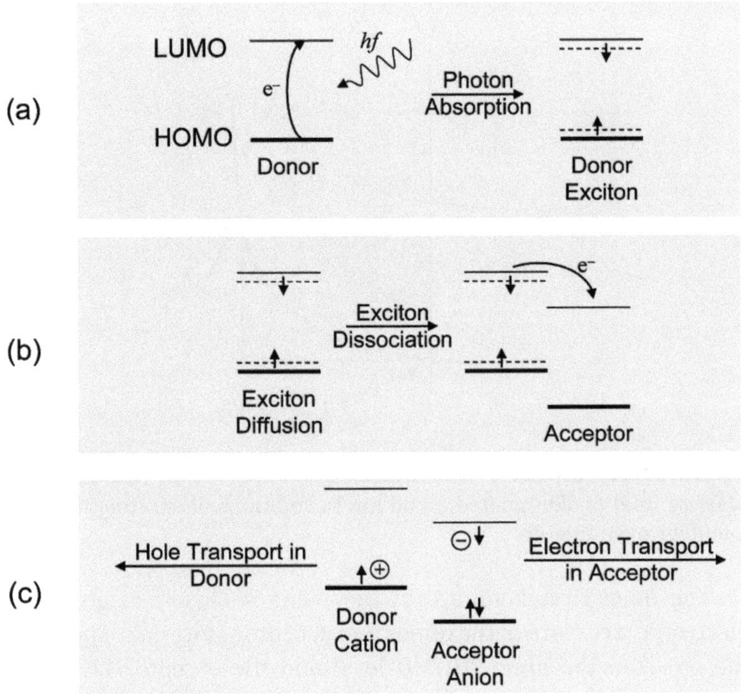

Figure 5.10. Operation of a photovoltaic cell.[18] (a) A photon is absorbed in the donor exciting an electron into its LUMO level to form an exciton. (b) The exciton diffuses to the donor–acceptor interface where the electron in the donor LUMO falls into the LUMO of the acceptor. (c) The charges are separated and transported through the donor or acceptor materials to their respective electrodes.

and the fill factor (FF). J_{SC} is the current density output when the load impedance is much smaller than the device impedance, V_{OC} is the voltage output when the load impedance is much greater than the device impedance, and FF is the ratio of the area of the largest rectangle that can fit within the device J–V curve (i.e., maximum output power) to that given by the product $J_{sc}V_{oc}$ (i.e., $FF = V_M J_M / J_{SC} V_{OC}$). This is summarized in Eq. 5.6 and Fig. 5.11.

$$\eta = \frac{P_{OUT}}{P_{IN}} = \frac{J_{SC}V_{OC}FF}{P_{IN}}. \tag{5.6}$$

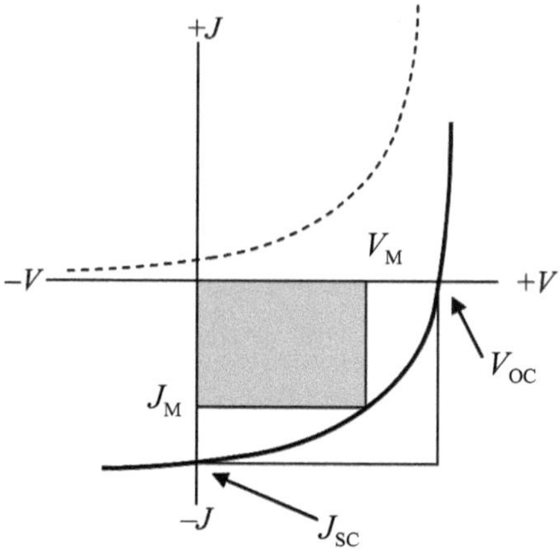

Figure 5.11. Typical *J*–*V* characteristics of a solar cell under dark (dashed line) or illuminated (solid line) conditions illustrating the important device parameters.

The open circuit voltage of OPV cells is closely related to the electronic structure of the donor and acceptor materials. Specifically, the offset in the donor HOMO level and the acceptor LUMO level determines V_{OC}.[83] To increase the internal efficiency of photon absorption in the donor it is preferable to use a material with a narrow gap between its HOMO and LUMO to enable absorption of the lowest energy photons of the solar spectrum. However, reducing the energy gap also reduces V_{OC} and hence the power conversion efficiency of the cell. Using an acceptor with higher electron affinity also has the same consequence. To obtain high η, devices are required to absorb a large fraction of the total flux of photons directed at the cell. This could be achieved by increasing the thickness of the cell. However, due to slow transport of charge carriers, thicker cells have an increased resistance which lowers the FF. Optimal device operation therefore depends on a compromise between exciton dissociation efficiency, photon absorption efficiency, and maximizing V_{OC}.

5.3.4 *Charge Transport in Organic Semiconductors*

The details of charge carrier transport in organic semiconductors generally differ from that of inorganic materials. The electron energy levels in inorganic atomic single crystals are sufficiently numerous to consider them part of a continuum known as an energy band. In this regime, the charge carriers are delocalized across the crystal and the speed at which they move through the material is limited by phonon scattering. This results in a charge carrier mobility that decreases with increasing temperature. In contrast, organic semiconductors form molecular crystals where electron energy levels are localized. This produces a regime in which carrier mobility is phonon assisted (increases with increasing temperature). This results in a higher hopping probability between transport sites upon increasing thermal energy.

Although the intermolecular bonding in organic crystals is dominated by weak van der Waals coupling, the absence of energy bands is not necessarily inherent to this class of materials. The characteristic temperature dependence of the mobility for band-like transport has been observed in organic single crystals. The origin of this dependence is a matter of current debate that will not be resolved here. In general, organic structures are disordered and don't display band-like behavior.

There are two broad classes of models that have been applied to charge transport in organic semiconductors in various ways. One of these is based on multiple trapping and release (MTR) of charge carriers and has been applied to transport in polycrystalline organic semiconducting films.[25,84,85] In this regime the charge carriers move through a series of localized trap states followed by thermally activated release into an extended transport state in the semiconductor. This model qualitatively describes the temperature dependence of the mobility in many organic semiconductors. MTR implies that increasing the temperature increases the probability of the thermally activated de-trapping process so carriers are more likely to have sufficient energy to reach the transport state and spend less time in traps. It also qualitatively describes the gate voltage dependence of the field-effect mobility. As V_G is increased, a higher density of charge carriers is introduced into the channel. These carriers fill the trap states first such that subsequent carriers are less likely to be trapped and are free to occupy the transport states.

The alternative models are based on hopping between polaron states. A polaron is the resultant charged state of an electron injected into the LUMO level or a hole into the HOMO level of a conjugated unit. The excitation produces an electrostatic distortion which results in negative or positive polaron levels within the LUMO-HOMO gap. An important example is based on Marcus Theory, originally developed to explain charge transfer in chemical reactions.[86–88] In this framework charges are transported only when the site energies of the initial and final states are equal. The theory predicts a hopping rate as a function of temperature (T) between sites i and j given by

$$k_{ij} = \frac{2\pi}{\hbar} \left|V_{ij}\right|^2 \sqrt{\frac{1}{4\pi k_B T \lambda}} \exp\left(-\frac{(\Delta G + \lambda)^2}{4\lambda k_B T}\right). \tag{5.7}$$

The change in Gibbs free energy is denoted by ΔG and the reorganization energy induced by the electron is given as λ. V_{ij} is the electronic coupling between the initial and final sites. This process depends on thermal fluctuation; hence, it is thermally activated even in the absence of disorder.

However, the polaronic nature of charge transport in organic semiconductors, particularly in solution processed systems, is generally hidden by energetic disorder. Bässler used Monte Carlo simulations based on a Millar-Abrahams[89,90] framework of hopping charge carriers to express the carrier mobility as a function of disorder in both site energy and intersite distance.[91] This was later modified by Novikov *et al.* to account for spatial correlation of charge-dipole interactions that dominated the disorder of site energies leading to improved low-field fits to experimental data.[92] This Correlated Disorder Model yielded the following expression for the charge carrier mobility:

$$\mu_{CDM}(E, T) = \mu_0 \exp\left[C_0 \sqrt{\frac{q_e a E}{\sigma}} \left(\left(\frac{\sigma}{k_B T}\right)^{3/2} - 2\right) - \left(\frac{3}{5}\frac{\sigma}{k_B T}\right)^2\right]. \tag{5.8}$$

In Eq. 5.8, $C_0 = 0.78$, a is the intersite distance, E is the electric field, T is the temperature, σ is the standard deviation of the Gaussian density of energy states, q_e is the charge of the electron, and k_B is Boltzmann's constant.

These disorder-based models are applicable for describing transport at low carrier concentrations but are unsuitable for describing charge transport through an OFET channel where the accumulated charge carrier densities are high. To describe transport in OFETs, Vissenberg and Matters developed a model based on percolation of charge carriers with variable range hopping.[93] Assuming charge carriers occupy an exponential density of states of width T_0, the authors suggested an expression for the mobility given by

$$\mu_{\mathrm{FE}} = \frac{\sigma_0}{q_{\mathrm{e}}}\left(\frac{(T_0/T)^4 \sin(\pi T/T_0)}{(2\alpha)^2 B_{\mathrm{c}}}\right)^{T_0/T}\left(\frac{(C_i V_{\mathrm{eff}})^2}{2k_{\mathrm{B}} T_0 \varepsilon_{\mathrm{S}} \varepsilon_0}\right)^{(T_0/T)-1}, \tag{5.9}$$

where α is an effective overlap parameter, σ_0 is the conductivity prefactor, B_{c} is the percolation criterion ($\approx$2.8 for a three-dimensional amorphous system), V_{eff} is the effective potential inducing charge at a given position along the channel, and ε_{S} is the dielectric constant of the semiconductor.

5.4 Electron Transport in Fullerene Thin-Film Transistors

This section will describe the operational considerations specific to OFETs based on C_{60}, C_{70}, C_{84}, and their derivatives as the semiconducting layer. A range of device architectures are presented and important results from the literature will be summarized with regard to electron transport. The environmental stability of n-channel behavior is discussed with a comparison to alternative semiconductors. The operation of low-voltage transistors implementing fullerenes is also discussed.

5.4.1 *Electron Injection*

The choice of electrode materials for injection and extraction of charge carriers to and from the semiconducting layer in OFETs is crucial for high-performance operation. In the simplest description, large offsets in metal work function and the molecular orbital energies of the semiconductor can hinder or even prevent injection and extraction of charge carriers. Common electrode metals include Ca, Al, and Au which have work functions of 2.9 eV, 4.1 eV, and 5.1 eV, respectively. These values should be compared with the LUMO levels of C_{60}, C_{70}, C_{84}, and their PCBM derivatives in Fig. 5.2. Despite the relatively high electron affinity of the fullerenes shown, few metals appear suitable for ohmic injection of electrons. Of those mentioned, Ca, although most energetically suitable, is unstable to oxidation on exposure to the atmosphere so is unsuitable for practical devices without encapsulation.

However, dipole states at the metal–organic interface can modify the simple Schottky picture and allow efficient injection despite apparent energetic offsets. Even Au, which should present a significant barrier, has been demonstrated to be capable of electron injection in n-channel fullerene devices.[39] A recent study[94] into the effect of introducing a thin C_{60} layer to the surface of Au has found that dipole formation at the interface pins the Fermi level of Au/C_{60} to the charge neutrality point of C_{60}. This gives an effective work function of $\sim$4.7 eV, reduced from 5.1 eV. Since the first layer of C_{60} acts as a modification layer, its molecular orbitals are bypassed for injection into the subsequent layer. This reduction in work function may lower the effective barrier for electron injection into the LUMO of fullerenes from Au electrodes. This also highlights the importance of the nature of the electrode/fullerene interface for a complete understanding of fullerene devices.

5.4.2 *Electron Transport in C_{60}, C_{70}, and C_{84} Devices*

Early work on fullerene thin-films in OFETs was first assessed in evaporated layers.[95] The authors reported films with randomly orientated polycrystalline grains of C_{60} with dimension $\sim$60 Å. The electron transporting transistors, which used an SiO_2 dielectric treated with tetrakis(dimethylamino)ethylene and Au/Cr contacts, recorded a respectable electron mobility of 0.3 cm^2/Vs. Although films deposited on the dielectric without treatment had reduced mobility, XRD studies indicated they were indistinguishable from films that were deposited on the treated dielectric. It was concluded

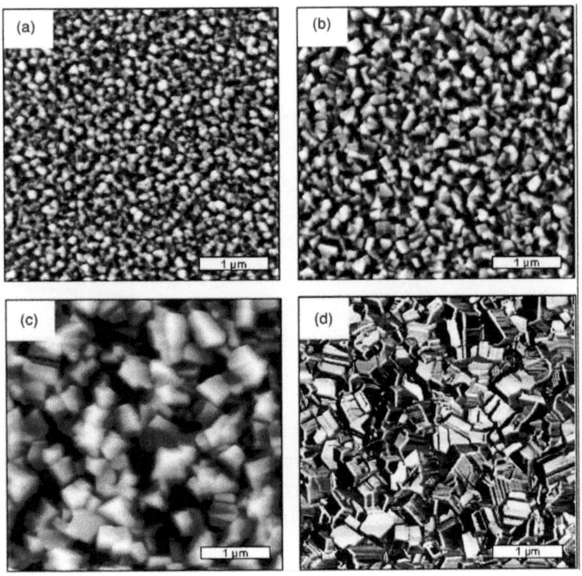

Figure 5.12. AFM images of C_{60} thin-films grown by hot wall epitaxy at different substrate temperatures: (a) 25°C, (b) 120°C, (c) unspecified temperature, and (d) 250°C. Image adapted from ref. 96. See also Color Insert.

that the treatment reduced the injection barrier to electron injection into the semiconductor.

The highest electron mobility values of any small molecule semiconductor have been obtained from C_{60} films grown by hot wall epitaxy using a polymeric divinyltetramethyldisiloxane-bis(benzocyclobutene) (BCB) dielectric.[13,96] The authors measured mobilities up to 6 cm^2/Vs. This was found to be highly dependent on the substrate temperature (see Fig. 5.13) during film growth where higher temperatures favored the growth of larger crystal domains as shown in Fig. 5.12. The crystallinity of the domains was determined by XRD and was suggested to be in agreement with the report by Kobayashi *et al.* of a face-centered cubic lattice.[64] Similar mobility values of 5 cm^2/Vs were subsequently reported by Zhang *et al.* following film growth by PVD also on a BCB dielectric.[14]

OFETs based on C_{70} have generally exhibited reduced mobility in comparison to C_{60}. Haddon, in an analogous report to ref. 95,

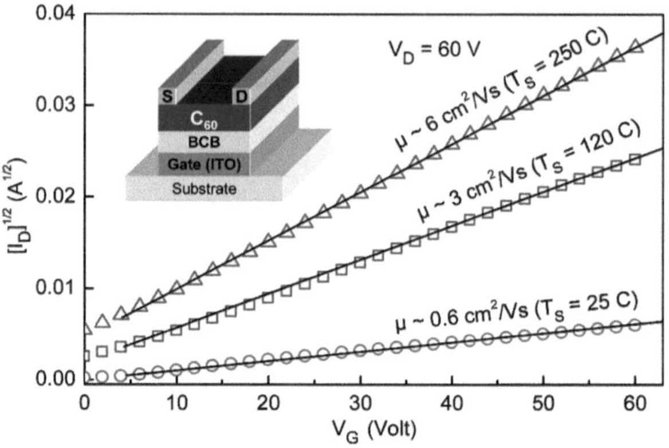

Figure 5.13. Square root of the drain current as a function of gate voltage for C_{60} devices grown by hot wall epitaxy at different substrate temperatures (T_S). Inset: Device structure used for measurements. Image adapted from ref. 13. See also Color Insert.

showed C_{70} transistors with a mobility of 2×10^{-3} cm^2/Vs using an untreated SiO_2 dielectric and Au/Cr contacts.[97] Although the film morphology was reported to display small disordered grains similar to C_{60} films produced using the same technique, the anisotropy of the fullerene cage was suggested to reduce the mobility in this case. This additional variable, not encountered with C_{60} films, apparently introduces further disorder into the film, modifying the solid state electronic structure. Haddock *et al.* reported a similar mobility discrepancy between devices based on C_{60} and C_{70} fabricated following the same procedure.[98]

The first example of an OFET based on a thermally evaporated thin-film of C_{84} showed electron mobility of 2.1×10^{-3} cm^2/Vs. The authors used a SiO_2 dielectric with Au bottom contacts and measured a normally on device that showed no saturation in the output curves at room-temperature. The drain current as a function of V_D was always > 0 A even at very negative V_G. This was attributed to bulk conductivity without clarification of the origin of the free carriers. Their analysis of the temperature dependence of the mobility suggested Arrhenius-type hopping transport of electrons with an activation energy of 0.13 eV.

5.4.3 Electron Transport in Solution Processed C_{60}-, C_{70}-, and C_{84}- PCBM Devices

The first successful demonstration of a solution processed fullerene layer for OFETs utilized the C_{60}-PCBM derivative, initially developed by Hummelen *et al.* as a soluble fullerene intermediate used in the preparation of a potential anti-HIV treatment.[99] C_{60}-PCBM has since become the most widely studied fullerene for molecular electronics.

One of the highest reported mobilities demonstrated with C_{60}-PCBM was recorded on a BCB dielectric with Ca top contact electrodes.[39] The authors reported an electron mobility of 0.21 cm^2/Vs as shown in Fig. 5.14a. They found that effective device mobilities decreased with increasing contact electrode work function. This is most likely due to the increased contact resistance to injection resulting from the increased barrier offset between the electrode work function and the fullerene LUMO level. Similar mobilities had been reported previously most notably by Singh *et al.* using a PVP dielectric and LiF/Al electrodes.[100]

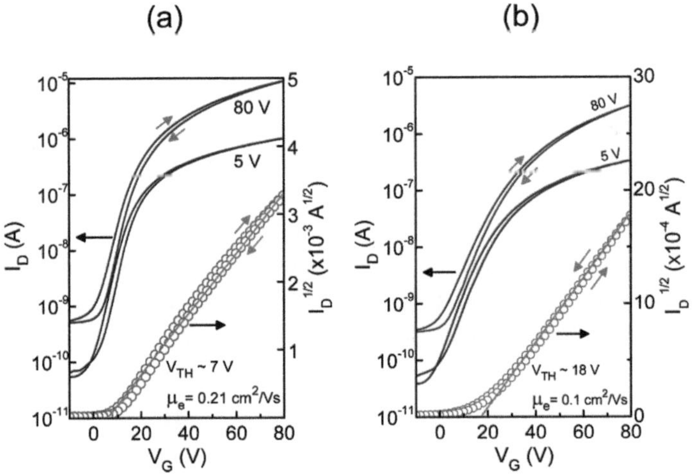

Figure 5.14. Transfer characteristics of bottom gate, top contact (a) C_{60}-PCBM ($W = 1$ mm and $L = 60$ μm) and (b) C_{70}-PCBM ($W = 1.5$ mm and $L = 60$ μm) transistors. Image adapted from ref. 39. See also Color Insert.

Wöbkenberg *et al.* also reported the highest recorded charge carrier mobility in C_{70}-PCBM OFETs[39] of ~0.1 cm^2/Vs shown in Fig. 5.14b. Previous work on OFETs based on C_{70} and its PCBM derivative suggested that it would have a lower mobility than C_{60}-PCBM. The authors demonstrate that this is not necessarily the case. Their reasoning for the observation was based on the increased solubility of C_{70}-PCBM compared to C_{60}-PCBM enabling formation of a more favorable interface with the dielectric for charge transport during spin-coating.

C_{84}-PCBM OFETs have also been demonstrated recently. Au bottom contact devices on an hexamethyldisilazane (HMDS)-treated SiO_2 dielectric yielded an electron mobility of 0.5×10^{-3} cm^2/Vs in films formed by drop casting the semiconductor from a chlorobenzene solution.[60] Thermal annealing of the semiconductor film under vacuum was found to increase the electron mobility by a factor of 6. This was attributed to an improvement of the injection interface as evidenced by the reduction of a superlinear I_{DS} increase at low V_D on the transistor output characteristics.

Interestingly, these devices were found to operate upon exposure to light and air for several months. OFETs based on lower PCBM analogues degrade rapidly upon atmospheric exposure without encapsulation. The enhanced lifetime upon atmospheric exposure was attributed to the lower lying LUMO level of C_{84} in comparison to C_{60} and C_{70} providing increased anionic stability. Anthopoulos *et al.* have suggested that the air stability of electron transporting small molecules depends on the position of their LUMO with respect to the reduction potential of H_2O as shown in Fig. 5.15. Alternatively, the inability of C_{84} to form a triplet state following optical excitation, in contrast to C_{60} and C_{70}, may prevent self-sensitized oxidative degradation following singlet-oxygen formation.

5.4.4 *Electron Transport in Devices with Alternative Fullerene Derivatives*

The freedom of chemical control over fullerene derivatives opens the door to a range of molecular structures tailored for purpose. This principle has been applied to specific molecular design for development of alternatives to PCBM for OFET applications.

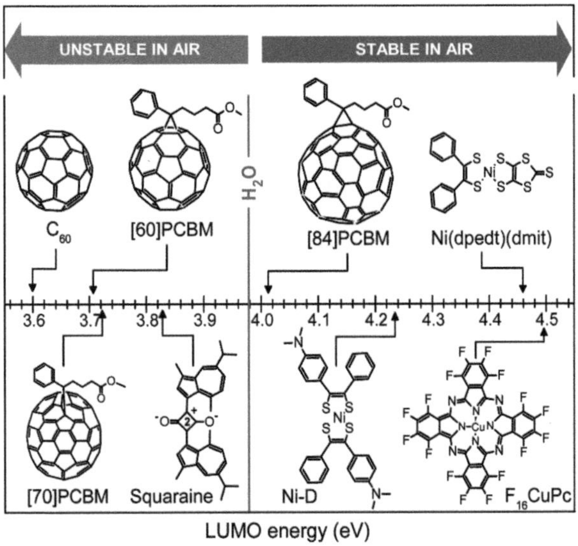

Figure 5.15. LUMO levels of small molecule electron transporting organic semiconductors compared to the reduction potential of H_2O. Image adapted from ref. 101.

To enhance air stability, electron withdrawing groups have been added to conjugated units to increase the electron affinity of several organic semiconductors in an attempt to circumvent the trap energy of atmospheric oxidants. An alternative, however, is to use side-chains that act as a structural barrier to the diffusion of oxidants into the transistor channel. One particular group of potential side-chains comprises perfluoroalkyl chains. These chains are chosen because they are hydrophobic, reducing the energetic favorability of water diffusion into the OFET channel.[37,69] Chikamatsu *et al.* reported the use of a perfluoroalkyl-substituted fulleropyrrolidine, illustrated in Fig. 5.16, which enabled n-channel transistor functionality for >140 hours under exposure to the atmosphere without significant modification to its electronic structure in comparison to C_{60}-PCBM.[37] Using an HMDS-treated SiO_2 dielectric and bottom contact Au electrodes, this fullerene also recorded an electron mobility of 0.25 cm^2/Vs under vacuum which reduced to 0.078 cm^2/Vs after exposure to air for five hours. The trend in film crystallinity and air stability of a series of fullerenes was found to correspond with

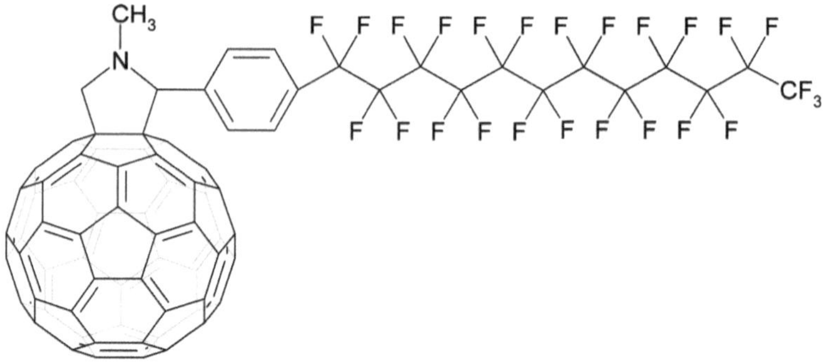

Figure 5.16. Molecular structure of a fluorinated fulleropyrrolidine used in ref. 37 with increased ambient stability compared to C_{60}-PCBM.

the length of perfluoroalkyl chain where longer chains formed more crystalline films with increased air stability. XRD results from these fullerene films concluded that high crystallinity was required for high mobility and air stability. The same report details additional derivatives with at least equivalent electron mobility to C_{60}-PCBM in the same device structure.

The surface energy of semiconductor solutions is an important parameter that should be taken into consideration when optimizing film morphology during solution processing. The choice of fullerene side-chain has been demonstrated to modify the liquid surface energy of semiconductor solutions. Fluorinated side-chains with a low surface energy were found to reduce the surface energy of a chlorobenzene solution in which they are the solute.[102] This enabled processing of a fluorinated fulleropyrrolidine on a low surface energy dielectric for fabrication of low-voltage transistors based on a self-assembled monolayer gate insulator.

5.5 Ambipolar Transport in Fullerene Thin-Film Transistors

The search for ambipolar organic semiconductors for exploiting the advantages of complementary-like logic has yielded few suitable examples. Here, we will discuss the advances made using fullerenes and their derivatives as ambipolar materials in OFETs. Device characteristics from the few literature examples will be presented and analyzed.

5.5.1 *Ambipolar Transport in Fullerene Transistors*

A rare example of ambipolar transport in a fullerene OFET has been presented in a solution processed C_{60}-PCBM device.[103] The authors reported electron and hole mobilities of 1×10^{-2} cm^2/Vs and 8×10^{-3} respectively. It was found that when using Au bottom contacts on an HMDS-treated SiO_2 dielectric, both species of carrier could be accumulated in the OFET channel despite significant apparent barriers to charge injection. The authors suggested that dipole formation at the contacts modified the injection barriers. However, superlinear increases in I_D when increasing V_D suggest injection was non-ohmic.

Following the same device fabrication procedure, C_{70}-PCBM has shown ambipolar transport albeit with more modest mobilities of 2×10^{-3} cm^2/Vs for electrons and 2×10^{-5} cm^2/Vs for holes.[104] This was again attributed to increased disorder resulting from the anisotropy of the fullerene. Additionally C_{70}-PCBM is processed as an isomeric mixture of the derivative which may compound the problem.

Ambipolar transport in OFETs based on the higher analogue, C_{84}-PCBM, has also been observed.[60] Both hole and electron transport could be explicitly shown in the same device but only at temperatures below 273 K. Although the p-channel was expected to remain at higher temperatures, the researchers were limited by the voltage range of their apparatus. A temperature-dependent threshold shift prevented observation of the hole current within the measurement window.

5.6 Fullerene-Based Microelectronics

As the final step toward practical applications of OFETs, the operation of fullerene-based microelectronics will be presented here as a brief review of important demonstrations in the literature.

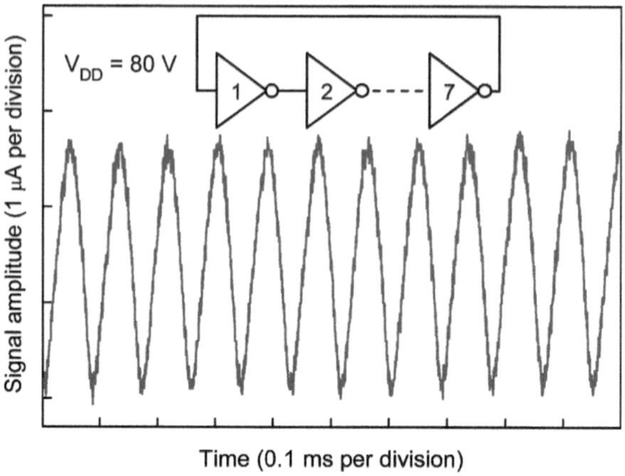

Figure 5.17. Signal output from a seven-stage ring oscillator based on C_{60} OFETs grown by hot wall epitaxy. Inset: circuit diagram for ring oscillator. Image adapted from ref. 13.

5.6.1 *Unipolar Logic Circuits*

C_{60} transistors grown by hot wall epitaxy have been integrated to fabricate unipolar seven-stage ring oscillators.[13] The output signal from this circuit is shown in Fig. 5.17. The authors reported peak oscillation frequencies of ~30.5 kHz corresponding to a stage delay of ~2.34 μs at $V_{DD} = 140$ V with transistors of length $L = 2.5$ μm. The oscillation frequency was found to depend strongly on V_{DD} in addition to the design constraints imposed by the widths of the load and driving transistors of each inverter stage.

Solution processed unipolar ring oscillators have also been fabricated with alternative fullerene derivatives. Based on a fluorinated fulleropyrrolidine, seven-stage oscillators were shown with a maximum oscillation frequency of 10.4 kHz corresponding to a mean stage delay of 6.86 μs.[69] However, this was achieved at a load voltage of 170 V. At the time of publication these were the fastest reported solution processed n-channel ring oscillators.

5.6.2 *Complementary Logic Circuits*

High-performance integrated complementary inverters based on evaporated layers of C_{60} and pentacene have been demonstrated.[105]

These transistors were able to operate at 5 V as a result of the thin polymer passivated Al_2O_3 dielectric. In addition, the devices were fabricated on flexible plastic substrates with no degradation to the inverter performance after bending. The high mobilities of the devices combined with optimized geometric device scaling enabled a high dc gain of 180 and noise margins > 80% of their maximum theoretical value. However, poor ambient stability of the C_{60} layer and the Ca contacts used for electron injection prevented operation of these circuits in air.

Solution processed complementary inverters have been shown based on C_{60}-PCBM and a polytriarylamine p-channel polymer.[39] Signal gains of 17 were demonstrated at $V_{DD} = 80$ V, limited by the mobility of the solution processed semiconductors. Respectable noise margins of 70% of their maximum value were obtained. Utilization of a higher performance solution processable p-channel material and optimization of geometric scaling could further improve the inverter characteristics.

Although the circuit was not integrated (transistors were fabricated on separate substrates) the report represents an important step towards high-performance solution processed complementary logic.

Complementary inverters based on solution processed C_{84}-PCBM transistors have also been reported with signal gain of 14. By combining the n-channel fullerene transistor with a p-channel device based on the hole transporting polymer poly[2-methoxy-5-(3′,7′-di-methyloctyloxy)]-p-phenylene vinylene (MDMO-PPV), the device displayed sinusoidal voltage inversion with an input frequency of 5 Hz. The authors note that the speed of the device is limited by parasitic resistances and capacitances as opposed to the intrinsic transistor performance. Although the example was not integrated, it is an important demonstration of a solution processed air-stable complementary inverter.

5.6.3 *Complementary-Like Logic Circuits*

Complementary-like inverters combine the advantages of both unipolar processability and complementary performance. However, few semiconductor materials display ambipolar characteristics

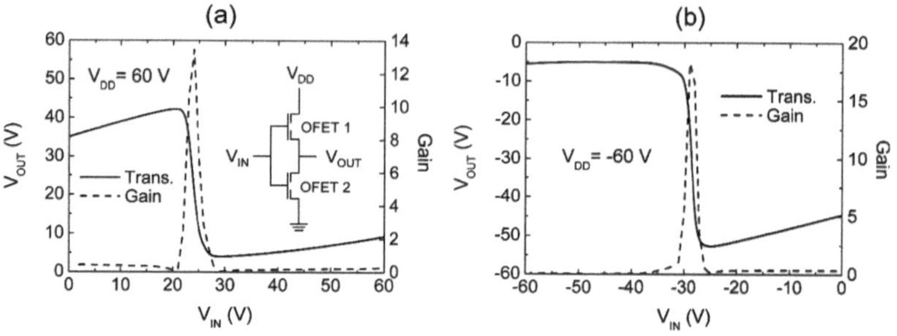

Figure 5.18. Transfer characteristics of a complementary-like inverter based on C_{60}-PCBM OFETs with V_{DD} and V_{IN} biased (a) positively and (b) negatively. Image taken from ref. 106.

from a single injection/extraction material with high performance. C_{60}-PCBM is a potential candidate with ambipolar characteristics as discussed in Section 5.5.1. Two OFETs described in ref. 103 were combined to form a complementary-like inverter[106] with a maximum signal gain of 18 as shown in Fig. 5.18.

C_{70}-PCBM OFETs, with lower electron and hole mobilities, have also been combined for fabrication of complementary-like inverters.[104] The inverters were able to reach a signal gain of ~6. The reduced gain in comparison to C_{60}-PCBM complementary-like inverters is a result of the mismatch in mobilities for electrons and holes.

Binary blends of MDMO-PPV with C_{60}-PCBM have been used to fabricate solution processed complementary-like inverters.[59] These devices combine the p-channel of the polymer with the n-channel of the fullerene in each transistor. This led to an inverter signal gain of 10 at $V_{DD} = 40$ V. Although more power is consumed in these circuits compared to their complementary counterparts, these devices are easier to integrate because both p- and n-channel semiconductors can be deposited in a single step.

5.7 Fullerene-Based Optoelectronics

This section will describe the operational considerations specific to BHJ OPV devices and phototransistors that incorporate fullerenes as the acceptor material. Fullerenes are used as an acceptor material in the majority of reports on BHJ OPV, so a brief summary of important advances will be given. Beyond this introduction, interested readers are pointed in the direction of more thorough reviews found in refs. 18, 19, and 107–109. The relatively new field of phototransistors based on fullerenes will also be explored.

5.7.1 *Fullerene-Based BHJ OPV*

The discovery of photoinduced charge transfer from a polymer to buckminsterfullerene on picosecond time scales in bilayer devices[110,111] led to the development of BHJ solar cells based on fullerenes in the mid-1990s.[112] Early studies concluded that charge transfer occurs on a time scale ~1000 times faster than radiative and non-radiative decay channels of the excited state leading to a quantum efficiency of near unity.[70] However, the power conversion efficiency of bilayer devices (Fig. 5.19a) was limited by the diffusion lengths of excitons.[113] Photoexcitations induced far from the heterojunction have enough time to recombine before reaching the donor–acceptor interface. An interpenetrating network of phase separated donor and acceptor material was proposed to enable a BHJ (Fig. 5.19b) with a larger D–A interface.[114] If the network is bicontinuous the efficiency of charge collection should also be high.

Fullerenes, which have a high electron affinity and high carrier mobility, are considered to be energetically almost ideal acceptor/electron transport materials for OPV. Much work on BHJ cells has therefore focused on optimizing the electronic structure

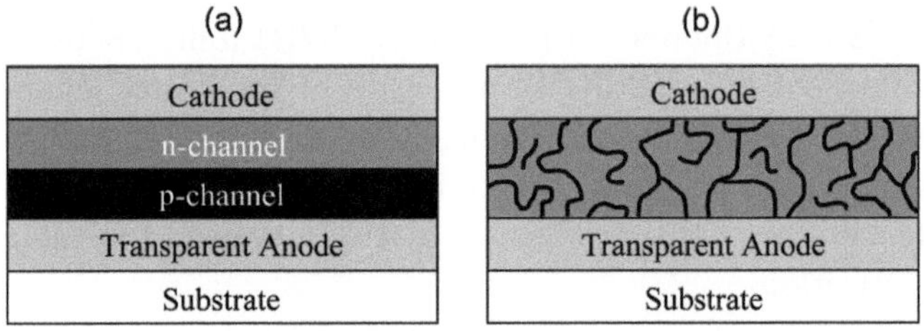

Figure 5.19. Cartoon profile illustrating the different photoactive layer morphology for (a) a bilayer and (b) a bulk heterojunction solar cell.

of polymer donor materials for efficient exciton dissociation and photon absorption. It should be noted that the power conversion efficiency of solar cells based on fullerenes is determined to a large extent by the overlap of the donor absorption spectrum with the solar emission spectrum. Although this is not the focus of the current chapter, the reader should bear this in mind in subsequent discussion.

The electronic structure of the two materials that comprise the BHJ is not the only factor that determines η. The nanostructure of the film morphology is extremely important in determining how much charge is collected by the electrodes. Several researchers have therefore strived to understand the thermodynamic properties of polymer-fullerene blends in an attempt to uncover the optimal thin-film processing conditions. The ideal film will be bicontinuous with domain sizes comparable to the exciton diffusion length (5–10 nm).[18] The two phases should also be well ordered to obtain fast charge transport thereby minimizing free carrier recombination. Forming ideal films is most easily achieved by solution processing so the most commonly utilized acceptor is C_{60}-PCBM.

One important factor influencing the film formation from a binary blend solution is the choice of solvent. Studies on MDMO-PPV:C_{60}-PCBM films spin-cast from toluene were found to exhibit lower η compared to films formed from a chlorobenzene solution.[115] This was attributed to the increased solubility of C_{60}-PCBM in chlorobenzene reducing the size of preformed clusters and allowing phase segregation on a smaller length scale compared to films formed from toluene. The solvent evaporation rate has also been found to affect film formation.[116] The power conversion efficiency of poly(3-hexylthiophene) (P3HT):C_{60}-PCBM films deposited from a dichlorobenzene (DCB) solution could be controlled to some extent by thermally controlling the solvent evaporation rate. Slower evaporation rates yielded improved efficiency where longer residence times for the solvent molecules allowed favorable phase reorganization.

The influence of the solubility of the fullerene on final film morphology and η has been explored for BHJ cells.[67] The solubility of a series of fullerenes in chlorobenzene was varied by modifying their side-chains and the morphology of films formed by blending them with P3HT was assessed. The authors concluded that maximizing the efficiency of devices required fullerene solubility in the range 30–80 mg/ml which is comparable to that of P3HT (50–70 mg/ml). Fullerenes with solubility below 20 mg/ml were found to form large aggregates in blend films which corresponded to low values of η. Films with phase domains with dimensions larger than the exciton diffusion length presumably exhibit a lower exciton dissociation efficiency. It was also found that fullerenes with solubility $>$ 90 mg/ml resulted in a reduced efficiency in P3HT blend solar cells. These films showed a homogeneous morphology implying intermixing of the two components without phase separation. This prevents the formation of a percolation pathway for charge carriers from the D–A interface to the electrodes.

Optimizing the blend ratio is also important for high-efficiency solar cells.[107] The fullerene content must be sufficiently high to allow percolation of electrons to the collecting electrodes. It must also be low enough to maximize photon absorption in the donor polymer. In general, this will depend on the miscibility of the donor and acceptor within the solution and the solid film after it has been formed. For example, the optimal blend ratio for MDMO-PPV:C_{60}-PCBM has been reported[117] to be 1:4 by weight compared to a ratio of around 1:1 for P3HT:C_{60}-PCBM films.[18] Additionally, the total concentration of the blend solution influences the solid film morphology.[107]

Post-deposition treatment has been found to be able to improve the morphology of polymer:fullerene films for high η. Treatments such as application of a large current, vapor annealing, and thermal annealing have all been shown to yield higher efficiencies in BHJ cells compared to cells without post-deposition treatment.[107] Thermal annealing above the glass transition temperature (T_G) of P3HT has been shown to enhance η in P3HT:C_{60}-PCBM films.[118,119] Thermal treatment was found to allow reorganization of the polymer chains and diffusion of the fullerene into a more ordered and thermodynamically favorable configuration of crystalline phase-separated domains of the two components. The phase behavior of this blend has been described in detail in a recent study.[45] Solvent annealing has been shown to provide a similar improvement in blend morphology for high-efficiency devices.[120] Exposure of

P3HT:C_{60}-PCBM to DCB was suggested to allow self-organization of the P3HT where subsequent thermal annealing allowed PCBM to diffuse and form aggregates. The general outcome of post-deposition treatment for improved efficiency is the formation of separated bicontinuous domains with high order to allow fast charge transport.[119]

Although most research reports have focused on C_{60} derivatives as acceptor materials, C_{70}-PCBM has also received some attention. The high symmetry of the C_{60} cage means that the lowest energy dipole transitions from the HOMO to the LUMO are forbidden to optical excitation. This results in low absorption coefficients for these materials. Lowering the symmetry of the acceptor molecules by moving to C_{70} derivatives allows these low energy transitions and increases absorption in the fullerene.[121] This has been exploited to increase the efficiency of solar cells using C_{70}-PCBM in comparison to the C_{60}-PCBM analogue.[121] Moving to C_{84}-PCBM, which has even stronger absorption in the visible spectrum, was actually found to reduce solar cell performance.[122] This is most likely a result of poor blend morphology.

Following the body of work focused on optimizing the film morphology and electronic structure since the suggestion of fullerene-based BHJ cells, the highest value of η reported in the scientific literature to date has reached 6.1%.[42] The authors utilized a blend of C_{70}-PCBM with the alternating copolymer poly[N-9″-hepta-decanyl-2,7-carbazole-alt-5,5-(4′,7′-di-2-thienyl-2′, 1′, 3′-benzothiadiazole)] (PCDTBT) without any post-deposition treatment. In comparison to the much more widely studied P3HT, PCDTBT has a narrower LUMO-HOMO gap with the implication that it can absorb more of the long wavelength end of the solar spectrum. Its relatively low-lying HOMO level compared to the LUMO of the fullerene also allows an increase in the device open circuit voltage. With an optimized nanomorphology the authors measured an internal quantum efficiency of nearly 100% implying that almost every photon absorbed results in a separated charged pair and that almost all carriers are collected by the electrodes. Additionally, an optical spacer was incorporated into the device structure to redistribute the incident light in the active layer to absorb a larger proportion of the incident photons.

The stability of solar cells is a crucial issue with regard to their successful commercialization. Electrical performance of OPV cells is typically found to degrade under exposure to atmospheric oxygen and water presumably by the action of these contaminants as charge traps. One route to circumvent this degradation is to encapsulate devices with a material that acts as a barrier to oxygen and water. This has been demonstrated recently with promising results. Hauch *et al.* have shown that a food package quality barrier film with a water vapor transmission rate of 0.2 g/(m^2 day) can enable P3HT:C_{60}-PCBM devices to survive for >1250 hours at 65°C under a relative humidity of 85%.[123] The authors suggest that this could allow an outdoor operational lifetime of two to three years. However, thermal instability of devices, relating to the phase behavior of the active layer, is an issue that requires different solutions.[18] Sivula *et al.* have reported that the addition of a diblock copolymer to P3HT:C_{60}-PCBM blends can reduce the interfacial energy between the polymer and fullerene and therefore attenuate the phase segregation induced by thermal annealing.[124] Alternatively, Drees *et al.* have shown that following cross-linking of a polymerizable fullerene derivative, the phase behavior of its blend with P3HT can be stabilized against thermal annealing.[125] The polymerization was found to hinder diffusion of the fullerene.

5.7.2 *Fullerene-Based Phototransistors and Electro-Optic Circuits*

The relatively recent development of organic phototransistors has opened the door for investigation into novel optoelectronic circuits based on bifunctional transistors. In particular, light-sensing OFETs (LS-OFETs)[126] are thought to be potential candidates for implementation in low-cost electro-optical transceivers and optical sensor arrays. Light-emitting OFETs (LE-OFETs),[127] with less obvious applications, are an interesting test bed for understanding recombination physics in organic semiconductors. Although light-emission was first observed in transistors based on a single component unipolar semiconductor,[127] reports on phototransistors based on fullerenes typically utilize ambipolar active layers containing a blend[126,128] or a bilayer[129] with p-channel materials.

Efficient photoinduced charge generation in LS-OFETs is required to allow identification of a photocurrent. One approach to achieving this is based on the well-known concept of the BHJ used in OPV. LS-OFETs that use this strategy are therefore based on polymer:fullerene blends. Marjanović *et al.* were the first to report a successful demonstration of photoresponsive BHJ OFETs.[126] They used a blend of MDMO-PPV:C_{60}-PCBM in a 1:4 weight ratio. The authors defined the photoresponsivity as

$$R = \frac{I_{\text{ph}}}{P_{\text{opt}}} = \frac{I_{\text{DS,light}} - I_{\text{DS,dark}}}{P_{\text{inc}}A}, \tag{5.10}$$

where I_{ph} is the photocurrent (the difference between the illuminated and dark drain–source currents) and P_{opt} is the incident optical power (incident power density multiplied by the effective device area, A). Their peak value of R was reported to be 5 A/W, suggesting that an increased I_{DS} upon illumination was a result of the additional contribution of photogenerated charge carriers in the bulk of the film. However, these devices only displayed n-channel behavior. This most likely results from the low work function LiF/Al contact electrodes limiting the change in current upon exposure to light.

The photovoltaic effect, in addition to ambipolar charge transport characteristics, was later shown in P3HT:C_{60}-PCBM BHJ OFETs.[128] The authors used asymmetric contacts (Au and Al) where the potential drop across the active layer created by the offset in work function between the metal electrodes was able to drive charge separation even under short circuit conditions, equivalent to an OPV cell. They found that with zero gate bias the device shows photovoltaic effects under illumination in addition to ambipolar OFET characteristics with gate bias. As an OPV cell the device was found to exhibit a modest η of 0.6%. However, this is achieved with an electrode spacing far exceeding the typical thickness of thin-film OPV device.

Logic functions such as OR and NOT gates have been fabricated with this class of active layer in OFET architecture where the input signals can be purely optical or a combination of electrical and optical.[130] The blend of MDMO-PPV:C_{60}-PCBM was used in a weight ratio of 1:15 to maximize electron mobility and photosensitivity (the ratio of illuminated to dark off-currents). The dynamic response of a LS-OFET was explored with a square wave optical input signal. Analysis of the V_{OUT} rise and fall times suggested a maximum operating frequency of ~15 kHz. Inverters incorporating a LS-OFET with a unipolar C_{60}-PCBM OFET were shown to produce a high V_{OUT} in the dark and a low V_{OUT} under illumination. This is a result of the ability to optically control the resistance of the LS-OFET channel. The characteristics of a LS-OFET and its use in an electro-optic NOT-gate are shown in Fig. 5.20. Similarly, by controlling the optical input as well as the electrical input to the gate of a single LS-OFET, an OR gate could also be realized.

To the best of our knowledge LE-OFETs based on fullerenes are yet to be demonstrated in the literature. However, the area of

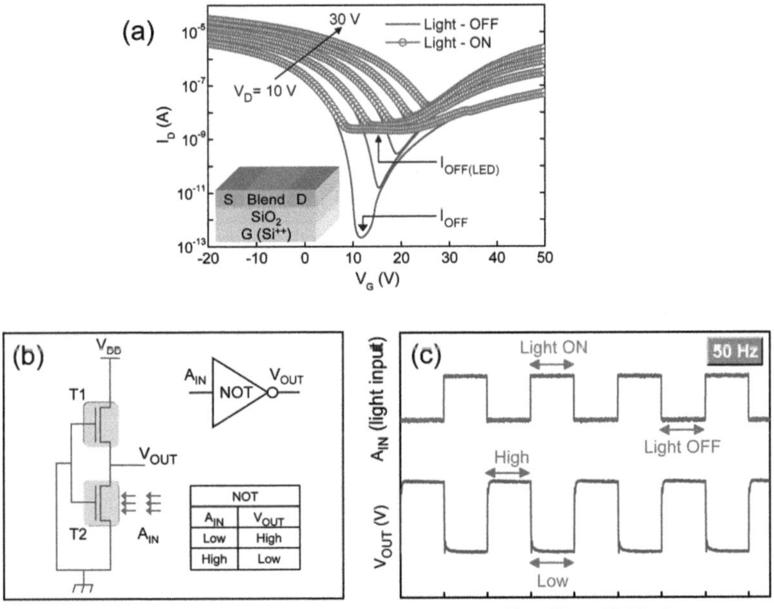

Figure 5.20. (a) Transfer characteristics of a LS-OFET using a C_{60}-PCBM:MDMO-PPV (15:1 by wt) blend as the active layer. (b) Circuit diagram of an electro-optic NOT gate with symbolic representation and truth table. (c) 50 Hz pulsed optical input (red line) and corresponding V_{OUT} (blue line) as a function of time. Image adapted from ref. 130. See also Color Insert.

organic phototransistors is still young and further developments in device and circuit design as well as fundamental understanding of organic semiconductors may yield novel applications inaccessible to otherwise alternative technologies.

5.8 Summary and Perspectives

We have seen in this chapter that fullerenes present an important class of semiconducting materials for the active layer in organic electronics and optoelectronics. Their unique electronic and structural properties combined with the ability to chemically tailor adducts allow them to be used as high-performance materials in a range of applications. Their high electron mobility in ordered films suggests their potential for implementation in integrated circuits for low-cost microelectronics. In addition, their favorable electronic properties and controllable phase behavior in polymer blends promises their use in commercial solar cell applications.

In transistors, further improvement in the field-effect mobility of solution cast fullerene films is still necessary to meet commercial requirements. Strategies to achieving this may lie in tailoring derivatives for self-assembly of highly ordered films. However, no extensive studies on solution preparation of fullerene thin-films for transistors have been published. A full understanding of the effects of processing conditions on pristine film morphology, and therefore electron mobility, is yet to be deduced.

The air stability of fullerene-based devices is currently a troubling issue particularly for transistors. Despite films of the higher fullerene C_{84} exhibiting air stability for several months, extraction of this material in large quantities has proved challenging and its widespread implementation has been hindered as a result. Although chemical tailoring of the cage can modify the electronic structure of a fullerene, this is yet to be exploited to sufficiently lower the LUMO level below the expected trap energy of atmospheric oxidants. Side-chain modification to provide a diffusion barrier has been shown to improve air stability but may not ultimately prevent long-term degradation without changing the electronic structure.

In solar cells fullerenes are well established as a promising candidate as the acceptor material for commercial devices. Further work on efficiency improvement is likely to be based on optimizing the electronic properties of the donor material with respect to the fullerene and on the morphology of the resulting blend. We therefore expect fullerenes to play a vital role in future device improvements towards achieving the goal of useful and cheap electrical conversion of solar energy.

Acknowledgments

The authors would like to thank the Engineering and Physical Sciences Research Council (EPSRC, grant numbers EP/C539516 and EP/E06455X) and Research Councils UK (RCUK) for financial support.

References

1. J. Koenigsberger and K. Schilling, *Ann. Phys.*, **337**, 179–230 (1910).
2. M. Pope, H. P. Kallmann, and P. Magnante, *J. Chem. Phys.*, **38**, 2042 (1963).
3. C. K. Chiang, C. R. Fincher, Y. W. Park, A. J. Heeger, H. Shirakawa, E. J. Louis, S. C. Gau, and A. G. MacDiarmid, *Phys. Rev. Lett.*, **39**, 1098–1101 (1977).
4. C. W. Tang, *Appl. Phys. Lett.*, **48**, 183 (1986).
5. H. Koezuka, A. Tsumura, and T. Ando, *Synth. Met.*, **18**, 699–704 (1987).
6. C. W. Tang and S. A. VanSlyke, *Appl. Phys. Lett.*, **51**, 913 (1987).
7. J. H. Burroughes, C. A. Jones, and R. H. Friend, *Nature*, **335**, 137–141 (1988).
8. J. H. Burroughes, D. D. C. Bradley, A. R. Brown, R. N. Marks, K. Mackay, R. H. Friend, P. L. Burns, and A. B. Holmes, *Nature*, **347**, 539–541 (1990).
9. A. Tsumura, H. Koezuka, and T. Ando, *Appl. Phys. Lett.*, **49**, 1210–1212 (1986).
10. S. R. Forrest, *Nature*, **428**, 911–918 (2004).

11. S. K. Park, D. A. Mourey, S. Subramanian, J. E. Anthony, and T. N. Jackson, *Appl. Phys. Lett.*, **93**, 043301 (2008).
12. R. Hamilton, J. Smith, S. Ogier, M. Heeney, J. E. Anthony, I. McCulloch, J. Veres, D. D. C. Bradley, and T. D. Anthopoulos, *Adv. Mater.*, **21**, 1166–1171 (2008).
13. T. D. Anthopoulos, B. Singh, N. Marjanovic, N. S. Sariciftci, A. M. Ramil, H. Sitter, M. Cölle, and D. M. de Leeuw, *Appl. Phys. Lett.*, **89**, 213504 (2006).
14. X.-H. Zhang, B. Domercq, and B. Kippelen, *Appl. Phys. Lett.*, **91**, 092114 (2007).
15. S. M. Sze and K. K. Ng, *Physics of Semiconductor Devices*, 3rd Edition, John Wiley & Sons, New Jersey, 2007.
16. L. Zhou, A. Wanga, S.-C. Wu, J. Sun, S. Park, and T. N. Jackson, *Appl. Phys. Lett.*, **88**, 083502 (2006).
17. Q. Cao, Z.-T. Zhu, M. G. Lemaitre, M.-G. Xia, M. Shim, and J. A. Rogers, *Appl. Phys. Lett.*, **88**, 113511 (2006).
18. B. C. Thompson and J. M. J. Fréchet, *Angew. Chem. Int. Ed. Engl.*, **47**, 58–77 (2008).
19. J. Peet, M. L. Senatore, A. J. Heeger, and G. C. Bazan, *Adv. Mater.*, **21**, 1521–1527 (2009).
20. S. Günes, H. Neugebauer, and N. S. Sariciftci, *Chem. Rev.*, **107**, 1324–1338 (2007).
21. B. W. D'Andrade and S. R. Forrest, *Adv. Mater.*, **16**, 1585–1595 (2004).
22. T. Agostinelli, M. Caironi, D. Natali, M. Sampietro, P. Biagioni, M. Finazzi, and L. Duò, *J. Appl. Phys.*, **101**, 114504 (2007).
23. M. Mushrush, A. Facchetti, M. Lefenfeld, H. E. Katz, and T. J. Marks, *J. Am. Chem. Soc.*, **125**, 9414–9423 (2003).
24. A. R. Brown, C. P. Jarrett, D. M. de Leeuw, and M. Matters, *Synth. Met.*, **88**, 37–55 (1997).
25. G. Horowitz, *Adv. Mater.*, **10**, 365–377 (1998).
26. C. Dimitrakopoulos and P. R. L. Malenfant, *Adv. Mater.*, **14**, 99–117 (2002).
27. B. Crone, A. Dodabalapur, Y.-Y. Lin, R. W. Filas, Z. Bao, A. La Duca, R. Sarpeshkar, H. E. Katz, and W. Li, *Nature*, **403**, 521–523 (2000).
28. T. N. Jackson, Y.-Y. Lin, D. J. Gundlach, and H. Klauk, *IEEE J. Sel. Top. Quantum Electron.*, **4**, 100–104 (1998).
29. Y. H. Kim, S. K. Park, D. G. Moon, W. K. Kim, and J. I. Han, *Jpn. J. Appl. Phys.*, **43**, 3605–3608 (2004).

30. P. F. Baude, D. A. Ender, M. A. Haase, T. W. Kelley, D. V. Muyres, and S. D. Theiss, *Appl. Phys. Lett.*, **82**, 3964–3966 (2003).
31. E. Cantatore, T. C. T. Geuns, G. H. Gelinck, E. van Veenendaal, A. F. A. Gruijthuijsen, L. Schrijnemakers, S. Drews, and D. M. de Leeuw, *IEEE J. Solid-State Circuits*, **42**, 84–92 (2007).
32. H. Sirringhaus, Nir Tessler, and R. H. Friend, *Science*, **280**, 1741–1744 (1998).
33. J. G. Laquindanum, H. E. Katz, A. Dodabalapur, and A. J. Lovinger, *J. Am. Chem. Soc.*, **118**, 11331–11332 (1996).
34. H. E. Katz, A. J. Lovinger, J. Johnson, C. Kloc, T. Siegrist, W. Li, Y.-Y. Lin, and A. Dodabalapur, *Nature*, **404**, 478–481 (2000).
35. C. R. Newman, C. D. Frisbie, D. A. da Silva Filho, J.-L. Brédas, P. C. Ewbank, and K. R. Mann, *Chem. Mater.*, **16**, 4436–4451 (2004).
36. H. Sirringhaus, T. Kawase, R. H. Friend, T. Shimoda, M. Inbasekaran, W. Wu, and E. P. Woo, **290**, 2123–2126 (2000).
37. M. Chikamatsu, A. Itakura, Y. Yoshida, R. Azumi, and K. Yase, *Chem. Mater.*, **20**, 7365–7367 (2008).
38. C. Piliego, D. Jarzab, G. Gigli, Z. Chen, A. Facchetti, and M. A. Loi, *Adv. Mater.*, **21**, 1573–1576 (2009).
39. P. H. Wöbkenberg, D. D. C. Bradley, D. Kronholm, J. C. Hummelen, D. M. de Leeuw, M. Cölle, and T. D. Anthopoulos, *Synth. Met.*, **158**, 468–472 (2008).
40. H. Yan, Z. Chen, Y. Zheng, C. Newman, J. R. Quinn, F. Dötz, M. Kastler, and A. Facchetti, *Nature*, **457**, 679–687 (2009).
41. J. Peet, J. Y. Kim, N. E. Coates, W. L. Ma, D. Moses, A. J. Heeger, and G. C. Bazan, *Nat. Mater.*, **6**, 497–500 (2007).
42. S. H. Park, A. Roy, S. Beaupré, S. Cho, N. Coates, J. S. Moon, D. Moses, M. Leclerc, K. Lee, and A. J. Heeger, *Nat. Photonics*, **3**, 297–303 (2009).
43. M. C. Scharber, D. Mühlbacher, M. Koppe, P. Denk, C. Waldauf, A. J. Heeger, and C. J. Brabec, *Adv. Mater.*, **18**, 789–794 (2007).
44. G. Dennler, M. C. Scharber, T. Ameri, P. Denk, K. Forberich, C. Waldauf, and C. J. Brabec, *Adv. Mater.*, **20**, 579–583 (2008).
45. C. Müller, T. A. M. Ferenczi, M. Campoy-Quiles, J. M. Frost, D. D. C. Bradley, P. Smith, N. Stingelin-Stutzmann, and J. Nelson, *Adv. Mater.*, **20**, 3510–3515 (2008).
46. H. W. Kroto, J. R. Heath, S. C. O'Brien, R. F. Curl, and R. E. Samlley, *Nature*, **318**, 162–163 (1985).

47. M. S. Dresselhaus, G. Dresselhaus, and P. C. Eklund, *J. Mater. Res.*, **8**, 2054–2097 (1993).
48. M. Bühl and A. Hirsch, *Chem. Rev.*, **101**, 1153–1183 (2001).
49. R. C. Haddon, *Acc. Chem. Res.*, **25**, 127–133 (1992).
50. Q. Xie, E. Pérez-Cordero, and L. Echegoyen, *J. Am. Chem. Soc.*, **114**, 3978–3980 (1992).
51. R. C. Haddon, L. E. Brus, and K. Raghavachari, *Chem. Phys. Lett.*, **125**, 459–464 (1986).
52. S. Saito, S. Okada, S. Sawada, and N. Hamada, *Phys. Rev. Lett.*, **75**, 685–688 (1995).
53. D. E. Manolopoulos and P. W. Fowler, *J. Chem. Phys.*, **96**, 7603–7614 (1992).
54. D. Bakowies, M. Kolb, W. Thiel, S. Richard, R. Ahlrichs, and M. M. Kappes, *Chem. Phys. Lett.*, **200**, 411–417 (1992).
55. K. Kikuchi, N. Nakahara, T. Wakabayashi, S. Suzuki, H. Shiromaru, Y. Miyake, K. Saito, I. Ikemoto, M. Kainosho, and Y. Achiba, *Nature*, **357**, 142–145 (1992).
56. F. Diederich, R. Ettl, Y. Rubin, R. L. Whetten, R. Beck, M. Alvarez, S. Anz, D. Sensharma, F. Wudl, K. C. Khemani, and A . Koch, *Science*, **252**, 548–551 (1991).
57. W. Krätschmer, L. D. Lamb, K. Fostiropoulos, and D. R. Huffman, *Nature*, **347**, 354–358 (1990).
58. Y. Yang, F. Arias, L. Echegoyen, L. P. F. Chibante, S. Flanagan, A. Robertson, and L. J. Wilson, *J. Am. Chem. Soc.*, **117**, 7801–7804 (1995).
59. E. J. Meijer, D. M. de Leeuw, S. Setayesh, E. van Veenendaal, B.-H. Huisman, P. W. M. Blom, J. C. Hummelen, U. Scherf, and T. M. Klapwijk, *Nat. Mater.*, **2**, 678–682 (2003).
60. T. D. Anthopoulos, F. B. Kooistra, H. J. Wondergem, D. Kronholm, J. C. Hummelen, and D. M. de Leeuw, *Adv. Mater.*, **18**, 1679–1684 (2006).
61. D. S. Bethune, R. D. Johnson, J. R. Salem, M. S. de Vries, and C. S. Yannoni, *Nature*, **366**, 123–128 (1993).
62. A. Y. Ganin, Y. Takabayashi, Y. Z. Khimyak, S. Margadonna, A. Tamai, M. J. Rosseinsky, and K. Prassides, *Nat. Mater.*, **7**, 367–371 (2008).
63. K. Tanigaki, S. Kuroshima, and T. W. Ebbesen, *Thin Solid Films*, **257**, 154–165 (1995).
64. S. Kobayashi, T. Takenobu, S. Mori, A. Fujiwara, and Y. Iwasa, *Appl. Phys. Lett.*, **82**, 4581–4583 (2003).

65. A. M. Ramil, Th. B. Singh, N. T. Haber, N. Marjanovic, S. Günes, A. Andreev, G. J. Matt, R. Resel, H. Sitter, and S. Sariciftci, *J. Cryst. Growth* **288**, 123–127 (2006).

66. M. T. Beck and G. Mándi, *Fullerenes Nanotubes Carbon Nanostruct.*, **5**, 291–310 (1997).

67. P. A. Troshin, H. Hoppe, J. Renz, M. Egginer, J. Y. Mayorova, A. E. Goryachev, A. S. Peregudov, R. N. Lyubovskaya, G. Gobsch, N. S. Sariciftci, and V. F. Razumov, *Adv. Funct. Mater.*, **19**, 779–788 (2009).

68. M. Chikamatsu, S. Nagamatsu, Y. Yoshida, K. Saito, K. Yase, and K. Kikuchi, *Appl. Phys. Lett.*, **87**, 203504 (2005).

69. P. H. Wöbkenberg, J. Ball, D. D. C. Bradley, T. D. Anthopoulos, F. Kooistra, J. C. Hummelen, and D. M. de Leeuw, *Appl. Phys. Lett.*, **92**, 143310 (2008).

70. B. Kraabel, C. H. Lee, D. McBranch, D. Moses, N. S. Sariciftci, and A. J. Heeger, *Chem. Phys. Lett.*, **213**, 389–394 (1993).

71. J. Zaumseil and H. Sirringhaus, *Chem. Rev.*, **107**, 1296–1323 (2007).

72. Z. Bao and J. Locklin, *Organic Field-Effect Transistors*, CRC Press, Boca Raton, 2007.

73. S. G. J. Mathijssen, M. Kemerink, A. Sharma, M. Cölle, P. A. Bobbert, R. A. J. Janssen, and D. M. de Leeuw, *Adv. Mater.*, **20**, 975–979 (2008).

74. S. Kobayashi, T. Nishikawa, T. Takenobu, S. Mori, T. Shimoda, T. Mitani, H. Shimotani, N. Yoshimoto, S. Ogawa, and Y. Iwasa, *Nat. Mater.*, **3**, 317–322 (2004).

75. A. Salleo, T. W. Chen, A. R. Völkel, Y. Wu, P. Liu, B. S. Ong, and R. A. Street, *Phys. Rev. B. Condens. Matter.*, **70**, 115311 (2004).

76. R. A. Street, J. E. Northrup, and A. Salleo, *Phys. Rev. B. Condens. Matter.*, **71**, 165202 (2005).

77. E. C. P. Smits, S. G. J. Mathijssen, P. A. van Hal, S. Setayesh, T. C. T. Geuns, K. A. H. A. Mutsaers, E. Cantatore, H. J. Wondergem, O. Werzer, R. Resel, M. Kemerink, S. Kirchmeyer, A. M. Muzafarov, S. A. Ponomarenko, B. de Boer, P. W. M. Blom, and D. M. de Leeuw, *Nature*, **455**, 956–959 (2008).

78. A. Dodabalapur, J. Laquindanum, H. E. Katz, and Z. Bao, *Appl. Phys. Lett.*, **69**, 4227–4229 (1996).

79. W. Fix, A. Ullmann, J. Ficker, and W. Clemens, *Appl. Phys. Lett.*, **81**, 1735–1737 (2002).

80. G. L. Gaines, M. P. O'Neil, W. A. Svec, M. P. Niemczyk, and M. R. Wasielewski, *J. Am. Chem. Soc.*, **113**, 719–721 (1991).

81. V. I. Arkhipov and H. Bässler, *Phys. Status. Solidi. A*, **201**, 1152–1187 (2004).
82. Y.-X. Liu, M. A. Summers, S. R. Scully, and M. D. McGehee, *J. Appl. Phys.*, **99**, 093521 (2006).
83. A. Gadisa, M. Svensson, M. R. Andersson, and O. Inganäs, *Appl. Phys. Lett.*, **84**, 1609–1611 (2004).
84. G. Horowitz, M. E. Hajlaoui, and R. Hajlaoui, *J. Appl. Phys.*, **87**, 4456–4463 (2000).
85. P. G. Le Comber and W. E. Spear, *Phys. Rev. Lett.*, **8**, 509–511 (1970).
86. R. A. Marcus, *J. Chem. Phys.*, **24**, 966–978 (1956).
87. J.-L. Brédas, D. Beljonne, V. Coropceanu, and J. Cornil, *Chem. Rev.*, **104**, 4971–5004 (2004).
88. J. L. Brédas, J. P. Calbert, D. A. da Silva Filho, and J. Cornil, *Proc. Natl. Acad. Sci. U.S.A.*, **99**, 5804–5809 (2002).
89. A. Miller and E. Abrahams, *Phys. Rev.*, **120**, 745–755 (1960).
90. V. Coropceanu, J. Cornil, D. A. da Silva Filho, Y. Olivier, R. Silbey, and J.-L. Brédas, *Chem. Rev.*, **107**, 926–952 (2007).
91. H. Bässler, *Phys. Status. Solidi. B*, **175**, 15–56 (1993).
92. S. V. Novikov, D. H. Dunlap, V. M. Kenkre, P. E. Parris, and A. V. Vannikov, *Phys. Rev. Lett.*, **81**, 4472–4475 (1998).
93. M. C. J. M. Vissenberg and M. Matters, *Phys. Rev. B. Condens. Matter.*, **57**, 12964–12967 (1998).
94. Z. B. Wang, M. G. Helander, M. T. Greiner, J. Qiu, and Z. H. Lu, *Appl. Phys. Lett.*, **95**, 043302 (2009).
95. R. C. Haddon, A. S. Perel, R. C. Morris, T. T. M. Palstra, A. F. Hebard, and R. M. Fleming, *Appl. Phys. Lett.*, **67**, 121–123 (1995).
96. Th. B. Singh, N. S. Sariciftci, H. Yang, L. Yang, B. Plochberger, and H. Sitter, *Appl. Phys. Lett.*, **90**, 213512 (2007).
97. R. C. Haddon, *J. Am. Chem. Soc.*, **118**, 3041–3042 (1996).
98. J. N. Haddock, X. Zhang, B. Domercq, and B. Kippelen, *Org. Electron.*, **6**, 182–187 (2005).
99. J. C. Hummelen, B. W. Knight, F. Lepeq, F. Wudl, J. Yao, and C. L. Wilkins, *J. Org. Chem.*, **60**, 532–538 (1995).
100. Th. B. Singh, N. Marjanovi , P. Stadler, M. Auinger, G. J. Matt, S. Günes, N. S. Sariciftci, R. Schwödiauer, and S. Bauer, *J. Appl. Phys.*, **97**, 083714 (2005).
101. T. D. Anthopoulos, G. C. Anyfantis, G. C. Papavassiliou, and D. M. de Leeuw, *Appl. Phys. Lett.*, **90**, 122105 (2007).

102. P. H. Wöbkenberg, J. Ball, F. B. Kooistra, J. C. Hummelen, D. M. de Leeuw, D. D. C. Bradley, and T. D. Anthopoulos, *Appl. Phys. Lett.*, **93**, 013303 (2008).

103. T. D. Anthopoulos, C. Tenase, S. Setayesh, E. J. Meijer, J. C. Hummelen, P. W. M. Blom, and D. M. de Leeuw, *Adv. Mater.*, **16**, 2174–2179 (2004).

104. T. D. Anthopoulos, D. M. de Leeuw, E. Cantatore, P. van 't Hof, J. Alma, and J. C. Hummelen, *J. Appl. Phys.*, **98**, 054503 (2005).

105. X.-H. Zhang, W. J. Potscavage Jr., S. Choi, and B. Kippelen, *Appl. Phys. Lett.*, **94**, 043313 (2009).

106. T. D. Anthopoulos, D. M. de Leeuw, E. Cantatore, S. Setayesh, E. J. Meijer, C. Tanase, J. Hummelen, and P. W. M. Blom, *Appl. Phys. Lett.*, **85**, 4205–4207 (2004).

107. H. Hoppe and N. S. Sariciftci, *J. Mater. Chem.*, **16**, 45–61 (2006).

108. X. Yang and J. Loos, *Macromolecules*, **40**, 1353–1362 (2007).

109. P. W. M. Blom, V. D. Mihailetchi, L. J. A. Koster, and D. E. Markov, *Adv. Mater.*, **19**, 1551–1566 (2007).

110. N. S. Sariciftci, L. Smilowitz, A. J. Heeger, and F. Wudl, *Science* **258**, 1474–1476 (1992).

111. L. Smilowitz, N. S. Sariciftci, R. Wu, C. Gettinger, A. J. Heeger, and F. Wudl, *Phys. Rev. B. Condens. Matter.*, **47**, 13835 (1993).

112. G. Yu, J. Gao, J. C. Hummelen, F. Wudl, and A. J. Heeger, *Science* **270**, 1789–1791 (1995).

113. N. S. Sariciftci, D. Braun, C. Zhang, V. I. Srdanov, A. J. Heeger, G. Stucky, and F. Wudl, *Appl. Phys. Lett.*, **62**, 585–587 (1993).

114. J. J. M. Halls, C. A. Walsh, N. C. Greenham, E. A. Marseglia, R. H. Friend, S. C. Moratti, and A. B. Holmes, *Nature* **376**, 498–500 (1995).

115. S. E. Shaheen, C. J. Brabec, N. S. Sariciftci, F. Padinger, T. Fromherz, and J. C. Hummelen, *Appl. Phys. Lett.*, **78**, 841–843 (2001).

116. G. Li, V. Shrotriya, J. Huang, Y. Yao, T. Moriarty, K. Emery, and Y. Yang, *Nat. Mater.*, **4**, 864–868 (2005).

117. J. K. J. van Duren, X. Yang, J. Loos, C. W. T. Bulle-Lieuwma, A. B. Sieval, J. C. Hummelen, and R. A. J. Janssen, *Adv. Funct. Mater.*, **14**, 425–434 (2004).

118. F. Padinger, R. S. Rittberger, and N. S. Sariciftci, *Adv. Funct. Mater.*, **13**, 85–88 (2003).

119. M. Campoy-Quiles, T. Ferenczi, T. Agostinelli, P. G. Etchegoin, Y. Kim, T. D. Anthopoulos, P. N. Stavrinou, D. D. C. Bradley, and J. Nelson, *Nat. Mater.*, **7**, 158–164 (2008).

120. Y. Zhao, Z. Xie, Y. Qu, Y. Geng, and L. Wang, *Appl. Phys. Lett.*, **90**, 043504 (2007).
121. M. M. Wienk, J. M. Kroon, W. J. H. Verhees, J. Know, J. C. Hummelen, P. A. van Hal, and R. A. J. Janssen, *Angew. Chem. Int. Ed., Engl.*, **42**, 3371–3375 (2003).
122. F. B. Kooistra, V. D. Mihailetchi, L. M. Popescu, D. Kronholm, P. W. M. Blom, and J. C. Hummelen, *Chem. Mater.*, **18**, 3068–3073 (2006).
123. J. A. Hauch, P. Schilinsky, S. A. Choulis, S. Rajoelson, and C. J. Brabec, *Appl. Phys. Lett.*, **93**, 103306 (2008).
124. K. Sivula, Z. T. Ball, N. Watanabe, and J. M. J. Fréchet, *Adv. Mater.*, **18**, 206–210 (2006).
125. M. Drees, H. Hoppe, C. Winder, H. Neugebauer, N. S. Sariciftci, W. Schwinger, F. Schäffler, C. Topf, M. C. Scharber, Z. Zhu, and R. Gaudiana, *J. Mater. Chem.*, **15**, 5158–5163 (2005).
126. N. Marjanović, Th. B. Singh, G. Dennler, S. Günes, H. Neugebauer, N. S. Sariciftci, R. Schwödiauer, and S. Bauer, *Org. Electron.*, **7**, 188–194 (2006).
127. A. Hepp, H. Heil, W. Weise, M. Ahles, R. Schmechel, and H. von Seggern, *Phys. Rev. Lett.*, **91**, 157406 (2003).
128. S. Cho, J. Yuen, J. Y. Kim, K. Lee, and A. J. Heeger, *Appl. Phys. Lett.*, **90**, 063511 (2007).
129. B. Park, P. Paoprasert, P. Gopalan, T. F. Kuech, and P. G. Evans, *Appl. Phys. Lett.*, **94**, 073302 (2009).
130. T. D. Anthopoulos, *Appl. Phys. Lett.*, **91**, 113513 (2007).

Chapter 6

Carbon Nanohorns Chemical Functionalization

Georgia Pagona and Nikos Tagmatarchis
Theoretical and Physical Chemistry Institute, National Hellenic Research Foundation, 48 Vass. Constantinou Avenue, Athens 11635, Greece
tagmatar@eie.gr

Carbon nanohorns (CNHs), an alternative of nanotubes with conical tips and high purity due to the absence of metal impurities, are assembled in a secondary spherical hyperstructure. Similar with nanotubes, CNHs are insoluble in all solvents. In this chapter, the most significant developments on the functionalization and solubilization of CNHs are presented. Selected examples from the recent literature have been collected and together with some original as well as established methodologies are discussed. Among these, 1,3-dipolar cycloadditions, aryl diazonium addition, Bingel cyclopropanation, amination, as well as oxidation and subsequent condensation reactions have been widely applied to covalently modify the outer skeleton or conical tips of CNHs. Furthermore, CNHs have been non-covalently functionalized with the aid of polymer wrapping and π–π stacking interactions with pyrenes or porphyrins. Finally, emphasis is placed on some potential applications of CNH-based hybrid materials, especially for drug delivery and photovoltaics.

Advances in Carbon Nanomaterials: Science and Applications
Edited by Nikos Tagmatarchis

ISBN 978-981-426-78-78 (Hardcover), 978-981-426-78-85 (eBook)
www.panstanford.com

6.1 Introduction

Carbon nanohorns (CNHs) fall within the large family of carbon nanostructures and more precisely are a promising alternative of carbon nanotubes with great potentiality for technological and biological applications. Although CNHs were observed earlier,[1,2] they were prepared in large quantities in 2004 from Iijima's group, using the technique of CO_2 laser ablation of graphite,[3,4] at room temperature, under an argon atmosphere. The structure of CNHs is similar to that of single-walled carbon nanotubes but with an irregular shape. Namely, they appear to be a cone-shaped graphitic aggregate, in which the direction of individual cones is radiated out from the center of the sphere, resembling the shape of a dahlia flower. The length of the conical tubes is 30–50 nm, their diameter is 2–5 nm, while the angle of the conical tip is calculated to 19°–20°. About 2000 of individual nanohorns assemble together to form a spherical aggregate with a diameter of about 100 nm. The overall size of this superstructure is compact, and only very recently, the separation and the isolation of an individual CNH was reported.[5]

An important advantage of CNHs that mainly discriminate them from carbon nanotubes is the absence of metal catalyst during their preparation. Thus, they are produced in clean form without the presence of impurities, contrary to carbon nanotubes which contain impurities of metal nanoparticles. It has also been reported that the type and the pressure of gas applied during the CNHs synthesis play an important role in the morphology and purity of the material. Thus, argon leads to CNHs aggregates with dahlia-like shape, helium results in CNHs aggregates with bud-like shape,[6,7] and there also exists a third morphology in which CNHs aggregate in a seed-like form.

The characteristic tubular structure and the conical tip of CNHs can be well observed under high-resolution transmission electron microscopy (HR-TEM). In Fig. 6.1, a graphical illustration and a real image of CNHs as obtained under the microscope are presented. The conical tips of CNHs have high energy due to the presence of the five five-membered rings — this is another characteristic that differentiates CNHs from carbon nanotubes. It has been reported that as-produced CNHs possess 70% structure of tube, 15% conical tip, 12% graphite, and 2.5% amorphous carbon.[8]

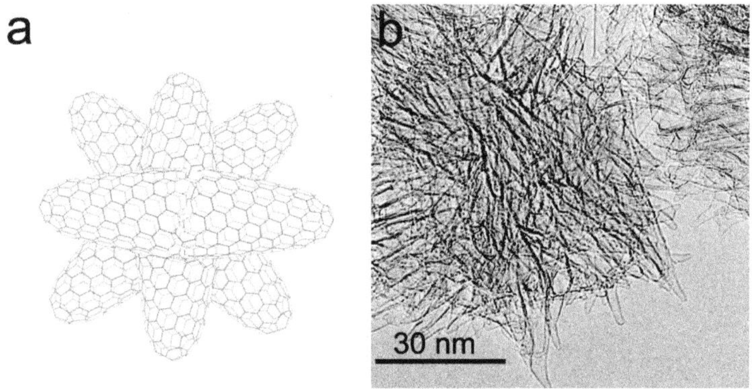

Figure 6.1. (a) Schematic illustration of CNHs aggregate and (b) HR-TEM image of CNHs.

Raman spectroscopy plays an important role on the structure determination of CNHs. Two characteristic Raman bands of almost equal intensity are observed in pristine CNHs, as it is shown in Fig. 6.2. The band at 1593 cm^{-1} is attributed to the E_{2g} vibrations of sp^2 carbon atoms (similarly with that of graphite; this is the so-called G-band), while the band at 1341 cm^{-1} (so-called D-band) is attributed to the A_{1g} vibrations of sp^3 carbon atoms that link each CNH forming the secondary spherical hyperstructure.[9–11]

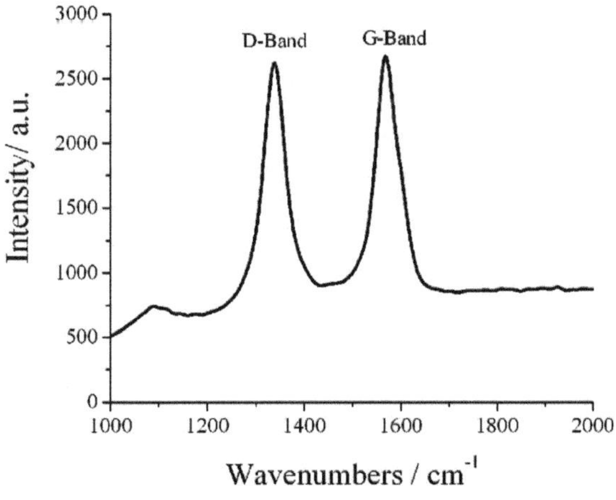

Figure 6.2. Raman spectrum of pristine CNHs (λ_{exc} = 488 nm) showing the characteristic D- and G-bands.

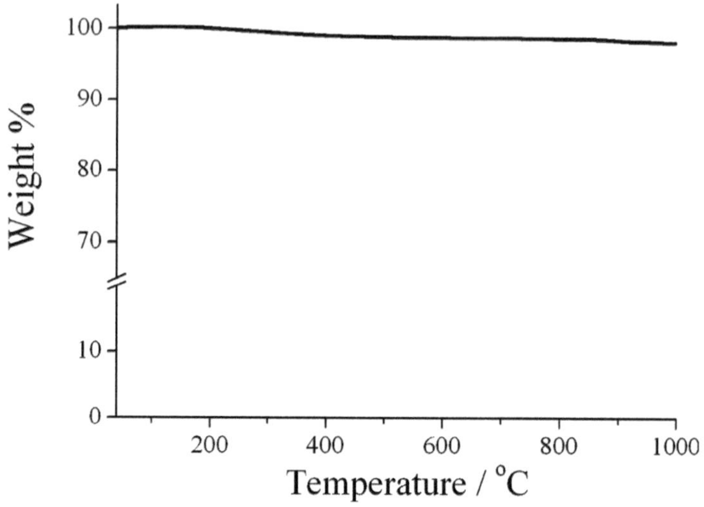

Figure 6.3. Thermogravimetric analysis graph of pristine CNHs, under nitrogen atmosphere.

The electronic absorption spectrum of CNHs is featureless, while the absorption monotonically decreases upon reaching the near infrared (NIR) region. As far as thermal stability concerns, CNHs are thermally stable up at least 900°C, under nitrogen atmosphere, as revealed by thermogravimetric analysis (TGA) studies. In Fig. 6.3 the thermograph of pristine CNHs is shown.

Potential applications of CNHs include gas adsorption and storage,[12–14] fuel cells,[15,16] catalytic nanoparticles support,[16–19] encapsulation of fullerenes[20–22] and metals,[23–25] and drug delivery.[26–37] However, a major obstacle that has to be overcome is their insolubility in all solvents and water, similarly like carbon nanotubes. In this context, chemical modification is the route that leads to solubilization enhancement, through the decoration of CNHs skeleton with a plethora of organic units. In general, functionalization of CNHs can occur either via covalent or supramolecular approaches. As far as introduction of organic moieties through stable bond formation onto CNHs skeleton concerns, two strategies are followed: (i) covalent bond formation at the sidewalls and (ii) oxidation of the conical tips, followed by condensation reactions with carboxylic acid units, as introduced during the oxidation process. On the other hand, in the context of supramolecular functionalization of CNHs based on non-covalent interactions, the major approaches followed are π–π stacking interactions with aromatic planar molecules as well as wrapping with polymers.

6.2 Chemical Functionalization of CNHS

Solubilization of CNHs is a major challenge since it enhances compatibility of CNHs with other materials, allows easier manipulation, enables comprehensive characterization via traditional spectroscopic techniques, and contributes to the better study and understanding of their solution properties.

Considering chemical functionalization of CNHs at the sidewalls, the following methodologies have been developed: (1) 1,3-dipolar cycloaddition reaction of *in situ* generated azomethine ylides,[38,39] (2) aryl addition via *in situ* generated aryl diazonium salts,[40] (3) Bingel cyclopropanation reaction,[41] (4) anionic polymerization,[42] (5) bulk free radical polymerization,[43] and (6) amine addition via sodium amide ($NaNH_2$) reaction.[44] Contrary, the chemical modification of CNHs at the conical tips is achieved by oxidation of CNHs,[45] introducing carboxylic groups which are used as grafting points for further condensation reactions with amines and alcohols, forming CNH-based amides and esters, respectively. Additionally, the carboxylic moieties at the conical tips of CNHs have also been utilized for metal complexation, thus introducing coordination chemistry as an alternative modification means.[46]

6.2.1 *Covalent Functionalization*

6.2.1.1 1,3-dipolar cycloaddition of *in situ* generated azomethine ylides

A versatile approach for the covalent functionalization and solubilization of CNHs is based on the 1,3-dipolar cycloaddition of *in situ* generated azomethine ylides, upon thermal condensation of aldehydes and α-amino acids. In this fashion, fused pyrrolidine rings are cycloadded onto the skeleton of CNHs, as it is shown in

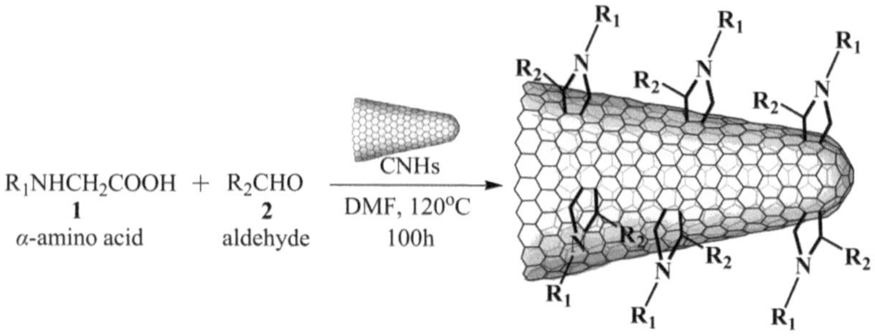

Figure 6.4. Functionalization of CNHs via 1,3-dipolar cycloaddition of azomethine ylides.

Fig. 6.4.[38,39] The novelty of the reaction is rationalized in the following two points: (i) modified aldehydes give rise to functionalized CNHs having substituted pyrrolidines on the α-carbon atom, while N-modified α-amino acids generate N-substituted pyrrolidines onto the skeleton of CNHs, as it is shown in Fig. 6.4, and (ii) plethora of commercially available aldehydes and α-amino acids, which may yield numerous and diversely modified CNHs. Therefore, in principle, any moiety can be successfully grafted to the graphitic network of CNHs, thus opening the way to the formation of diverse hybrid nanostructur

In a typical experimental procedure, an excess of modified glycines **1** and aldehydes **2** (Fig. 6.4) were added to a suspension of CNHs in N,N-dimethylformamide (DMF), and the mixture was heated at 120°C for 100 h. After centrifugation, the dense black supernatant DMF solution was passed through a PTFE filter and the functionalized CNHs were collected on top of the filter. As an immediate result of the functionalization reaction, the resulting modified CNHs **3** were rendered soluble in several organic solvents, depending on the functional group introduced. In this context, when the polar ethylene glycol chain was introduced on the α-amino acid part, the nanohorns produced were rendered soluble in polar solvents, such as dichloromethane, chloroform, and acetone. However, the presence of apolar or with medium polarity alkyl chains on the α-amino acid resulted in solubility only in toluene and DMF.

For the characterization of the modified CNHs, complementary spectroscopic and microscopy techniques were used. Microscopy analysis verified the presence of nanohorns on the functionalized material. Briefly, when analyzing a typical TEM image of pyrrolidine-modified CNHs, it is evident that both the unique structure and dahlia-like morphology of CNH aggregates are preserved. On the other hand, from the spectroscopic point of view, Raman spectroscopy of the pyrrolidines-modified CNHs shows a significant increase of the D band as compared with the intact CNHs material. This is attributed to the covalent chemical modification of CNHs and the introduction of sp^3 hybridized carbon atoms at positions where pyrrolidine rings are fused onto the graphitic skeleton of CNHs.

Although CNHs do not possess any well-resolved electronic absorption spectrum, ultraviolet-visible spectroscopy (UV-Vis) of modified CNHs with chromophore moieties allows the estimation of organic groups attached onto the skeleton of CNHs. For example, when pyrene aldehyde utilized as a reactant in a typical 1,3-dipolar azomethine ylides cycloaddition reaction with CNHs, the characteristic absorption profile of pyrene, which was incorporated as substituent of the α-carbon of the pyrrolidine rings on the modified CNHs, can be used to calculate the number of pyrrolidines in the hybrid material.

Having introduced the 1,3-dipolar cycloaddition methodology as a powerful functionalization means, the next step was the preparation of CNH-based donor–acceptor ensembles. In this direction, the covalent linkage of photo- or electro-active moieties, as substituents of pyrrolidines grafted onto the nanostructured network of CNHs, yielded some novel hybrid materials potential useful in energy conversion systems, such as photovoltaic and/or photoelectrochemical cells. In more detail, ferrocene-modified CNHs were synthesized (Fig. 6.5) utilizing ferrocene aldehyde in the 1,3-dipolar cycloaddition reaction. Alternatively, the same ferrocene-modified CNHs were also prepared by a typical condensation reaction between ferrocene acid and the free amino-functionalized CNHs, as derived from the corresponding N-Boc-protected material.[47] The plethora of ferrocene units all around the skeleton of CNH is expected to significantly contribute towards managing intramolecular charge-transfer reactions. In the same context, following the 1,3-dipolar

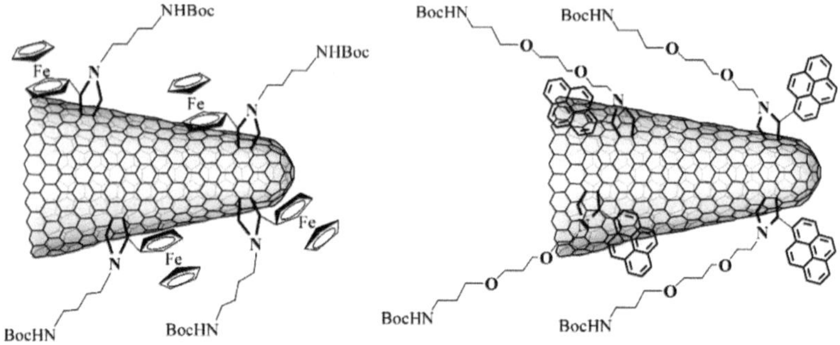

Figure 6.5. Representative CNH-based hybrid materials prepared via the 1,3-dipolar cycloaddition reaction of azomethine ylides.

cycloaddition reaction of azomethine ylides, pyrene groups were also covalently attached onto CNHs (Fig. 6.5), showing an efficiency on electron-transfer processes.[48]

Recently, the 1,3-dipolar cycloaddition reaction of azomethine ylides onto CNHs was performed with the aid of microwaves.[49] Under solvent-free conditions, the microwave-assisted introduction of pyrrolidine rings on the surface of CNHs was achieved very fast, thus significantly extending the strategies available for the covalent functionalization of CNHs.

Finally, theoretical calculations, based on the AM1, DFT, and ONIOM methods, on modified CNHs with the 1,3-dipolar cycloaddition reaction of azomethine ylides, indicated that greater binding energy and reactivity occurs at the conical tips of CNHs.[50] This was related with the higher strain of the conical ends of CNHs due to the presence of the five-membered rings, in sharp contrast for areas remotely located, where the presence of only six-membered rings with reduced reactivity exists.

6.2.1.2 Aryl addition via *in situ* generated aryl diazonium salts

Another efficient and simple strategy for the covalent sidewall functionalization of CNHs developed was based on their reaction with *in situ* generated aryl diazonium salts. The original methodology was successfully applied for the functionalization of carbon nanotubes.[51–56] Briefly, aryl diazonium salts were *in situ* generated

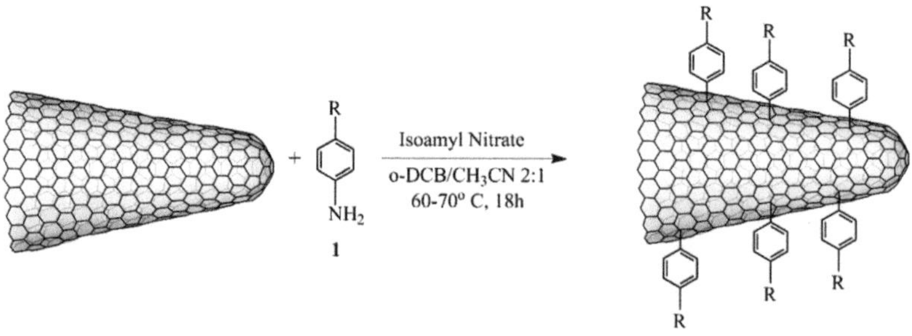

Figure 6.6. Schematic illustration of aryl-functionalized CNHs by *in situ* generated aryl diazonium salts.

by substituted anilines, and reacted with CNHs. The existence of a plethora of commercially available substituted anilines as well as the possibility of customized synthesis of more sophisticated aniline derivatives opened a new chapter in the chemical modification of CNHs (Fig. 6.6).[40] The direct result from the aryl functionalization of CNHs was the solubility achieved either in organic solvents or even in water.

A typical example was that of a Boc amino-protected aniline derivative, shown in Fig. 6.7a. Initially, functionalization of CNHs yielded the aryl-modified material, having a terminal Boc unit, while being well dispersed in common organic solvent. Then, deprotection under acidic conditions furnished the corresponding ammonium-modified hybrid material, shown in Fig. 6.7b, which in turn was soluble in aqueous media due to the presence of the cationic ammonium species. At this point it should be mentioned that further modification of the material can occur by exploiting the free amino groups, through coupling with suitable organic moieties, generating advanced CNH-based hybrid materials.

6.2.1.3 Bingel cyclopropanation reaction

The Bingel cyclopropanation reaction,[57] first employed in fullerene chemistry, with the bromo derivative of diethyl malonate in the presence of a base such as sodium hydride or DBU,[58–60] leading to methanofullerenes, allows the incorporation of a diverse selection of functional groups to the fullerene cage and has been employed

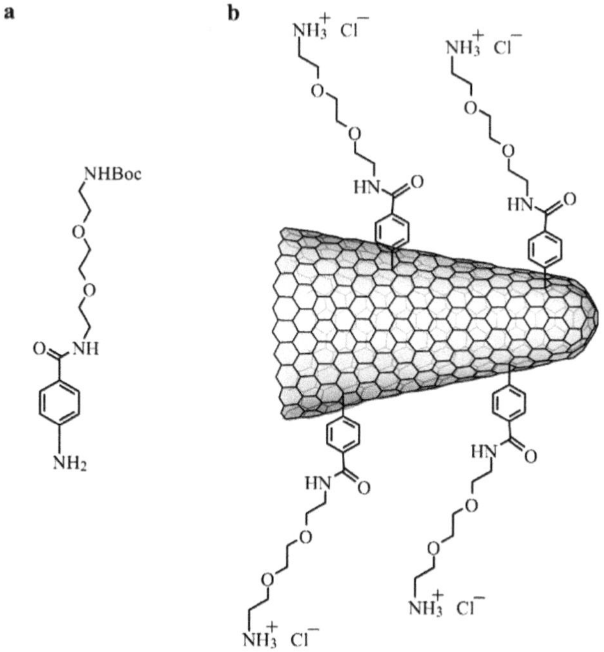

Figure 6.7. (a) Custom-synthesized aniline derivative and b) water-soluble CNHs-based hybrid material, prepared after aryl functionalization with the aniline derivative shown in (a) and followed deprotection under acidic conditions of the Boc-group.

successfully also in the functionalization of singlewall carbon nanotubes.[61] However, among the modification strategies for carbon nanotubes,[62] the Bingel reaction is the least applied. Moreover, when tested to CNHs, difficulties were encountered, thus not allowing the successful functionalization and solubilization of the material.[41] To overcome this obstacle, the Bingel cyclopropanation reaction was explored with the aid of microwaves.

Microwave-assisted chemistry is an extremely attractive synthetic route that allows the synthesis of the desired product in a fraction of time and in many cases without the use of organic solvents. Thus, following the Bingel reaction conditions, functionalized CNHs bearing malonate units along their skeleton were synthesized, with the aid of microwave irradiation and without the use of solvent,

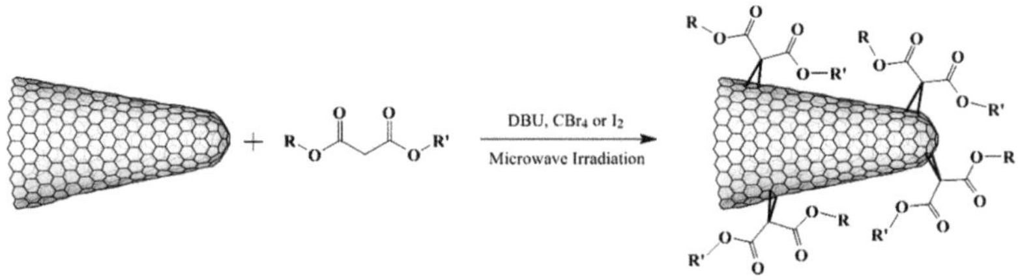

Figure 6.8. Microwave-assisted chemical functionalization of CNHs with malonate derivatives via Bingel reaction.

as shown in Fig. 6.8.[41] By comparison to conventional synthetic attempts, the microwave-assisted Bingelfunctionalized CNHs exhibit a high degree of functionalization, proving that this synthetic attempt is a viable alternative for the preparation of Bingelmodified CNHs. Importantly, the modified CNHs exhibited various degrees of functionalization, depending on the microwave irradiation duration as evidenced by Raman and TGA measurements. Furthermore, synthetic attempts to produce appropriate malonate derivatives bearing lightharvesting molecules were also successful and the resulting functionalized CNHs bearing pyrene and anthracene were further characterized by optical and electrochemical methods.

6.2.1.4 Anionic polymerization

Polymer functionalization of CNHs is also a promising approach toward homogeneous distribution of CNHs in polymer matrixes. Therefore, it is not surprising that polyisoprene as well as a diblock copolymer of polystyrene-b-polyisoprene were covalently grafted onto the sidewalls of CNHs through the grafting-to approach.[42] Briefly, the anion at the end of the polymer chains, synthesized by anionic polymerization high vacuum techniques, reacted with pristine CNHs. The immediate result of the reaction was the solubilization enhancement achieved. Similarly with the other covalently functionalized CNH materials, the characterization was confirmed by diverse spectroscopic techniques, as well as TGA, TEM, and dynamic light scattering.

6.2.1.5 Bulk free radical polymerization

In the grafting-to methodology previously described, polymer chains already prepared were attached to CNHs. However, the polymers used to decorate the surface of CNHs contribute only to the solubilization of the carbon nanostructure, since they do not contain functional sites for further exploitation. To overcome such deficiencies, a quick and facile protocol for the covalent functionalization of CNHs, using *in situ* bulk free radical polymerization of methacrylic acid, was followed. The formed polymer was a polyelectrolyte, offering a large number of ionic groups all around the skeleton of CNHs, thus facilitating water solubility. Moreover, these ionic sites were utilized to direct the synthesis of gold nanoparticles on the surface of the polymer decorated hybrid material.[43] In this context, gold nanoparticles were localized at the periphery of polymer decorated CNHs, as a result of complexation between negatively charged polymer chains and gold ions. Finally, the hybrid material was soluble in aqueous media, facilitating its processability, and was fully characterized by a wide gamut of complementary analytical techniques, microscopy, and thermal analysis.

6.2.1.6 $NaNH_2$ addition and amination reactions

$NaNH_2$ is a strong base and was found effective to introduce amine functions to CNHs. In this frame, when pristine CNHs treated in liquid ammonia with $NaNH_2$, a water-soluble material was obtained.[44] The amine-modified CNHs were satisfactorily characterized through a variety of analytical techniques as well as microscopy, while their aqueous solubility allowed to perform biological studies. Thus, fluoresceine moieties were conjugated with the aminemodified CNHs (Fig. 6.9) and the hybrid material was incubated with mammalian cells.[44] With the aid of confocal fluorescence microscope, it was proved that CNHs were inserted into mammalian cells, while at the same time, studies of cytotoxicity revealed that the material possess low values.[33] This low cytotoxicity of CNHs and modified CNHs was rationalized to the absence of transition metal particle impurities.

Moving a step forward, the amine-functionalized CNHs were further reacted with a biotinylated diamide material. In such a way,

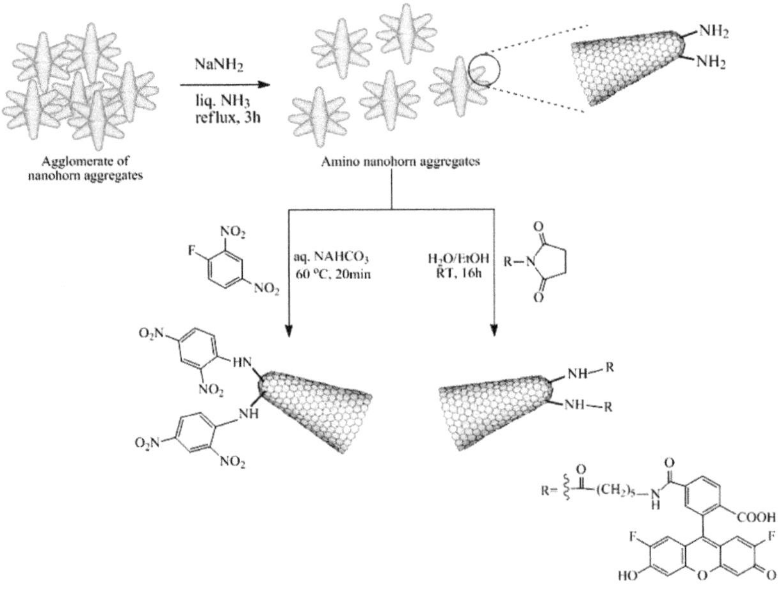

Figure 6.9. Sodium amide addition to CNHs, followed by derivatization with fluorescein.

the formation of CNHs-based conjugate was possible, in which a long biotinylated chain was grafted to the skeleton of CNHs through a stable and rigid amide bond.[63] Careful examination by HR-TEM allowed the identification of conformational changes observed, thus opening the way for possible future developments in the imaging of modified CNHs and other similar materials.

Using the chemical modification of sidewalls of CNHs with diamines and further chemical reaction of the free amino function with fluorescent molecule, new biocompatible CNHs hybrids were also prepared.[64] These derivatives were incubated with phagocytes (defensive cells of pathogenic viruses) and CNHs penetrated them without influencing the life of the cell. These results gave a new dimension in drug delivery systems, by introducing the use of modified CNHs as carriers of biological activated phagocytes, for strengthening the defensive system of organisms.

The amine-functionalized CNHs were also utilized to conjugate porphyrins with carboxylic acid moieties as light-harvesting

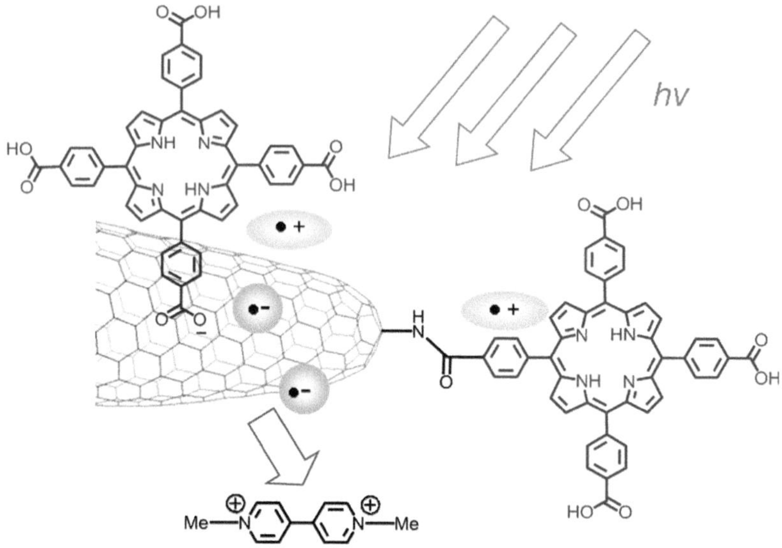

Figure 6.10. Representative illustration of amine-functionalized CNHs conjugated with porphyrins moieties, as light-harvesting antennae. See also Color Insert.

antennae. Recently, a CNHs–porphyrin hybrid material was synthesized (Fig. 6.10) which was characterized by spectroscopy and microscopy. Photoinduced electron-transfer processes of the nanohybrids of CNHs in aqueous environment were revealed with the aid of time-resolved absorption and fluorescence measurements. From the observed fluorescence quenching of free porphyrin acid moieties by the amine-modified CNHs material, chargeseparation via the excited singlet state of the porphyrin units, generating radical cations localized in the porphyrins and electrons trapped in CNHs, were suggested.[65]

6.2.1.7 Oxidation

Covalent functionalization of CNHs can also be performed at the conical end of the material. However, prior of this, oxidation of CNHs must be performed to introduce the appropriate carboxylic moieties which are utilized as starting points for the functionalization. Oxidation of CNHs was achieved either through (i) a mild but powerful oxidative treatment, during which shortening of nanohorns and

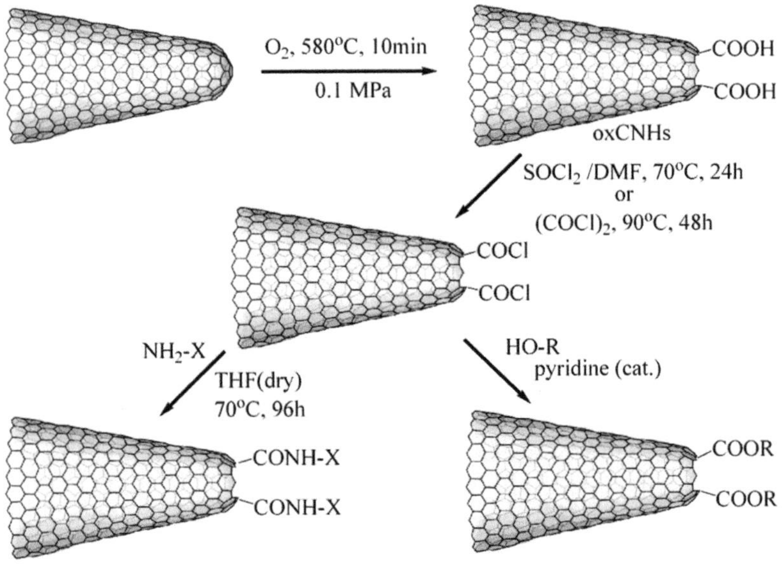

Figure 6.11. Oxidation and cone-end functionalization of CNHs.

formation of any type of impurities is excluded[45] or (ii) a light-assisted oxidation with hydrogen peroxide.[66]

As it is shown in Fig. 6.11, the as-generated carboxylic acid terminated nanohorns were converted to the corresponding acyl chlorides (CNH–COCl) upon treatment with either thionyl chloride, in the presence of a catalytic amount of DMF, or simply in refluxing oxalyl chloride. Treatment of CNH–COCl, in completely anaerobic and dry conditions, with a variety of amines and alcohols possessing either short or long hydrophobic alkyl chains, polar oligoethylenic units, aromatic chromophores such as pyrene or anthracene groups, or even masked active groups suitable for further organic exploitations, gave the corresponding CNH-based amides and esters, respectively.[45]

The first indication for the covalent conical-tip modification of CNHs was delivered by infrared (IR) spectroscopy, due to the presence of the characteristic carbonyl moiety. Additionally, electronic absorption spectroscopy as well as fluorescence emission were also important tools for the characterization of the functionalized

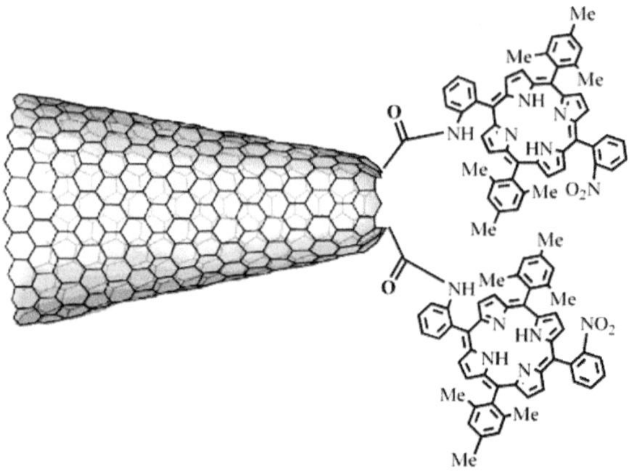

Figure 6.12. Representative example of CNHs–H_2P hybrid material synthesized via cone-end functionalization.

CNHs when a chromophore was introduced. Similarly like the case when functionalization occurs at the sidewalls of CNHs, morphological characterization of the material was delivered by electron microscopy. Importantly, during the oxidation process and the introduction of carboxylic units, the conical tips of CNHs were broken and holes were pierced on their skeleton.

A very well-established example of such carboxylated-modified CNHs concerned their condensation with an amino-modified porphyrin material. Thus, activation of the carboxylic acids with oxalyl chloride, followed by a typical coupling reaction with the aminoporphyrin, resulted in the formation of a novel hybrid material in which the porphyrin unit was connected to the CNHs tips through a robust amide bond (Fig. 6.12).[67] Spectroscopic and photophysical studies revealed that CNHs served as electron acceptors while the photoexcited porhyrine moieties were the electron donor. The formation of a charge-separated state $CNH^{\bullet -}$–$H_2P^{\bullet +}$ in polar solvents was also identified and the dynamics of the system were evaluated with the aid of time-resolved fluorescence studies as well as transient absorption spectroscopy. Thus, in non-polar solvents, intramolecular energy-transfer quenching of the photoexcited H_2P singlet excited state by CNHs was shown to occur on a pico-second time scale.

Moving a step forward, photoelectrochemical electrodes with the CNHs–H_2P hybrid material were constructed.[68] The film of CNHs–H_2P onto the nanostructured SnO_2 electrode exhibited an incident photon-to-photocurrent efficiency of 5.8% in a three-compartment electrochemical cell. Fluorescence lifetime measurements revealed that electron transfer from the singlet excited state of porphyrin to CNHs takes place. In addition, direct electron injection from reduced nanohorns to the conduction band of SnO_2 electrode occurs. Overall, these results demonstrated the potentiality and applied utility of CNHs in directing efficient charge transport in photoelectrochemical devices, such as solar cells.

In another report, the covalent fixation of a polyethylene oxide (PEO) through a stable ester bond formation to oxidized CNHs was also performed.[69] The synthesis of CNHs–PEO material initially involved the oxidation of pristine CNHs, the activation of the introduced carboxylic groups and their esterification with the hydroxyl group of the polymer chains. The CNHs–PEO material was soluble in a variety of solvents, which are thermodynamically compatible for PEO, like water, tetrahydrofuran, $CHCl_3$ and DMF (*ca.* 0.5–0.7 mg/mL). The grafting of the macromolecules on the surface of CNHs was identified by UV-Vis and attenuated total reflection IR spectroscopy, as well as by TGA. Moreover, the size of the functionalized nanostructure in water was determined by dynamic light scattering. Finally, the incorporation of CNHs–PEO in poly(hydroxyl styrene) was studied by means of optical microscopy, indicating the miscibility of the components at certain compositions.

Additional studies with oxidized CNHs were carried out, where peptides and proteins were attached covalently via the carboxylic function of oxidized CNHs, thus giving rise to some interesting and novel hybrid materials suitable for biotechnological applications.[27,28] For example, the Alexa Fluoro 488-labeled bovine serum albumin (BSA) protein was coupled to carboxylic units of oxCNHs, thus obtaining a new watersoluble CNHsbased material, as it is shown in Fig. 6.13. Upon incubation of the hybrid material with cells, uptake of CHNs through an endocytosis pathway was observed.[34]

In another important study, functionalized CNHs were employed as components for photodynamic therapy (PDT) of cancer. Moreover,

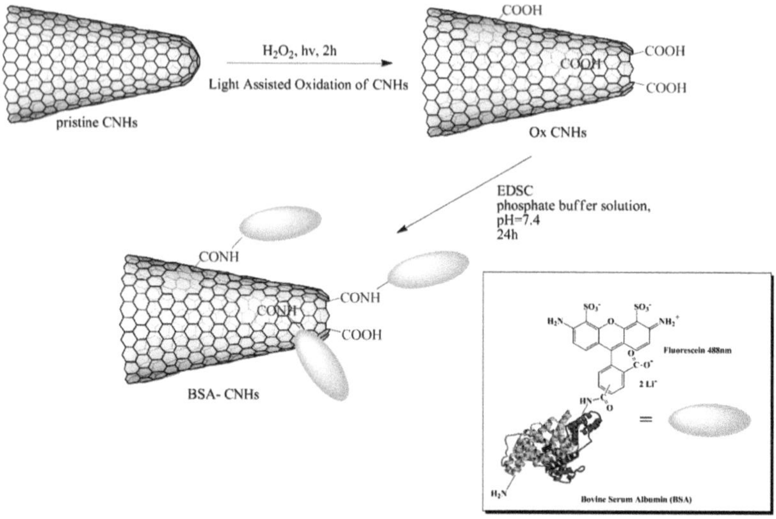

Figure 6.13. Light-assisted oxidation of CNHs followed by BSA coupling to the so-formed carboxylic groups.

as PDT is a noninvasive phototherapy, it can be combined with hypothermia to induce tumor cells death. Thus, oxidized CNHs were condensed with BSA for biocompatibility, while from the holes pierced on the skeleton of CNHs during oxidation, zinc phthalocyanine (ZnPc) was loaded (Fig. 6.14).[70] In that multifuctional CNH-based material, the ZnPc was acting as PDT agent, while the oxidized CNHs, due to their ability to absorb light in the near-IR region, can cause cell death by localized photothermal or photohypothermia effect. Additionally, the photophysical properties of the ZnPc/CNHs–BSA hybrid material were also examined.[71] Thus, conditions for electron- and/or energy-transfer mechanisms, useful not only for the PDT application but also for the photosynthetic model and photovoltaics, were revealed.

Finally, a sandwich-type hybrid of oxidized CNHs with TiO_2 and porphyrin acid was prepared via the dentate binding of TiO_2 nanoparticles to the carboxylates.[72] The resulting nanohybrid showed excellent electrocatalysis toward reduction of chloramphenicol (CAP), leading to a sensitive amperometric biosensor for CAP, which can be further extended for applications in photovoltaics and photocatalysis.

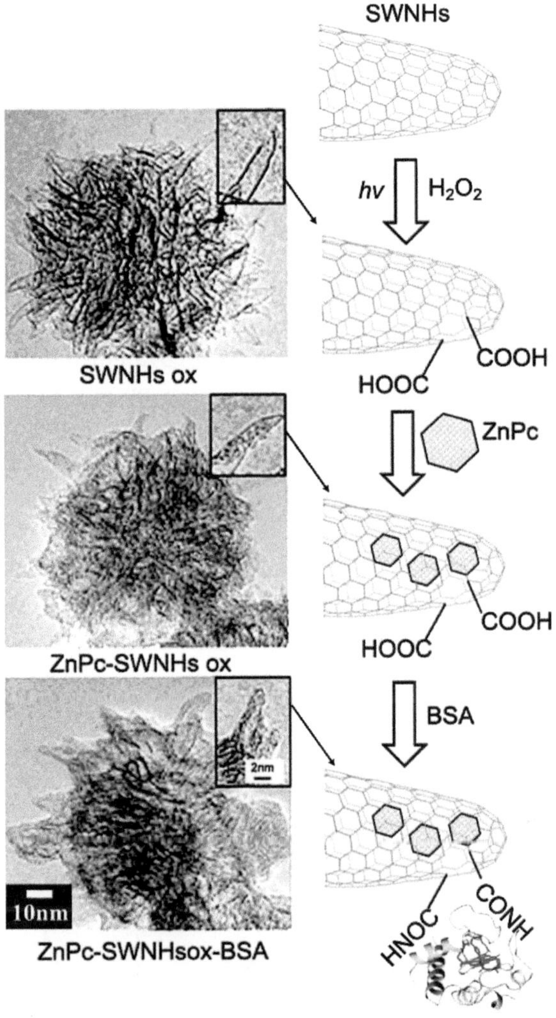

Figure 6.14. Preparation of ZnPc–oxCNHs–BSA. Right panel: Diagram showing synthetic steps of ZnPc–CNHs–BSA. Left panel: TEM visualization of ZnPc–CNHs–BSA at each stage of synthesis. Insets: Magnified images. See also Color Insert.

6.2.2 *Non-Covalent Functionalization*

Although covalent attachment of various addends either onto the graphite-like sidewalls of CNHs (e.g., via pyrrolidine moieties) or at the conical-shaped tip (e.g., via formation of amides, esters) leads to significant solubilization and dispersion of the functionalized material, the resulting perturbation of the continuous π-electronic network of CNHs is a significant implication, especially when applications based on nanoelectronics are considered. Therefore, to overcome drawbacks arising from such issues, supramolecular approaches utilizing either non-covalent π–π stacking interactions between the skeleton of CNH with aromatic organic materials and synergistic electrostatic interactions or polymer wrapping were developed.

The very first report on the non-covalent functionalization of CNHs deals with the interaction of a pyrene derivative with the surface of CNHs.[25] More specifically, 1-pyrenebutanoic acid succinimidyl ester was used for the solubilization of CNHs, with the aid of π–π interactions between the pyrene unit and the sidewalls of CNHs, while in the following step the free group of succinimidyl ester reacted with amino-modified surfaces (Fig. 6.15). Thus, peptides

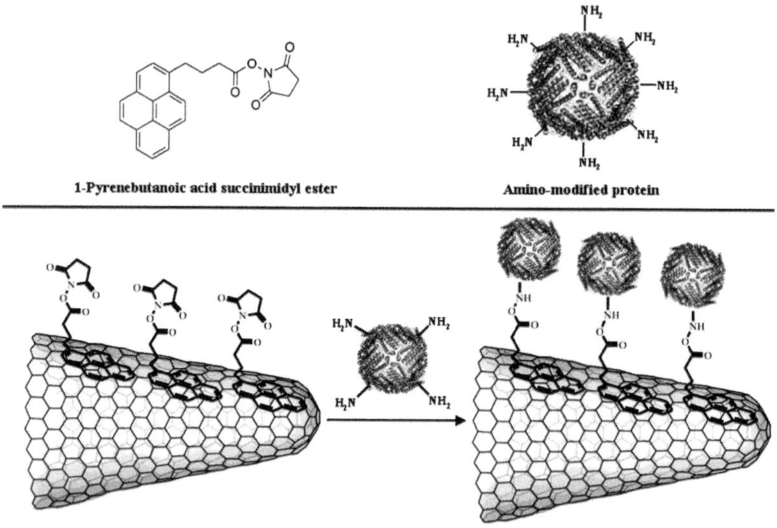

Figure 6.15. 1-pyrenebutanoic acid succinimidyl ester adsorbed onto CNHs via π–π stacking interactions. Protein is immobilized through formation of amide bond between free amine groups on the protein and the succinimidyl ester. See also Color Insert.

found immobilized onto the sidewalls of CNHs, through conjugation with the pyrene derivative, and opened up new fields such as bioassembly and biosensors with CNHs-based materials.[26]

In another typical example, a tetracationic water-soluble porphyrin (H_2P^{4+}) was immobilized by π–π stacking interactions onto the skeleton of CNHs, without disrupting the continuous π-electronic network of the nanomaterial (Fig. 6.16).[73] The stable aqueous solution of the CNHs–H_2P^{4+} nanoensemble was examined both by electron microscopy and spectroscopic techniques. The

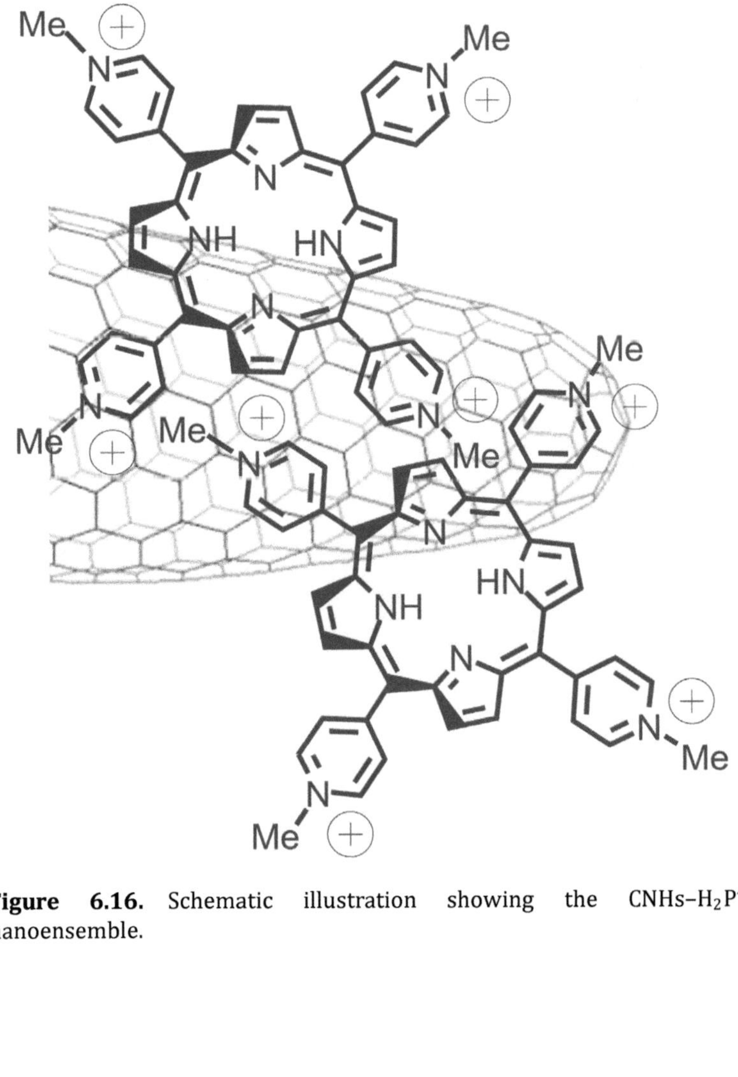

Figure 6.16. Schematic illustration showing the CNHs–H_2P^{4+} nanoensemble.

efficient emission quenching of the H_2P^{4+} moiety in the CNH–H_2P^{4+} nanoensemble was probed by steady-state as well as time-resolved photoluminescence, suggesting charge-separation from the photoexcited H_2P^{4+} to CNHs. Additionally, transient absorption spectroscopy, with the aid of methyl viologen dication (MV^{2+}) and a hole trap, verified the presence of charge-separated state of $(CNHs)^{\bullet-}-(H_2P^{4+})^{\bullet+}$.

Moreover, such nanohybrids possessing cationic charges were utilized for the electrostatic association of negatively charged molecules, leading to more complex and advanced materials. In this frame, the coulombic association of the negatively charged tetrathiafulvalene carboxylate (TTF^-) units with the positively charged pyrene (pyr^+) noncovalently immobilized on the surface of CNHs gave rise to the watersoluble CNH–pyr^+–TTF^- nanosized architecture (Fig. 6.17).[74] The three-component nanoensemble was

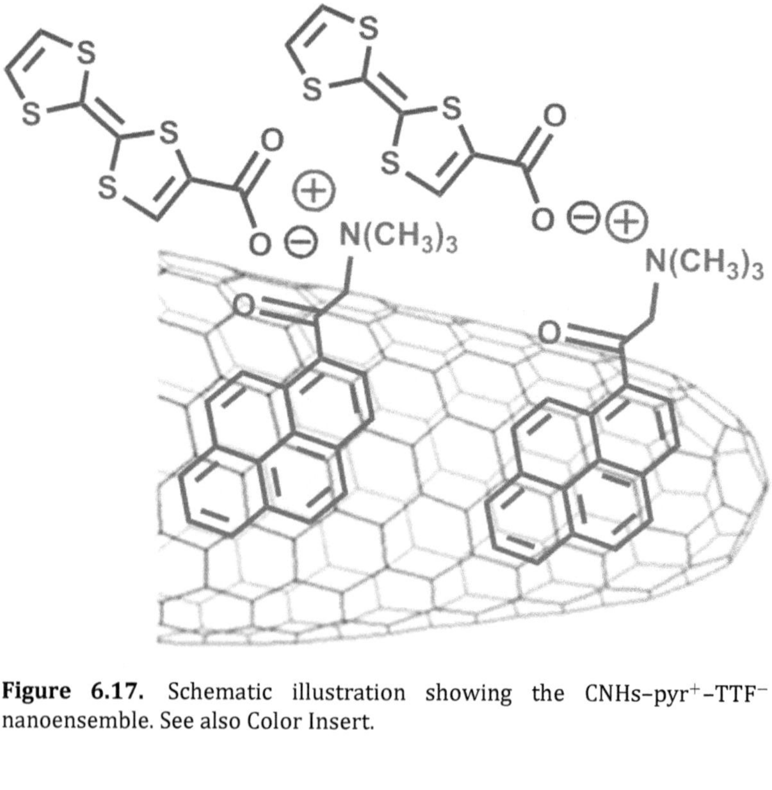

Figure 6.17. Schematic illustration showing the CNHs–pyr^+–TTF^- nanoensemble. See also Color Insert.

structurally and morphologically characterized. The one-electron reduced and oxidized species such as $(CNHs)^{\bullet -}-pyr^{+}-(TTF^{-})^{\bullet +}$ and $(CNHs)^{\bullet -}-(pyr^{+})^{\bullet +}-(TTF^{-})$ were identified directly by the transient spectral measurements and indirectly by the accumulation of electron on methyl viologen dication (MV^{2+}). Kinetic analyses of the time profiles of the fluorescence and transient absorptions gave information regarding charge-separation rate and quantum yields through the excited singlet sate of pyr^{+} and lifetimes for the charge-separated state, respectively. In addition, the photoexcitation of CNHs also afforded the accumulation of MV^{+}, suggesting the photoinduced charge-separation through the CNHs.

The results from the non-covalent functionalization of CNHs for applications in the fields of biotechnology and medicine are also very important. Drastic substances such as drugs, enzymes, and proteins were adsorbed onto the surface of CNHs, or even encapsulated inside the empty space of CNHs, thus creating some novel hybrid materials. For example, the anti-inflammatory glucocorticoid dexamethasone was adsorbed on CNHs and the drug's release rate was studied both in neutral solutions and in solutions of growth of cells.[75]

Moreover, the well-known anticancer drug cisplatin (CDDP) was encapsulated in oxidized CNHs possessing nanosized holes. During that study, it was observed that the rate of disengagement in neutral solution was smaller, concerning the solubilization of free medicine in the solution, while experiments in cancer cells showed anticancer activity.[29–31] Changing the solvent from DMF to water, better adsorption of CDDP on modified CNHs was achieved, while the rate of disengagement remained the same. The CDDP@CNHs hybrid showed high anticancer activity, both *in vitro* and *in vivo* experiments. This hybrid material actually increased the concentration of CDDP which is released in the cells, leading to the death of cancer cells.[32] The slow rate of release substantially maintains the concentration of drug constant and acts drastically for more time at the cancer cell. In the same concept, CDDP was adsorbed in oxidized CNHs which followed by non-covalent modification with polyethylene glycol (PEG)-modified peptides to enhance water solubility. The new CNHs-based hybrid material was indeed very well soluble in water. Experiments with cancer cells showed significant anticancer activity, proving that the interaction with peptide did not influence the ability of the adsorbed CDDP to be released. Therefore, it was concluded that such material is a suitable candidate for clinical experimental studies and applications of chemotherapy.[76]

Oxidized CNHs were used for the adsorption of drugs such as the antibiotic vancomycin hydrochloride and the study of drug release from CNHs. The prerequisite of water solubilization was achieved by introducing PEG polymeric chains in the nanostructured material. The release rate of the drug in the CNHs-based hybrid material was slow and constant, confirming that CNHs can be used as drug delivery systems.[77] In the same frame, the non-covalent modification of oxidized CNHs with the anticancer drug doxoroubisin (DXR), which had a PEG-modified amino end, led to the formation of water-soluble material, namely PEG–DXR–CNHs (Fig. 6.18). Dynamic light scattering measurements used to calculate the average size of the nanohybrid material, which was found to be 160 nm. *In vitro* experiments showed that in such range the PEG–DXR–CNHs cannot be removed from blood through liver or spleen, so it is expected to act against cancer tumors. Moreover at the same time,[36,78] *in vivo* experiments, with injection of PEG–DXR–CNHs in cancer tumors, showed effective tumor reduction, indicating that such water-soluble hybrid systems can be used in chemotherapy.

Finally, apart from drugs, non-covalent attachment of anti-viruses in oxidized CNHs was recently reported. The water solubility of these hybrid CNHs-based material was enhanced via the introduction of a PEG chain. At the one end of the PEG polymeric chain, the anti-virus was attached, while at the other end a phospholipid was incorporated contributing to solubility enhancement. The release of the anti-virus from CNHs hybrid material was performed by NIR laser excitation (1064 nm), opening new avenues for the confrontation of harmful viruses.[79]

6.3 Conclusions and Outlook

Based on the existing methodologies for the functionalization of CNHs, a series of novel hybrid materials can be obtained. One can choose from the introduction of conventional chemical bond

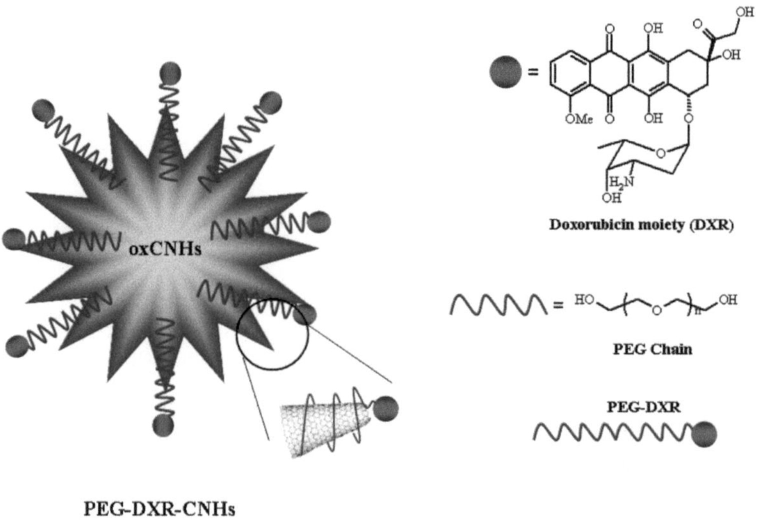

Figure 6.18. Schematic illustration of the PEG–DXR–CNHs hybrid material.

formation to supramolecular interactions to solubilize the otherwise insoluble CNHs. Moreover, based on the 1,3-dipolar cycloaddition of azomethine ylides, the *in situ* generated aryl diazonium salt functionalization, the Bingel cyclopropanation, polymer functionalization, amine addition, oxidation of CNHs, along with the π–π stacking interactions, as well as the polymer wrapping, CNHs-based hybrid materials potentially suitable for applications in solar cells and drug delivery have been synthesized.

Keeping in mind that only recently CNHs have been started to become available in bulk quantities, an even higher amount of research dedicated to the functionalization of CNHs and a plethora of hybrid materials are envisioned. Therefore, it is expected that in the near future, modified CNHs will play a major role in diverse technological fields.

Acknowledgments

Partial financial support from the EU FP7, Capacities Program, NANOHOST project (GA 201729) is acknowledged. We are also indebted to our collaborators, whose names appear in the reference section, for the fruitful cooperation on the chemical functionalization and properties evaluation of some CNHs-based hybrid materials.

References

1. P. J. F. Harris, S. C. Tsang, J. B. Claridge, and M. L. H. Green, *J. Chem. Soc. Faraday Trans.* **90**, 2799 (1994).
2. S. Iijima, T. Wakabayashi, and Y. Achiba, *J. Phys. Chem.* **100**, 5839 (1996).
3. S. Iijima, M. Yudasaka, R. Yamada, S. Bandow, K. Suenaga, F. Kokai, and K. Takahashi, *Chem. Phys. Lett.* **309**, 165 (1999).
4. T. Azami, D. Kasuya, R. Yuge, M. Yudasaka, S. Iijima, T. Yoshitake, and Y. Yoshimi Kubo, *J. Phys. Chem. C* **112**, 1330 (2008).
5. M. Zhang, T. Yamaguchi, S. Iijima, and M. Yudasaka, *J. Phys. Chem. C* **113**, 11184 (2009).
6. F. Kokai, K. Takahashi, D. Kasuya, M. Yudasaka, and S. Iijima, *Appl. Surf. Sci.* **197–198**, 650 (2002).
7. D. Kasuya, M. Yudasaka, K. Takahashi, F. Kokai, and S. Iijima, *J. Phys. Chem. B* **106**, 4947 (2002).
8. S. Iijima, *Physica* **323**, 1 (2002).
9. K. Murata, K. Kaneko, W. A. Steele, F. Kokai, K. Takahashi, D. Kasuya, K. Hirahara, M. Yudasaka, and S. Iijima, *J. Phys. Chem. B* **105**, 10210 (2001).
10. S. Bandow, A. M. Rao, G. U. Sumanasekera, P. C. Eklund, F. Kokai, K. Takahashi, M. Yudasaka, and S. Iijima, *Appl. Phys. A* **71**, 561 (2000).
11. T. Fujimori, K. Urita, Y. Aoki, H. Kanoh, T. Ohba, M. Yudasaka, S. Iijima, and K. Kaneko, *J. Phys. Chem. C* **112**, 7552 (2008).
12. K. Murata, K. Kaneko, F. Kokai, K. Takahashi, M. Yudasaka, and S. Iijima, *Chem. Phys. Lett.* **331**, 14 (2002).
13. T. Ohba, T. Omori, H. Kanoh, M. Yudasaka, S. Iijima, and K. Kaneko, *Chem. Phys. Lett.* **389**, 332 (2004).
14. K. Murata, K. Kaneko, H. Kanoh, D. Kasuya, K. Takahashi, K. Kokai, M. Yudasaka, and S. Iijima, *J. Phys. Chem. B* **106**, 11132 (2002).
15. E. Bekyarova, K. Murata, M. Yudasaka, D. Kasuya, S. Iijima, H. Tanaka, H. Kahoh, and K. Kaneko, *J. Phys. Chem. B* **107**, 4681 (2003).

16. N. Sano and S. Ukita, *Mater. Chem. Phys.* **99**, 447 (2006).
17. E. Bekyarova, A. Hashimoto, M. Yudasaka, Y. Hattori, K. Murata, H. Kanoh, D. Kasuya, S. Iijima, and K. Kaneko, *J. Phys. Chem. B* **109**, 3711 (2005).
18. T. Yoshitake, Y. Shimakawaa, S. Kuroshima, H. Kimura, T. Ichihashi, Y. Kubo, D. Kasuya, K. Takahashi, F. Kokai, M. Yudasaka, and S. Iijima, *Physica B* **323**, 124 (2002).
19. T. Itoh, H. Danjo, W. Sasaki, K. Urita, E. Bekyarova, M. Arai, T. Imamoto, M. Yudasaka, S. Iijima, H. Kanoh, and K. Kaneko, *Carbon* **46**, 172 (2008).
20. K. Murata, K. Hirahara, M. Yudasaka, S. Iijima, D. Kasuya, and K. Kaneko, *J. Phys. Chem. B* **106**, 12668 (2002).
21. K. Ajima, M. Yudasaka, K. Suenaga, D. Kasuya, T. Azami, and S. Iijima, *Adv. Mater.* **16**, 397 (2004).
22. R. Yuge, M. Yudasaka, J. Miyawaki, Y. Kubo, T. Ichihashi, H. Imai, E. Nakamura, H. Isobe, H. Yorimitsu, and S. Iijima, *J. Phys. Chem. B* **109**, 17861 (2005).
23. A. Hashimoto, H. Yorimitsu, K. Ajima, K. Suenaga, H. Isobe, J. Miyawaki, M. Yudasaka, S. Iijima, and E. Nakamura, *Proc. Natl. Acad. Sci. U.S.A.* **101**, 8527 (2004).
24. J. Miyawaki, M. Yudasaka, H. Imai, H. Yorimitsu, H. Isobe, E. Nakamura, and S. Iijima, *Adv. Mater.* **18**, 1010 (2006).
25. E. Bekyarova, A. Hashimoto, M. Yudasaka, Y. Hattori, K. Murata, H. Kanoh, D. Kasuya, S. Iijima, and K. Kaneko, *J. Phys. Chem. B* **109**, 3711 (2005).
26. J. Zhu, D. Kase, K. Shiba, D. Kasuya, M. Yudasaka, and S. Iijima, *NanoLett.* **3**, 1033 (2003).
27. K. Shiba, *J. Drug Target.* **14**, 512 (2006).
28. D. Kase, J. L. Kulp, M. Yudasaka, J. S. Evans, S. Iijima, and K. Shiba, *Langmuir* **20**, 8939 (2004).
29. T. Matsui, N. Matsukawa, K. Iwahori, K. I. Sano, K. Shiba, and I. Yamashita, *Langmuir* **23**, 1615 (2007).
30. K. Ajima, M. Yudasaka, T. Murakami, A. Maigne, K. Shiba, and S. Iijima, *Mol. Pharm.* **2**, 475 (2005).
31. K. Ajima, M. Yudasaka, A. Maigne, J. Miyawaki, and S. Iijima, *J. Phys. Chem. B* **110**, 5773 (2006).
32. K. Ajima, A. Maigne, M. Yudasaka, and S. Iijima, *J. Phys. Chem. B* **110**, 19097 (2006).

33. K. Ajima, T. Murakami, Y. Mizoguchi, K. Tsuchida, T. Ichihashi, S. Iijima, and M. Yudasaka, *ACS Nano* **2**, 2057 (2008).
34. J. Miyawaki, M. Yudasaka, T. Azami, Y. Kubo, and S. Iijima, *ACS Nano* **2**, 213 (2008).
35. M. Zhang, M. Yudasaka, K. Ajima, J. Miyawaki, S., and Iijima, *ACS Nano* **1**, 265 (2007).
36. S. Lacotte, A. Garcia, M. Decossas, W. T. Al-Jamal, S. Li, K. Kostarelos, S. Muller, M. Prato, H. Dumortier, and A. Bianco, *Adv. Mater.* **20**, 2421 (2008).
37. T. Murakami, J. Fan, M. Yudasaka, S. Iijima, and K. Shiba, *Mol. Pharm.* **3**, 407 (2006).
38. N. Tagmatarchis, A. Maigne, M. Yudasaka, and S. Iijima, *Small* **2**, 490 (2006).
39. C. Cioffi, S. Campidelli, F. G. Brunetti, M. Meneghetti, and M. Prato, *Chem. Commun.* 2129 (2006).
40. G. Pagona, N. Karousis, and N. Tagmatarchis, *Carbon* **46**, 604 (2008).
41. S. P. Economopoulos, G. Pagona, M. Yudasaka, S. Iijima, and N. Tagmatarchis, *J. Mater. Chem.* **19**, 7326 (2009).
42. G. Mountrichas, S. Pispas, and N. Tagmatarchis, *Chem. Eur. J.* **13**, 7595 (2007).
43. G. Mountrichas, S. Pispas, and N. Tagmatarchis, *Chem. Eur. J.* **16**, 5927 (2010).
44. H. Isobe, T. Tanaka, R. Maeda, E. Noiri, N. Solin, M. Yudasaka, S. Iijima, and E. Nakamura, *Angew. Chem. Int. Ed.* **45**, 6676 (2006).
45. G. Pagona, N. Tagmatarchis, J. Fan, M. Yudasaka, and S. Iijima, *Chem. Mater.* **18**, 3918 (2006).
46. G. Rotas, A. S. D. Sandanayaka, N. Tagmatarchis, T. Ichihashi, M. Yudasaka, S. Iijima, and O. Ito, *J. Am. Chem. Soc.* **130**, 4725 (2008).
47. G. Pagona, G. Rotas, I. D. Petsalakis, G. Theodorakopoulos, J. Fan, A. Maigné, M. Yudasaka, S. Iijima, and N. Tagmatarchis, *J. Nanosci. Nanotechnol.* **7**, 3468 (2007).
48. A. S. D. Sandanayaka, G. Pagona, J. Fan, N. Tagmatarchis, M. Yudasaka, S. Iijima, Y. Araki, and O. Ito, *J. Mater. Chem.* **17**, 2540 (2007).
49. N. Rubio, M. A. Herrero, M. Meneghetti, A. Diaz-Ortiz, M. Schiavon, M. Prato, and E. Vazquez, *J. Mater. Chem.* **19**, 4407 (2009).
50. I. D. Petsalakis, G. Pagona, N. Tagmatarchis, and G. Theodorakopoulos, *Chem. Phys. Lett.* **448**, 115 (2007).
51. J. L. Bahr and J. M. Tour, *Chem. Mater.* **13**, 3823 (2001).

52. J. L. Bahr, J. Yang, D. V. Kosynkin, M. J. Bronikowski, R. E. Smalley, and J. M. Tour, *J. Am. Chem. Soc.* **123**, 6536 (2001).
53. C. A. Dyke and J. M. Tour, *J. Am. Chem. Soc.* **125**, 1156 (2003).
54. C. A. Dyke and J. M. Tour, *J. Phys. Chem. A* **108**, 11151 (2004).
55. Z. Chen, K. Kobashi, U. Rauwald, R. Booker, H. Fan, and W.-F. Hwang, *J. Am. Chem. Soc.* **128**, 10568 (2006).
56. Z. Guo, F. Du, D. Ren, Y. Chen, J. Zheng, and Z. Liu, *J. Mater. Chem.* **16**, 3021 (2006).
57. C. Bingel, *Chem. Ber.* **126**, 1957 (1993).
58. J.-F. Nierengarten, V. Gramlich, F. Cardullo, and F. Diederich, *Angew. Chem. Int. Ed.* **35**, 2101 (1996).
59. Y. Nakamura, S. Minami, K. Iizuka, and J. Nishimura, *Angew. Chem. Int. Ed.* **42**, 3158 (2003).
60. S. Gonzalez, N. Martin, and D. M. Guldi, *J. Org. Chem.* **68**, 779 (2003).
61. K. Coleman, S. Bailey, S. Fodgen, and M. Green, *J. Am. Chem. Soc.* **125**, 8722 (2003).
62. D. Tasis, N. Tagmatarchis, A. Bianco, and M. Prato, *Chem. Rev.* **106**, 1105 (2006).
63. E. Nakamura, M. Koshino, T. Tanaka, Y. Niimi, K. Harano, Y. Nakamura, and H. Isobe, *J. Am. Chem. Soc.* **130**, 7808 (2008).
64. S. Lacotte, A. Garcia, M. Decossas, W. T. Al-Jamal, S. Li, K. Kostarelos, S. Muller, M. Prato, H. Dumortier, and A. Bianco, *Adv. Mater.* **20**, 2421 (2008).
65. A. S. D. Sandanayaka, O. Ito, T. Tanaka, H. Isobe, E. Nakamura, M. Yudasaka, and S. Iijima, *New J. Chem.* **33**, 2261 (2009).
66. M. Zhang, M. Yudasaka, K. Ajima, J. Miyawaki, and S. Iijima, *ACS Nano* **1**, 265 (2007).
67. G. Pagona, A. S. D. Sandanayaka, Y. Araki, J. Fan, N. Tagmatarchis, G. Charalambidis, A. G. Coutsolelos, B. Boitrel, M. Yudasaka, S. Iijima, and O. Ito, *Adv. Funct. Mater.* **17**, 1705 (2007).
68. G. Pagona, A. S. D. Sandanayaka, T. Hasobe, G. Charalambidis, A. G. Coutsolelos, M. Yudasaka, S. Iijima, and N. Tagmatarchis, *J. Phys. Chem. C* **112**, 15735 (2008).
69. G. Mountrichas, N. Tagmatarchis, and S. Pispas, *J. Nanosci. Nanotechnol.* **9**, 3775 (2009).
70. M. Zhang, T. Murakami, K. Ajima, K. Tsuchida, A. S. D. Sandanayaka, O. Ito, S. Iijima, and M. Yudasaka, *Proc. Natl. Acad. Sci. U.S.A.* **105**, 14773 (2008).

71. A. S. D. Sandanayaka, O. Ito, M. Zhang, K. Ajima, S. Iijima, M. Yudasaka, T. Murakami, and K. Tsuchida, *Adv. Mater.* **21**, 4366 (2009).
72. W. Tu, J. Lei, L. Ding, and H. Ju, *Chem. Commun.* 4227 (2009).
73. G. Pagona, A. S. D. Sandanayaka, Y. Araki, J. Fan, N. Tagmatarchis, M. Yudasaka, S. Iijima, and O. Ito, *J. Phys. Chem. B.* **110**, 20729 (2006).
74. G. Pagona, A. S. D. Sandanayaka, A. Maigne, J. Fan, G. C. Papavassiliou, I. D. Petsalakis, B. R. Steele, M. Yudasaka, S. Iijima, N. Tagmatarchis, and O. Ito, *Chem. Eur. J.* **13**, 7600 (2007).
75. T. Murakami, K. Ajima, J. Miyawaki, M. Yudasaka, S. Iijima, and K. Shiba, *Mol. Pharm.* **1**, 399 (2004).
76. S. Matsumura, K. Ajima, M. Yudasaka, S. Iijima, and K. Shiba, *Mol. Pharm.* **4**, 723 (2007).
77. J. Xu, M. Yudasaka, S. Kouraba, M. Sekido, Y. Yamamoto, and S. Iijima, *Chem. Phys. Lett.* **461**, 189 (2008).
78. T. Murakami, H. Sawada, G. Tamura, M. Yudasaka, S. Iijima, and K. Tuchida, *Nanomedicine* **3**, 453 (2008).
79. M. Miyako, H. Nagata, K. Hirano, K. Sakamoto, Y. Makita, K. Nakayama, and T. Hirotsu, *Nanotechnology* **19**, 075106 (2008).

Chapter 7

Endohedral Metallofullerene Functionalization

Yutaka Maeda[1], Takeshi Akasaka[2], and Shigeru Nagase[3]

[1]*Department of Chemistry, Tokyo Gakugei University*
Koganei, Tokyo 184-8501, Japan
PRESTO, Japan Science and Technology Agency
Chiyoda, Tokyo 102-0075, Japan
[2]*Life Science Center for Tsukuba Advanced Research Alliance, University of Tsukuba*
Tsukuba, Ibaraki 305-8577, Japan
[3]*Department of Theoretical and Computational Molecular Science,*
Institute for Molecular Science Myodaiji,
Okazaki 444-8585, Japan
akasaka@tara.tsukuba.ac.jp; nagase@ims.ac.jp

Endohedral fullerenes have attracted special interest since the first proposal of their existence in 1985. They are a new type of carbon cluster containing one or more atoms inside the hollow fullerene cage. Particularly, endohedral metallofullerenes have attracted broad attention because of their properties resulting from an intramolecular metal–fullerene cage interaction. Their recent production and isolation has enabled detailed characterizations of metallofullerenes' chemical reactions. Endohedral metallofullerenes' unique chemical properties and structures have been revealed through studies of functionalization.

Advances in Carbon Nanomaterials: Science and Applications
Edited by Nikos Tagmatarchis

ISBN 978-981-426-78-78 (Hardcover), 978-981-426-78-85 (eBook)
www.panstanford.com

7.1 Introduction

Endohedral metallofullerenes are created by trapping metal atoms or metal clusters into a fullerene cage, which naturally combines the properties of fullerenes and metals. This novel hybrid molecule was indicated by mass spectrometry as early as 1985.[1] Six years later, successful synthesis and isolation of La@C_{82} were reported by Smalley and coworkers.[2] In subsequent years, great efforts have been undertaken for the synthesis of various endohedral metallofullerenes. Recent production and isolation of endohedral metallofullerenes have made it possible to investigate their chemical properties.[3,4]

Therefore, this chapter presents a description of recent progress made in the field of endohedral metallofullerenes, which involves their structural characterization and chemical functionalization. Importantly, we attempt to understand their structural and chemical features imparted on them by the encapsulated metallic species.

7.2 Reduction and Oxidation

Because of the odd-numbered electron transfer from encapsulated metal to the fullerene cage, trivalent mono-metallofullerenes (M^{3+}@C_{2n}^{3-}) have an unpaired electron on the fullerene cage.[4–8] Their paramagnetic nature has prevented detailed experimental characterization of them. Recently, preparation and isolation of the M@C_{82} (M = Y, La, Ce, Pr) anion have been attained using an electrochemical[9–12] and chemical method,[13] which were also used to generate the metallofullerenes in its cationic form. They show diamagnetic properties; those anions show extraordinary stability even under ambient conditions, making them suitable for NMR spectroscopic studies. The La@C_{82} anion is stable at 170°C, under photoirradiation (cutoff $<$ 300 nm) at 20°C, or in an acidic solution ($pK_a \geq 4$).

The C_{82} fullerene has nine distinct isomers (C_{3V}(a), C_{3V} (b), C_{2V}, C_2 (a), C_2 (b), C_2 (c), C_S(a), C_S (b), and C_S (c)) that satisfy the so-called isolated pentagon rule.[14] Because of three-electron transfer from La to C_{82}, it was recently predicted that encapsulation of La inside the C_{2V}, C_{3V}(b), and Cs (C) isomers is energetically much more favorable, and that this engenders endohedral structures with C_{2V}, C_{3V}(b), and C_S(c) symmetry, respectively.[5] These structures are mutually similar in energy and respectively have 24 [17(4) + 7(2)], 17 [11(6) + 5(3) + 1(1)], and 44 [38(2) + 6(1)] nonequivalent carbons, where the values in parentheses denote the relative intensities. Actually, ^{13}C NMR measurement of M@C_{82} in its anionic form was performed, revealing clearly that M@C_{82}−A (M = Y,[12] La,[9,15] Ce,[11,15] Pr[16]) and La@C_{82}−B[10], respectively have C_{2V} and C*s* symmetry.

Reduction can occur even in some solvents such as DMF and pyridine.[17] In other reports, azacrown[18] or unsaturated thiacrown[19] having proper size was observed to form 1:1 complex with La@C_{82}, in which La@C_{82} accepted one electron and converted to anion. The guest and host molecular interaction of La@C_{82} with those crown ethers was believed to have facilitated the electron transfer process. Recently, reversible intermolecular spin-site exchange systems at complete equilibrium in solution were achieved using La@C_{82} and *N, N,N',N'*-tetramethyl-*p*-phenylenedamine, which respectively form stable diamagnetic anion and radical cations (Scheme 7.1)[20] It is noteworthy that the systems show thermochromism and solvatochromism.

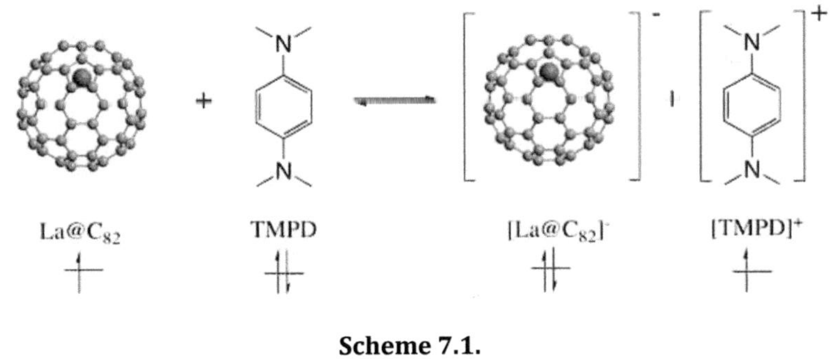

Scheme 7.1.

$M_3N@C_{2n}$ has a closed-shell structure, although their anions and cations usually have open-shell structures and lower stabilities; $[Sc_3N@C_{68}]^+$ is the first electro-synthesized cation. It was characterized by *in situ* ESR and absorption spectroscopic studies.[21] The 22 lines in its ESR spectrum originate from three equivalent Sc hyperfine splittings of 1.289 *g*. No observable N hyperfine splitting was detected under identical experimental conditions. The $M@C_{82}$ anions show high stability and high solubility. That is, they are soluble in polar solvents such as mixed acetone and CS_2 and insoluble in nonpolar solvents such as toluene or CS_2. This character contrasts with that exhibited by neutral fullerenes, which are insoluble in polar solvents and soluble in nonpolar solvents. Several groups have reported convenient methods for separation of endohedral metallofullerenes from carbon soot by chemical reduction or solvent extraction of carbon soot by electrochemical reduction, in which selective reduction of the endohedral metallofullerenes with low redox potentials occurs.[22–24] Selective chemical oxidation is also applied for separation of two isomers of $Sc_3N@C_{80}$ (D_{5h} and I_h).[25]

7.3 Disilylation

Numerous experimental studies have been performed to functionalize empty fullerenes such as C_{60} and C_{70} to elucidate the basic chemical properties and obtain new derivatives with interesting material, catalytic, or biological properties. A new procedure to functionalize C_{60}, C_{70}, and higher fullerenes by the addition of silicon[26–31] and germanium compounds[32] has been developed. It is an interesting challenge to disclose how reactivities of empty fullerenes are modified upon endohedral metal-doping. Consequently, the first exohedral functionalization was conducted for $La@C_{82}$ with 1,1,2,2-tetra-mesityl-1,2-disilirane[33,34] (Scheme 7.2).

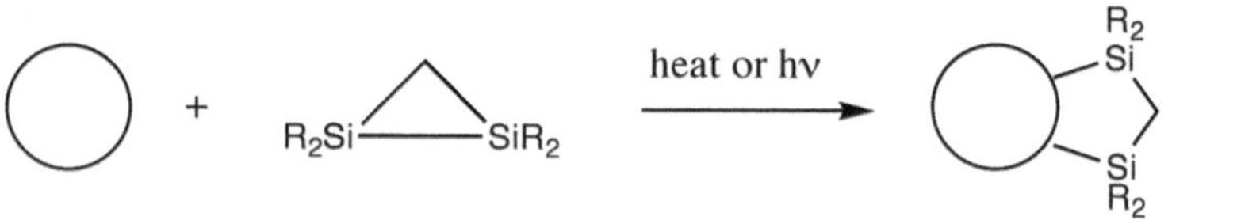

$La@C_{82}$ $\quad$ $Si_2R_4CH_2$ $\quad$ $La@C_{82}Si_2R_4CH_2$

R: a = 2,4,6-trimetylphenyl
b = 2, 6-diethylphenyl

Scheme 7.2.

Table 7.1. Reactivity of fullerenes toward disilirane and redox potentials of fullerenes

	Reactivity[a]			
Compound	$h\nu$	Heat	${}^{OX}E_1^b$	${}^{Red}E_1^b$
C_{60}	Yes	No (80°C)	+1.21	−1.12
C_{70}	Yes	No (80°C)	+1.19	−1.09
C_{82}	Yes	No (80°C)	+0.72	−0.69
$Y@C_{82}$	—	Yes (80°C)	+0.10	−0.37
$La@C_{82}(C_{2v})$	Yes	Yes (80°C)	+0.07	−0.42
$La@C_{82}(C_S)$	Yes	Yes (80°C)	−0.07	−0.47
$Ce@C_{82}$	Yes	Yes (80°C)	+0.08	−0.41
$Pr@C_{82}(C_{2v})$	Yes	Yes (80°C)	+0.07	−0.39
$Pr@C_{82}(C_S)$	Yes	Yes (80°C)	−0.07	−0.48
$La_2@C_{80}$	Yes	Yes (80°C)	+0.56	−0.31
$Ce_2@C_{80}$	Yes	Yes (80°C)	+0.57	−0.39
$Sc_3N@C_{80}$	Yes	No (80°C)	+0.34	−1.24
$Sc_3C_2@C_{80}$	Yes	Yes (80°C)	−0.03	−0.50

[a]"Yes" signifies the formation of a 1:1 adduct of fullerene and disilirane; "No" signifies that a 1:1 adduct was not formed, and no change in the starting fullerene was observed.
[b]Half-wave potentials unless otherwise stated. Values are relative to the ferrocene–ferrocenium couple.

The photochemical reaction was first tested. An interesting finding is that $La@C_{82}$ reacts thermally with disilirane, affording the 1:1 adduct. This contrasts sharply against the fact that empty fullerenes — C_{60}, C_{70}, and so on — react with disilirane only in a photochemical manner. Apparently, the facile thermal addition of disilirane to $La@C_{82}$ is attributable to the stronger electron acceptor and donor properties (Table 7.1).[11,16,33,35] The ESR spectra measured during the reaction mainly reveal formation of two regioisomers, which suggests that the regioselectivity as well as the reactivity of empty fullerenes can be controlled by endohedral metal-doping. Under both conditions, digermylation of $La@C_{82}$ with digermirane was achieved and the mono-adduct was characterized using mass spectrometry and ESR measurement.[36]

For observation of the degree to which the reactivity changes when a different metal is inside the cage, the respective reactions of disilirane with $M@C_{82}$ (M = La,[33,34] Y,[34] Pr,[37] Ce,[11] Gd[38]), $M_2@C_{80}$(La,[39] Ce[40]), $Sc_3C_2@C_{80}$,[41] and $Sc_3N@C_{80}$[42] were also investigated. These results reveal that $Sc_3N@C_{80}$ reacts only photochemically with disilirane, which contrasts with the fact that other metallofullerenes react both photochemically and thermally. This difference is not surprising because the reduction potential of $Sc_3N@C_{80}$ is comparable to those of empty fullerenes, as shown in Table 7.1.

Reduction and oxidation can change chemical properties of endohedral metallofullerenes.[13] The reaction of $[La@C_{82}]^+SbCl_6^-$ with disilirane at room temperature in the dark caused formation of the corresponding 1:1 adduct, as confirmed based on the FAB mass spectrum. Under the same conditions investigated, the reactions of $[Y@C_{82}]^+SbCl_6^-$, $[La@C_{82}\text{-}B]^+SbCl_6^-$, and $[Ce@C_{82}]^+SbCl_6^-$ with disilirane resulted in formation of the corresponding 1:1 adduct. These results indicate that oxidation is an effective method to control the reactivity of endohedral metallofullerenes with disilirane. The reaction of $M@C_{82}$ anions (M = Y, La, Ce) with disilirane was also investigated. However, no adduct was formed either thermally (80°C) or photochemically (400 nm <). This behavior of metallofullerene anions differs greatly from that of the neutral forms: the latter reacts with the disilirane thermally and photochemically. Redox properties are extremely important in determining the reactivity of fullerenes and endohedral metallofullerenes. $M@C_{82}$ cation (M = Y, La, Ce) react readily with the nucleophilic disilirane. Meanwhile, $M@C_{82}$ anion (M = Y, La, Ce) do not react with disilirane. This difference might result from the electrophilicity of fullerenes. In other words, the reactivity of fullerene toward disilirane increases by oxidation and decreases by reduction. It is notable that the fullerene reactivity can be tuned by ionization.

In 1997, Akasaka *et al.* reported the three-dimensional random motion of two La atoms in $La_2@C_{80}$.[43] Nagase *et al.* found that the three-dimensional random motion of two La atoms in $La_2@C_{80}$ can be restricted to the circular motion in a plane by attaching an electron-donating molecule, such as disilirane, on the outer surface of the C_{80} cage.[44] Disilylation of $M_2@C_{80}$ (M = La and Ce) yielded only one mono-adduct.[39,40] Structures of their mono-adducts, especially the motion of the encapsulated metals, were elucidated through NMR spectroscopic studies and the use of X-ray crystallographic method (Fig. 7.1). The addend is bridging on the 1,4-position

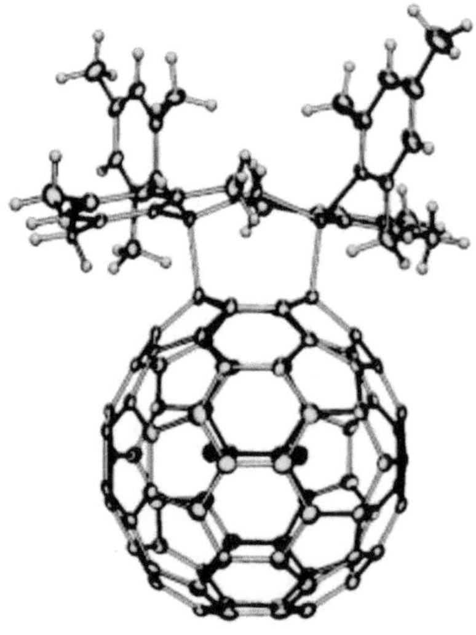

Figure 7.1. An ORTEP drawing of $Ce_2@C_{80}Si_2Mes_4CH_2$.

of a six-membered ring, affording two dynamically exchanged conformers. The motion of the encapsulated metals was found to depend on their own characteristics. In the case of disilylated $Ce_2@C_{80}$, the two Ce atoms are localized at the pole–pole plane inside the cage, which is parallel to the addition sites.[39,40] However, regarding disilylated $La_2@C_{80}$, the two La atoms are in a two-dimensional hopping motion between two addition sites along the equatorial plane. These two results markedly contrast with their three-dimensional motion in pristine $Ce_2@C_{80}$ or $La_2@C_{80}$, and their almost fixed positions in carbene-[45] or Prato-adducts[46] of $Ce_2@C_{80}$ or $La_2@C_{80}$.

7.4 Reaction with Nitrogen Compounds

The first synthesis of methanofullerene derivatives of $La@C_{82}$ was reported by Suzuki *et al.*[47] The reaction of $La@C_{82}$ with diphenyldiazomethane at 60°C engendered formation of the corresponding 1:1 adduct (Scheme 7.3). Results of subsequent ESR analyses of the reaction mixture suggest that four or more position isomers are generated.

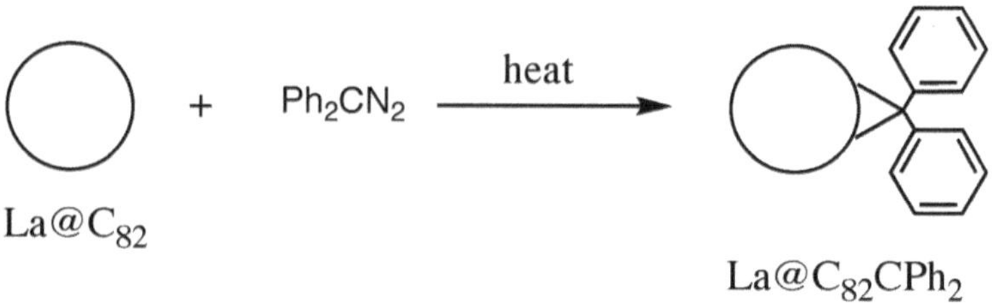

Scheme 7.3.

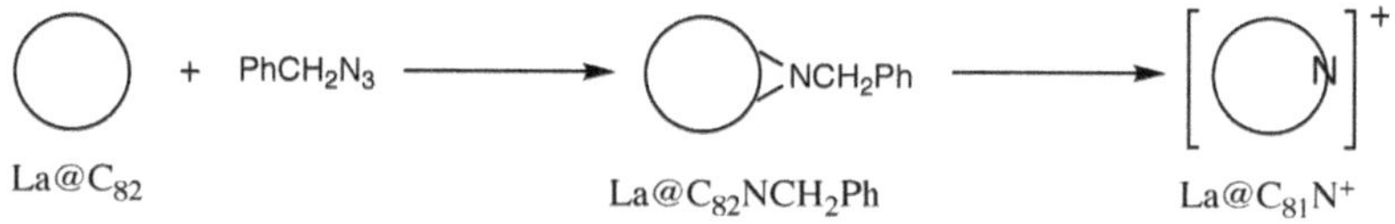

Scheme 7.4.

Organic azides can act as 1,3-dipoles and undergo [3+2] cycloaddition; then thermal nitrogen extrusion of the adducts engenders aza-bridged fullerenes.[48–52] Azafulleroids are known to be suitable precursors for the formation of heterofullerenes such as $C_{59}N^+$.[53,54] Akasaka *et al.* reported evidence of the formation of azametallofullerene ions in gas phase.[55] The reaction of benzyl azide with La@C_{82} and La_2C_{80} at 170°C was conducted (Scheme 7.4). The FAB mass analysis of adducts of La@C_{82} or La_2@C_{80} with benzyl azide shows ion peaks of La@$C_{81}N^+$ or La_2@$C_{79}N^+$. To confirm the formation of azafullerene ions, the ^{15}N-labeled analogues were used for reaction. Intense fragmentation signals for the La@$C_{81}{}^{15}N^+$ and La_2@$C_{79}{}^{15}N^+$ were observed. In 2008, M_2@$C_{79}N$ (M = Y and Tb) was prepared by conducting electric-arc processes. The unique structures of azametallofullerenes were revealed using X-ray structure analyses by the Dorn group.[56]

7.5 Prato Reaction

The Prato reaction is the reaction between fullerene and azomethine ylide with 1,3-dipole character (Scheme 7.5). Azomethine ylides can be generated *in situ* from various readily accessible chemicals. The great popularity of this reaction in fullerene chemistry is attributable to its good selectivity on [6,6]-bond and the general tolerance of widely various functional groups.[57,58]

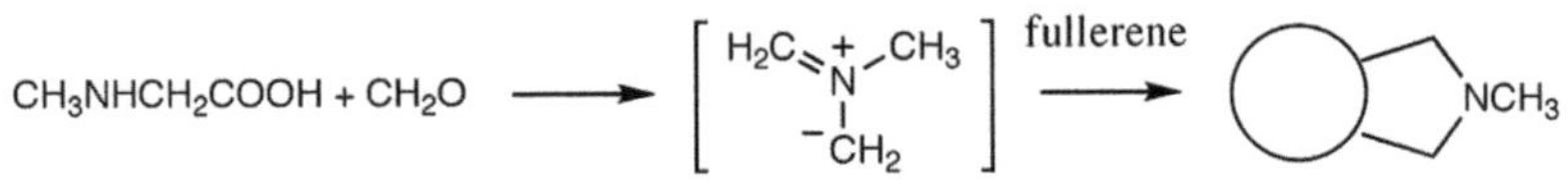

Scheme 7.5.

The reactivities of metallofullerenes with azomethine ylide have been studied recently.[59–61] The reaction of M@C_{82} (M = Y, La, Gd) with azomethine ylide in toluene solution formed the corresponding adducts. It is particularly interesting that the addition to La@C_{82} is extremely efficient and, to some extent, regioselective. The precipitation of adducts during reaction takes place because of the low solubility in toluene. This inhibits further addition of azomethine ylide. On the other hand, three isomers of tris-adduct of Y@C_{82} were found under similar reaction conditions.

The regioselective reaction of Dy@C_{82} with twitterion, which is formed by reaction of phosphine and electro-withdrawing acetylene, was achieved. The structure of the mono-adduct of Dy@C_{82} with phosphorus yilde was determined using X-ray crystallography. The addition position of phosphorus yilde is [6,6]-double bond, which is the same as the addition position of carbene to M@C_{82} (Scheme 7.6).[62]

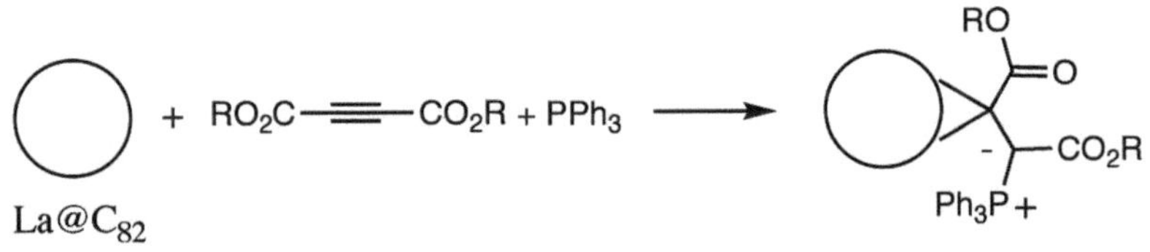

Scheme 7.6.

The Prato reactions involving M_3N@C_{80} have attracted much interest. Results show that the addition of *N*-alkylazomethine occurred on the [5,6]-bond of Sc_3N@C_{80},[63,64] but mainly on the [6,6]-bond of Er_3N@C_{80}[65] and Gd_3N@C_{80}.[66,67] Results of further studies suggest that the [6,6]-pyrrolidine adducts of Y_3N@C_{80} and Er_3N@C_{80} were thermally isomerized to a [5,6]-adduct, although

[5,6]-adduct of $Sc_3N@C_{80}$ and [6,6]-adduct of $Gd_3N@C_{80}$ are unchanged. Additional theoretical studies also confirmed their different stabilities. The density functional theory-calculated formation energy of the [5,6]-adduct of $Sc_3N@C_{80}$ is 11.7 kcal/mol lower than that of the [6,6]-adduct. In the case of $Gd_3N@C_{80}$, the formation energy of its [5,6]-adduct is 0.4 kcal/mol higher than that of its [6,6]-adduct.[67] Consequently, it appears that [5,6]-adducts of $M_3N@C_{80}$ (M = Sc, Y) are thermodynamically favored products and [6,6]-adducts of $Y_3N@C_{80}$ are kinetically favored ones; [6,6]-adducts of $Gd_3N@C_{80}$ are both thermodynamically and kinetically favored products. Such different regioselectivities and stabilities are putatively attributable to their different internal metal ions, which is considered a feature of metallofullerene chemistry. Only a thermodynamically favored [5,6]-adduct was observed in the reaction of $Sc_3N@C_{80}$ with *N*-alkylazomethine ylide. However, using less reactive 1,3-dipolar ylide (*N*-tritylazomethine ylide), [6,6]-pyrrolidino-adduct of $Sc_3N@C_{80}$ was detected together with its [5,6]-adduct (Scheme 7.7).[68] This [6,6]-adduct can also be

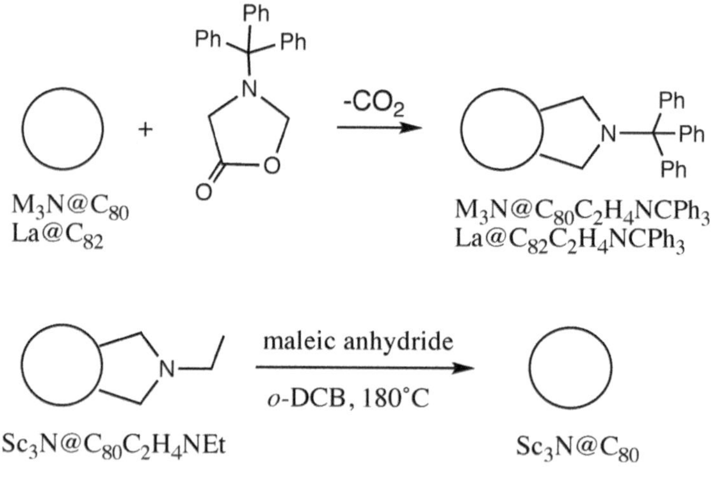

Scheme 7.7.

isomerized thermally to [5,6]-adduct. Similarly, $La_2@C_{80}(I_h)$ reacted with *N*-tritylazomethine ylide, yielding both [5,6]- and [6,6]-adduct.[46] The [6,6]-adduct is separable from its [5,6]-isomer by crystallization (Fig. 7.2). Results of both experimental and

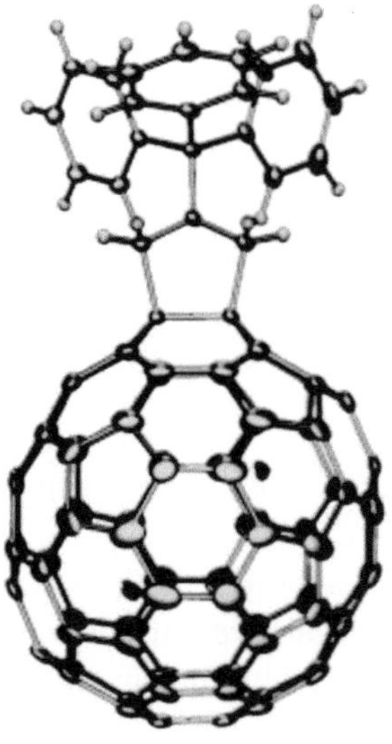

Figure 7.2. An ORTEP drawing of $La_2@C_{80}(CH_2)_2NCPh_3$.

theoretical studies suggest that the two La atoms are highly localized in the C_{80} cage of the [6,6]-adduct. In comparison, $Sc_3N@C_{80}(D_{5h})$ and $Sc_3N@C_{78}(D_{3h})$ each exhibited higher reactivity than $Sc_3N@C_{80}(I_h)$. Their mono-adducts yielded from Prato reactions were characterized as kinetically favored [6,6]-adducts.[69]

It was reported previously that pyrrolidines undergo a retro-cycloaddition reaction that engenders alkene and the azomethine ylide. Actually, *N*-ethylpyrrolidino-$Sc_3N@C_{80}$, heated in the presence of maleic anhydride in sealed tubes in the dark, reveals that the retro-reaction occurs even for metallofullerene derivatives in high yield.[70]

7.6 Cycloaddition of Diene and Benzyne

The [4+2] cycloaddition of metallofullerenes was first achieved on $Sc_3N@C_{80}$ with ^{13}C-labeled 6,7-dimethoxyisochroman-3-one, which forms an *o*-quinone under heating (Scheme 7.8). The addition site occurred at a [5,6]-bond using both NMR and single-crystallographic results.[71,72] The same reaction was also performed on $Gd_3N@C_{80}$

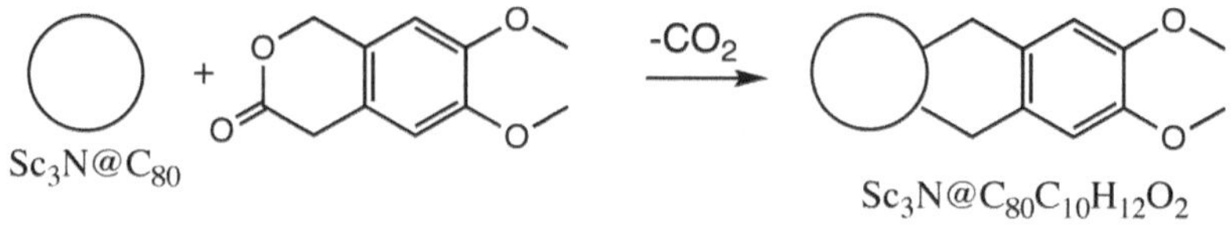

Scheme 7.8.

and a bis-adduct was isolated, but no structural information was reported.[73]

Addition of cyclopentadiene (Cp) to La@C_{82} shows a surprisingly high selectivity because only one regioisomer is formed. Activation energy of the retro-reaction of LaC_{82}Cp that is even lower than that of the retro-reaction of C_{60}Cp was revealed. The La@C_{82}Cp undergoes a retro-reaction, even at ambient conditions, which disables the determination of its molecular structure.[74] On the other hand, the adduct of La@C_{82} with 1,2,3,4,5-pentamethyl cyclopentadiene (Cp*) shows higher stability than that with Cp (Scheme 7.9). A

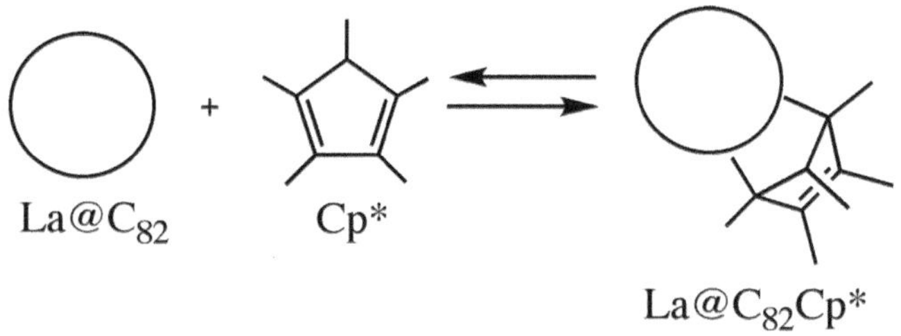

Scheme 7.9.

single-crystallographic result of La@C_{82}Cp* reveals that the addition position of cyclopentadiene moiety is the most positively charged and higher π-orbital axis vector (POAV) angle carbon atoms.[75] Advanced techniques for separation of metallofullerenes from fullerene mixtures including chemical separation methods using a reactive cyclopentadienyl resin or aminosilica to immobilize fullerene contaminants on solid supports have been established.[76–78]

Benzyne, generated by the diazotization of anthranilic acid with isoamyl nitrite, added to Gd@C_{82} forming two isolable isomers of mono-adduct and electrochemical measurements, disclosed that the electronic structure of pristine Gd@C_{82} is changed dramatically

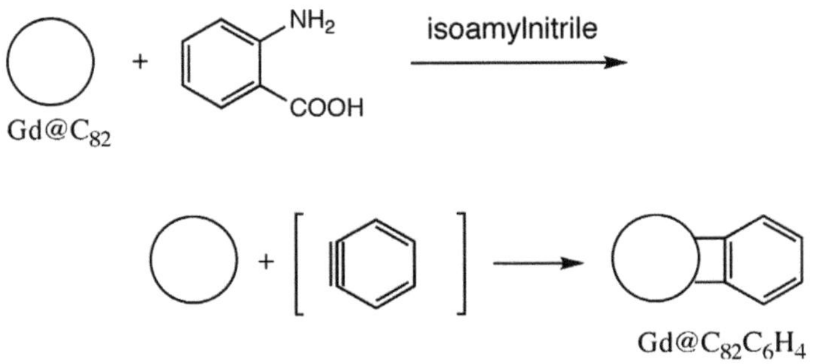

Scheme 7.10.

(Scheme 7.10).[79] For ($Gd@C_{82}$: ${}^{Ox}E$: 0.20 V ${}^{Red}E$: −0.25 V, $Gd@C_{82}(C_6H_4)$: ${}^{Ox}E$: 0.26 V ${}^{Red}E$: −0.97 V), because of the high reactivity of benzyne, multiple adducts are not avoidable, even at lower temperatures.

7.7 Addition of Carbene

The first reported carbene reaction of metallofullerene was conducted by irradiation of 2-adamantane-2,3-[3H]-diazirine and $La@C_{82}(C_{2V})$ in degassed solvent.[80] Because 24 nonequivalent carbons exist in $La@C_{82}$, addition might take place at several sites to afford numerous possible mono-adduct isomers. It is particularly interesting that this reaction yielded only two mono-adducts, indicating the high regioselectivity of carbene towards $La@C_{82}(C_{2V})$ (Scheme 7.11). The two mono-adducts were determined either

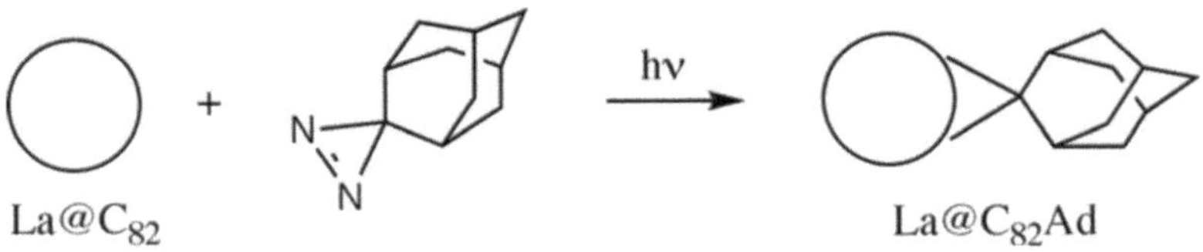

Scheme 7.11.

using X-ray crystallography or NMR spectroscopy in addition to theoretical calculations (Fig. 7.3).[80,81] Addition sites involved two [6,6]-bonds adjacent to the La^{3+} ion. Theoretical calculations of $La@C_{82}(C_{2V})$ disclosed that one carbon on the six-membered ring

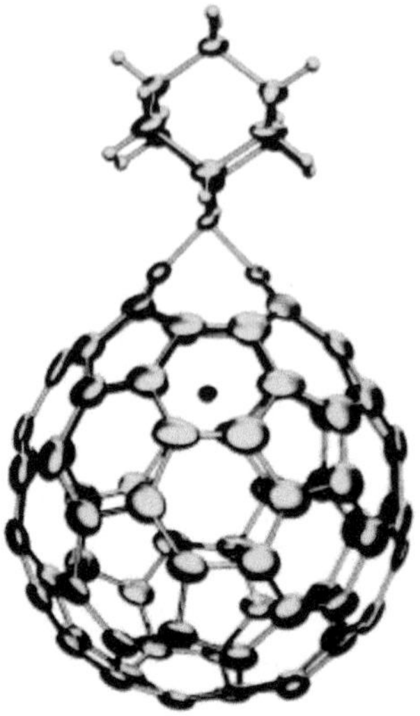

Figure 7.3. An ORTEP drawing of La@C_{82}Ad.

adjacent to the metal atom has both higher POAV angle and charge density value than others (Table 7.2). Accordingly, it is more reactive toward adamantylidene, which acts as an electrophile in this reaction. For the first time, a thermal isomerization from major mono-adduct to minor nomo-adduct of metallofullerene adducts was reported. Theoretical calculations revealed that the major mono-adduct is 2.5 kcal/mol less stable than the minor mono-adduct.[81] The addition positions of and minor isomer of Ce@C_{82}Ad was determined using X-ray crystallography.[82]

Unlike the closed cyclopropane structure of a typical [6,6]-adduct of C_{60},[83] the two [6,6]-adducts of La@C_{82} each have a broken [6,6]-bond because of the carbene addition, forming methanofulleroids instead of methanoadducts of La@C_{82}(C_{2V}). This open cage structure was found to be a common feature of all reported carbene derivatives of metallofullerenes as La@C_{82}(C_s)[84] and Gd@C_{82}.[85]

This carbene reaction is quite clean and simple. For that reason, it has been used frequently as a probe to examine the chemical reactivities and structures of various metallofullerenes. The reactions of adamantylidene carbene with M_2@C_{80}(I_h)[45] (M = La, Ce), La_2@C_{78},[86]Sc_2C_2@C_{82},[87] Sc_3C_2@C_{80},[88] and non-IPR

Table 7.2. Charge density, spin density, and POAV angles of carbon atoms in $La@C_{82}$, as calculated at the B3LYP level

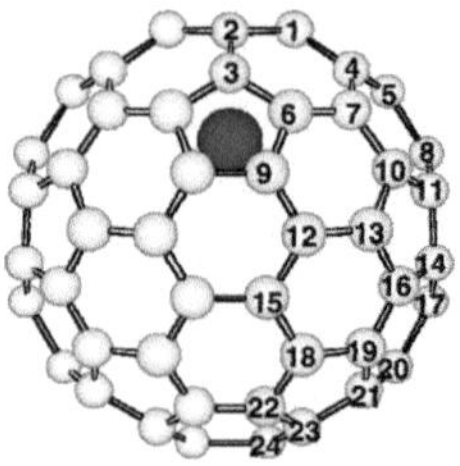

Carbon number	Change density	Spin density	POAV
1	−0.136	−0.012	11.2
2	−1.170	0.023	11.3
3	−0.099	0.031	8.8
4	−0.071	0.035	9.6
5	−0.045	−0.019	9.2
6	−0.037	−0.004	10.5
7	−0.092	−0.242	9.1
8	−0.061	0.008	8.9
9	−0.022	0.037	9.9
10	−0.020	0.066	10.7
11	−0.021	−0.016	10.6
12	−0.047	−0.003	7.7
13	−0.006	−0.001	10.9
14	0	0.057	11.0
15	−0.027	0.014	7.4
16	−0.012	−0.001	10.6
17	−0.036	−0.031	8.2
18	0.004	0.045	11.0
19	−0.006	0.046	10.9
20	0.026	0.011	8.4
21	0.002	0.029	10.5
22	−0.006	−0.012	10.7
23	0.006	0.063	10.7
24	−0.025	−0.026	8.3

$La_2@C_{72}$[89,90] were reported recently (Fig. 7.4). For $M_2@C_{80}$ ($M = La$, Ce) and $Sc_3C_2@C_{80}$, irrespective of the encapsulated metal or cluster, the additions occur exclusively on the [6,6]-bond. In contrast, the carbene additions on non-IPR $La_2@C_{72}$[89] select either [5,6]-bonds or a [6,6]-bond that is adjacent to the fused-pentagon pair.[90] The second addition also preferably occurred on a [5,6]-bond on the other pentagon pair. Even at longer reaction time, no multi-addition was observed, indicating the higher reactivities of the fused-pentagon regions. Regarding $La_2@C_{78}$, the selectivity of carbene addition is lower. Both [5,6]-bond and [6,6]-bond are involved in the additions.

Akasaka *et al.* reported the FET properties of metallofullerene derivative $La@C_{82}Ad$.[91] It is particularly interesting that pristine

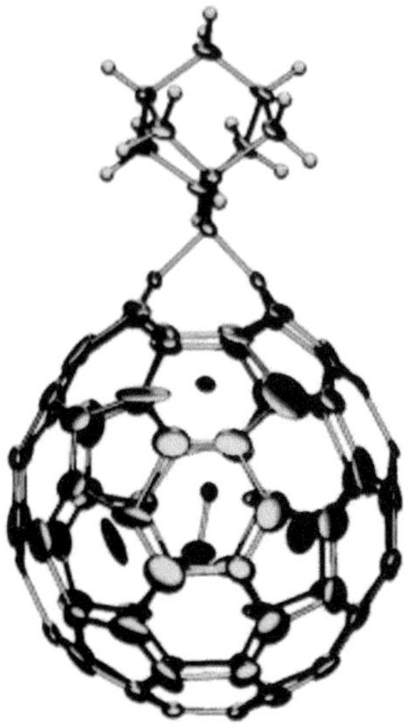

Figure 7.4. An ORTEP drawing of $Sc_3C_2N@C_{80}Ad$.

La@C_{82}Ad and La@C_{82}Ad film show the *n*-type action, but the La@C_{82}Ad nanorod shows the *p*-type action. The thin films and whiskers of other empty fullerenes are well known to show *n*-type semiconductivity. Consequently, the *p*-type behavior of the La@C_{82}Ad nanorod is unique within the fullerene FETs. A magnetic orientation of nanorods was also observed by scanning electron microscope (SEM) observation. The suspension of the La@C_{82}Ad nanorods in isopropyl alcohol was placed in a magnetic field at 12 T, and the solvent was vaporized slowly at room temperature. In fact, SEM images show that the La@C_{82}Ad nanorods orient perpendicularly to the magnetic field.

7.8 Nucleophilic Addition

Another efficient chemical modification method used in fullerene chemistry is the Bingel–Hirsch reaction.[66,92–94] Its mechanism involves a nucleophilic attack of a carbon anion that is produced *in situ* by deprotonation of α-halo esters or α-halo ketones. This method gives easy access to versatile fullerene derivatives as well as water-soluble fullerenes.

Bingel–Hirsch reaction was also performed on metallofullerenes with the aim of obtaining various methanoadducts. Because of multiple electron transfers from encapsulated metals or clusters to fullerene cages, metallofullerenes are not such good electron-deficient species as empty fullerenes. Nevertheless, the Bingel–Hirsch reaction of $M_3N@C_{80}$($M = Y$,[95] Gd[96]), $Sc_3N@C_{78}$,[97] and $Gd_3N@C_{84}$[98] can proceed smoothly at room temperature (Scheme 7.12). The mono-adduct of $Y_3N@C_{80}$ was characterized as a

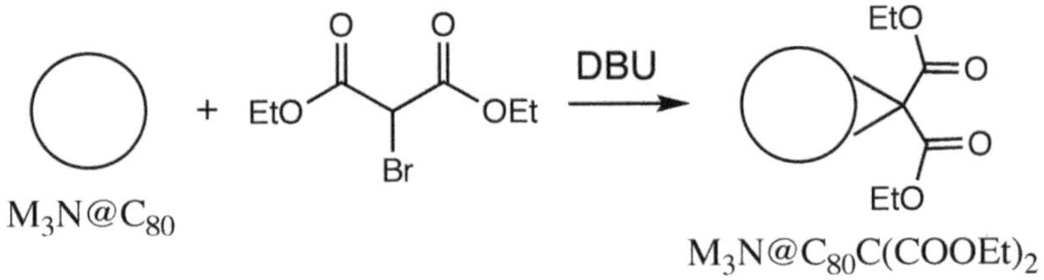

Scheme 7.12.

[6,6]-methanofulleroid adduct,[95] different from the methano-adducts of C_{60} or C_{70}. In fact, $Sc_3N@C_{78}$ readily afforded a symmetric bis-adduct with high regioselectivity.[97] $Sc_3N@C_{80}$ and $Gd_3N@C_{88}$[98] do not undergo Bingel–Hirsch reaction under identical experimental conditions. Such inertness is attributed to the fact that a smaller encapsulated cluster or larger cage size induced a lower degree of pyramidalization of cage carbons.

Reaction of $La@C_{82}$ and diethyl bromomalonate in the presence of 1,8-diazabicyclo[5.4.0]-undec-7-ene (DBU) afforded four isomers with singly bonded structures and one cycloadduct, which is the typical product in Bingel-Hirsch reaction for empty fullerenes (Scheme 7.13).[99,100] Four show diamagnetic properties, in contrast to paramagnetic properties of $La@C_{82}$ and another minor product. The X-ray structure of its major diamagnetic product is shown in Fig. 7.5. The addition site is far from the La^{3+} ion and has the most positive charge density. The reaction mechanism is proposed as mainly involving a nucleophilic attack of carbonanion on

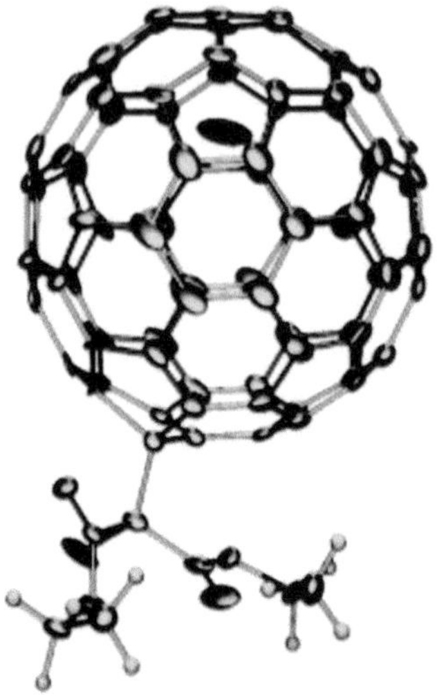

Figure 7.5. An ORTEP drawing of La@C_{82}CBr(CO_2Et)$_2$.

La@C_{82}. The following — otherwise slowly proceeding — bromo-leaving and cyclopropanation processes are possibly replaced by an unidentified rapid oxidation of the intermediate. Compared to pristine La@C_{82}, its diamagnetic mono-adducts have negatively shifted first reduction potentials and positively shifted first oxidation potentials, suggesting their larger HOMO–LUMO gaps. On the other hand, its minor paramagnetic mono-adduct was characterized as a methanofulleroid La@C_{82}C(CO_2Et)$_2$ by NMR spectroscopic studies of its anion.[100] The La@C_{82}C(CO_2Et)$_2$ and above described Y_3N@C_{82}C(CO_2Et)$_2$ were found to have common features in their redox behaviors. Both exhibited high stabilities in their one-electron reductive states, which is in contrast with the previously reported retro-cycloaddition of [C_{60}C(CO_2Et)$_2$]$^-$.

The Bingel–Hirsch reaction is not restricted to highly acidic carbonyl compound such as bromomalonate. Actually, La@C_{82} even reacted with malonate in the presence of DBU at elevated temperature.[101] This reaction affords a 7,13-bismalomate derivative of La@C_{82} with high regioselectivity, which dimerizes during the crystallization process, thereby indicating its more reactive radical character (Scheme 7.13).

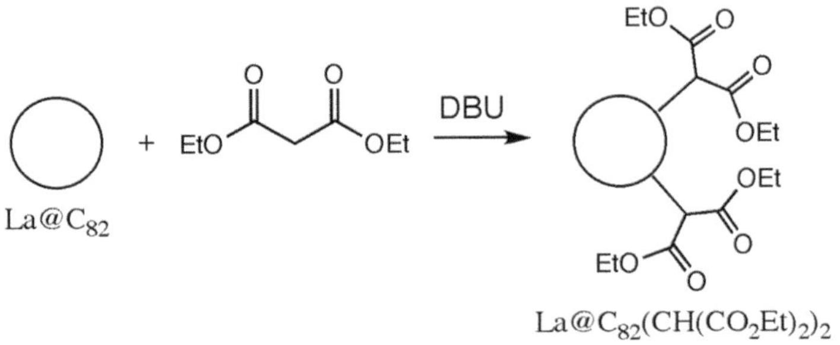

Scheme 7.13.

7.9 Radical Addition

Endohedral metallofullerenes behave as either a radical sponge or a radical. Additions of perfluoroalkyl groups generated from perfluoroalkyl iodides (R_fI) have been conducted on a mixture of $Sc_3N@C_{80}$ and $Sc_3N@C_{80}$ under vacuum and over 500°C, yielding $Sc_3N@C_{80}(CF_3)_{2n}$ ($n \leq 6$) (Scheme 7.14). For their bis-CF_3

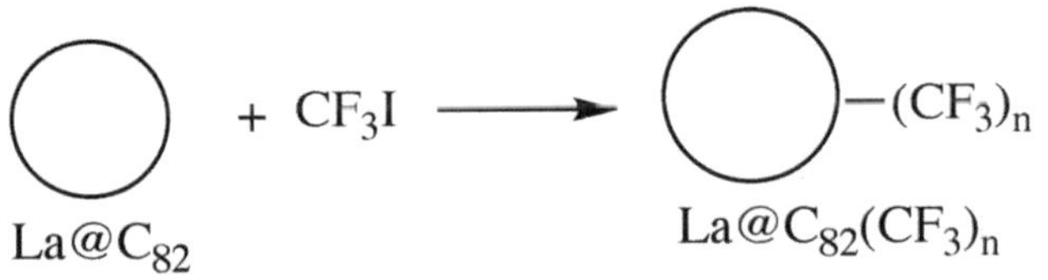

Scheme 7.14.

derivatives, two CF_3 groups were shown to be added equivalently on either isomer according to ^{19}F NMR spectroscopic studies.[102] A 1,4-addition was proposed based on results of theoretical studies, which possibly engenders the derivatives with minimum formation energies. In addition, $La@C_{82}$ was reported to undergo multiple additions of fluoroalkyl radicals (Scheme 7.15).[103] Attachment of a fluorous label group changes the fullerene solubility. The fluorous-phase partitioning method (liquid–liquid extraction) aided by multi-stage recycling high-performance liquid chromatography (HPLC) caused isolation of an adduct, $La@C_{82}(C_8F_{17})_2$, in isomer-free form. Dorn and Gibson *et al.* reported similar photoreaction of $Sc_3N@C_{80}$ with benzyl bromide. Photochemically generated benzyl radicals

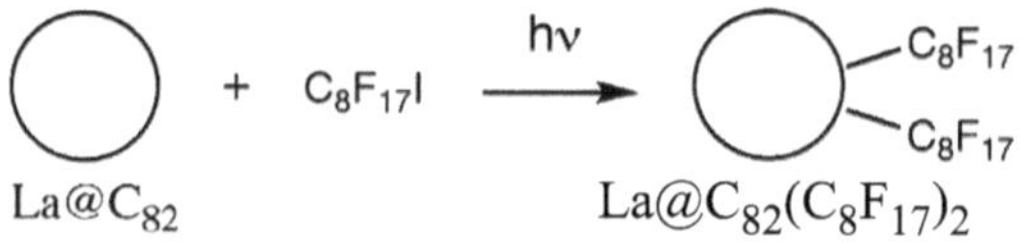

Scheme 7.15.

react with $Sc_3N@C_{80}$ to produce a dibenzyl adduct in high yield and high regioselectvity (Scheme 7.16).[104] Radical additions of $Y@C_{82}$ with perfluoroalkyl groups generated from $AgCF_3CO_2$ were performed similarly.[105] Series mono-adducts and multi-adducts, such as $Y@C_{82}(CF_3)$, $Y@C_{82}(CF_3)_3$, and $Y@C_{82}(CF_3)_5$, were isolated and characterized. Only odd number of CF_3 groups was added to $Y@C_{82}$, resulting in derivatives with closed-shell structures, which are apparently thermodynamically favored products. The two isomers of $Y@C_{82}(CF_3)_5$ were proposed by the authors as having an addition pattern with 1,4-additions across four contiguous six-membered rings.

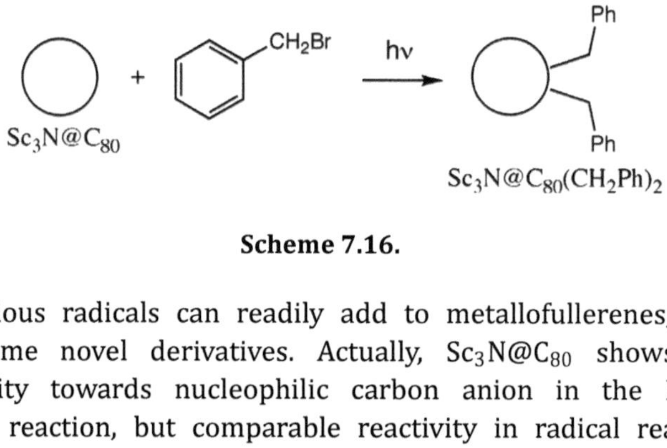

Scheme 7.16.

Various radicals can readily add to metallofullerenes, forming some novel derivatives. Actually, $Sc_3N@C_{80}$ shows inert reactivity towards nucleophilic carbon anion in the Bingel–Hirsch reaction, but comparable reactivity in radical reactions. Its reaction-generated malonate radical proceeds smoothly in refluxed chlorobenzene, yielding two methanofulleroids: $Sc_3N@C_{80}C(CO_2Et)_2$ and $Sc_3N@C_{80}CH(CO_2Et)$.[106] Further studies of $M^{3+}@C_{2n}^{3-}$ type metallofullerenes definitively revealed their unique radical character, which was believed arising from the unpaired electron on their SOMO. Results showed that $La@C_{82}$ even thermally reacted with toluene in presence of 3-triphenylmethyl-5-oxazolidinone, leading to four mono-adducts commonly described as $La@C_{82}CH_2C_6H_5$ (Scheme 7.17).[107] This result indicates that $La@C_{82}$ is more reactive with even unstable radical than with azomethine ylide. Alternatively, under photoirradiation conditions, $La@C_{82}$ can react directly not only with toluene, but also α,α,2,4-tetrachlorotoluene, yielding $La@C_{82}$-$CHClC_6H_3Cl_2$. One isomer was fully determined. The addition site was revealed to possess the highest spin density and high POAV angle by theoretical calculations, thereby confirming the proposed radical–radical reaction.

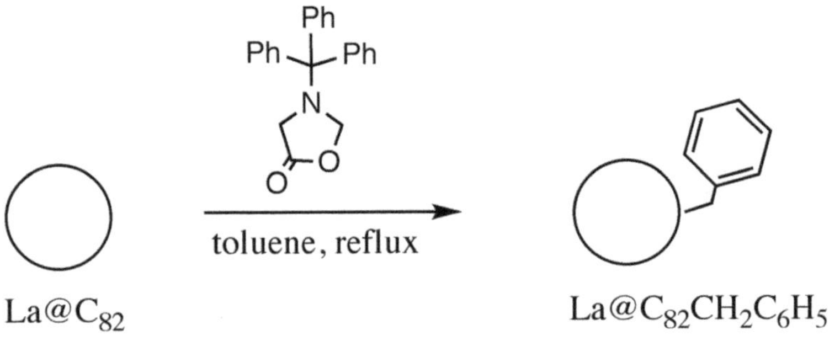

Scheme 7.17.

In 1991, Smalley and coworkers reported that $La@C_{60}$, $La@C_{74}$, and $La@C_{82}$ were produced especially abundantly in soot, but only $La@C_{82}$ was extracted with toluene.[2] Since then, the chemistry of soluble metallofullerenes has been started. To date, many soluble metallofullerenes have been separated and characterized. However, insoluble metallofullerenes, such as $La@C_{60}$ and $La@C_{74}$, have not yet been isolated, although they are regularly observed in raw soot using mass spectrometry. Recently, $La@C_{72}$[108] and $La@C_{74}$[109] have been isolated in forms of their derivatives as adducts of the dichlorophenyl group by HPLC of extract from soot using 1,2,4-trichlorobenzene, and subsequently characterized. Their X-ray structures exhibit a non-IPR $C_{72}(C_2)$ cage for $La@C_{72}$ and an IPR $C_{74}(D_{3h})$ cage for $La@C_{74}$ (Fig. 7.6). The addition position of the dichlorophenyl group has high SOMO spin-density and the high POAV value on fullerene cages. This result indicates that dichlorophenyl radical, which might be produced by the reaction of 1,2,4-trichlorobenzene with reductant, such as lanthanum carbide in the raw soot, adds to one of these carbons to produce the sta-

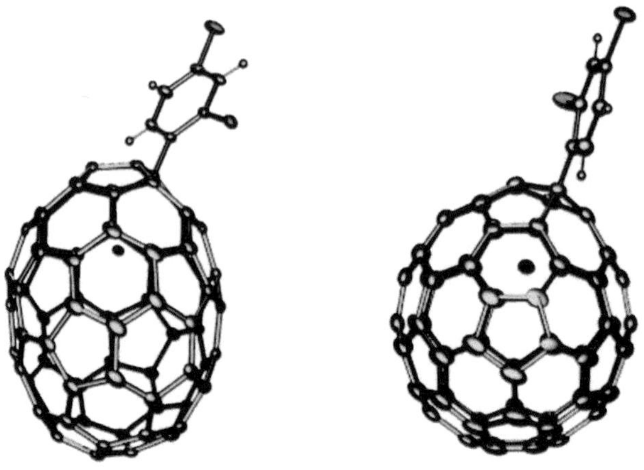

Figure 7.6. An ORTEP drawing of $La@C_{72}C_6H_3Cl_2$ and $La@C_{74}C_6H_3Cl_2$.

ble and soluble adduct. In $La@C_{72}$, La^{3+} ion is localized close to the fused [5,5]-junction, indicating their strong interaction. The respective distances between La and two carbons of the [5,5]-junction are 2.615 and 2.606 Å. These are somewhat shorter than the calculated values of 2.714 and 2.680 Å. In the case of $La@C_{74}$, the encapsulated La atom is localized mainly at a site near the dichlorophenyl group, which slightly deviated from the calculated optimal site along the C_2 axis on the σ_h plane. This site shift might result from introduction of the dichlorophenyl group.

7.10 Conclusion

By recent development of synthesis and separation techniques, the yields of endohedral metallofullerenes have been greatly increased. These advances make the macro-quantities of metallofullerenes presently available in laboratory production. Hereby, functionalization of endohedral metallofullerenes has been greatly progressed in last decade. Reactivity, stability, and regioselectivity of endohedral metallofullerenes are well controlled by intramolecular metal–fullerene cage interaction. Functionalization of endohedral metallofullerenes which have novel characteristics different from empty fullerenes opens new material, catalytic, and biological science and applications.

Acknowledgments

This work was supported in part by a Grant-in-Aid for Scientific Research on Innovation Areas (No. 20108001, "pi-Space"), a Grant-in-Aid for Scientific Research (A)(No. 20245006), a Grant-in-Aid for Young Scientists (B)(No. 23750035), the 21st Century COE Program, The Next Generation Super Computing Project (Nanoscience Project), the Nanotechnology Support Project, and a Grant-in Aid for Scientific Research on Priority Area (Nos. 20036008, 20038007) from the Ministry of Education, Culture, Sports, Science, and Technology of Japan.

References

1. J. R. Heath, S. C. O'Brien, Q. Zhang, Y. Liu, R. F. Curl, H. W. Kroto, F. K. Tittel, and R. E. Smalley, *J. Am. Chem. Soc.* **107**, 7779–7780 (1985).
2. Y. Chai, T. Guo, C. Jin, R. E. Haufler, L. P. F. Chibante, J. Fure, L. Wang, J. M. Alford, and R. E. Smalley, *J. Phys. Chem.* **95**, 7564–7568 (1991).
3. H. Shinohara, *Rep. Prog. Phys.* **63**, 843–892 (2000).
4. *Endofullerenes: A New Family of Carbon Clusters*, Kluwer Academic Publishers, Dordrecht, 2002.
5. K. Kobayashi and Nagase. S., *Chem. Phys. Lett.* **282**, 325–329 (1998).
6. T. Kato, S. Suzuki, K. Kikuchi, and Y. Achiba, *J. Phys. Chem.* **97**, 13425–13428 (1993).
7. K. Yamamoto, H. Funasaka, T. Takahashi, and T. Akasaka, *J. Phys. Chem.* **98**, 2008–2011 (1994).
8. J. Ding, L. T. Weng, and S. Yang, *J. Phys. Chem.* **100**, 11120–11121 (1996).
9. T. Akasaka, T. Wakahara, S. Nagase, K. Kobayashi, M. Waelchli, K. Yamamoto, M. Kondo, S. Shirakura, S. Okubo, Y. Maeda, T. Kato, M. Kako, Y. Nakadaira, R. Nagahata, X. Gao, E. Van Caemelbecke, and K. M. Kadish, *J. Am. Chem. Soc.* **122**, 9316–9317 (2000).

10. T. Akasaka, T. Wakahara, S. Nagase, K. Kobayashi, M. Waelchli, K. Yamamoto, M. Kondo, S. Shirakura, Y. Maeda, T. Kato, M. Kako, Y. Nakadaira, X. Gao, E. Van Caemelbecke, and K. M. Kadish, *J. Phys. Chem. B* **105**, 2971–2974 (2001).
11. T. Wakahara, J. I. Kobayashi, M. Yamada, Y. Maeda, T. Tsuchiya, M. Okamura, T. Akasaka, M. Waelchli, K. Kobayashi, S. Nagase, T. Kato, M. Kako, K. Yamamoto, and K. M. Kadish, *J. Am. Chem. Soc.* **126**, 4883–4887 (2004).
12. L. Feng, T. Wakahara, T. Tsuchiya, Y. Maeda, Y. Lian, T. Akasaka, N. Mizorogi, K. Kobayashi, S. Nagase, and K. M. Kadish, *Chem. Phys. Lett.* **405**, 274–277 (2005).
13. Y. Maeda, J. Miyashita, T. Hasegawa, T. Wakahara, T. Tsuchiya, L. Feng, Y. Lian, T. Akasaka, K. Kobayashi, S. Nagase, M. Kako, K. Yamamoto, and K. M. Kadish, *J. Am. Chem. Soc.* **127**, 2143–2146 (2005).
14. D. E. M. P. W. Fowler, *An Atlas of Fullerenes*, Clarendon, Oxford, 1995.
15. T. Tsuchiya, T. Wakahara, Y. Maeda, T. Akasaka, M. Waelchli, T. Kato, H. Okubo, N. Mizorogi, K. Kobayashi, and S. Nagase, *Angew. Chem. Int. Ed. Engl.* **44**, 3282–3285 (2005).
16. T. Wakahara, S. Okubo, M. Kondo, Y. Maeda, T. Akasaka, M. Waelchli, M. Kako, K. Kobayashi, S. Nagase, T. Kato, K. Yamamoto, X. Gao, E. Van Caemelbecke, and K. M. Kadish, *Chem. Phys. Lett.* **360**, 235–239 (2002).
17. S. P. Solodovnikov and S. F. Lebedkin, *Russ. Chem. Bull.* **52**, 1111–1113 (2003).
18. T. Tsuchiya, K. Sato, H. Kurihara, T. Wakahara, T. Nakahodo, Y. Maeda, T. Akasaka, K. Ohkubo, S. Fukuzumi, T. Kato, N. Mizorogi, K. Kobayashi, and S. Nagase, *J. Am. Chem. Soc.* **128**, 6699–6703 (2006).
19. T. Tsuchiya, H. Kurihara, K. Sato, T. Wakahara, T. Akasaka, T. Shimizu, N. Kamigata, N. Mizorogi, and S. Nagase, *Chem. Commun.* 3585–3587 (2006).
20. T. Tsuchiya, K. Sato, H. Kurihara, T. Wakahara, Y. Maeda, T. Akasaka, K. Ohkubo, S. Fukuzumi, T. Kato, and S. Nagase, *J. Am. Chem. Soc.* **128**, 14418–14419 (2006).
21. S. Yang, P. Rapta, and L. Dunsch, *Chem. Commun.* 189–191 (2007).
22. R. D. Bolskar and J. M. Alford, *Chem. Commun.* 1292–1293 (2003).
23. T. Tsuchiya, T. Wakahara, S. Shirakura, Y. Maeda, T. Akasaka, K. Kobayashi, S. Nagase, T. Kato, and K. M. Kadish, *Chem. Mater.* **16**, 4343–4346 (2004).

24. T. Kodama, K. Higashi, T. Ichikawa, S. Suzuki, H. Nishikawa, I. Ikemoto, K. Kikuchi, and Y. Achiba, *Chem. Lett.* **34**, 464–465 (2005).

25. B. Elliott, L. Yu, and L. Echegoyen, *J. Am. Chem. Soc.* **127**, 10885–10888 (2005).

26. T. Akasaka, W. Ando, K. Kobayashi, and S. Nagase, *J. Am. Chem. Soc.* **115**, 10364–10365 (1993).

27. T. Akasaka, E. Mitsuhida, W. Ando, K. Kobayashi, and S. Nagase, *J. Am. Chem. Soc.* **116**, 2627–2628 (1994).

28. T. Wakahara, A. Han, Y. Niino, Y. Maeda, T. Akasaka, T. Suzuki, K. Yamamoto, M. Kako, Y. Nakadaira, K. Kobayashi, and S. Nagase, *J. Mater. Chem.* **12**, 2061–2064 (2002).

29. A. Han, T. Wakahara, Y. Maeda, Y. Niino, T. Akasaka, K. Yamamoto, M. Kako, Y. Nakadaira, K. Kobayashi, and S. Nagase, *Chem. Lett.* 974–975 (2001).

30. Y. Maeda, G. M. Aminur Rahman, T. Wakahara, M. Kako, M. Okamura, S. Sato, T. Akasaka, K. Kobayashi, and S. Nagase, *J. Org. Chem.* **68**, 6791–6794 (2003).

31. T. Akasaka, T. Suzuki, Y. Maeda, M. Ara, T. Wakahara, K. Kobayashi, S. Nagase, M. Kako, Y. Nakadaira, M. Fujitsuka, and O. Ito, *J. Org. Chem.* **64**, 566–569 (1999).

32. T. Akasaka, Y. Maeda, T. Wakahara, T. Mizushima, W. Ando, M. Wälchli, T. Suzuki, K. Kobayashi, S. Nagase, M. Kako, Y. Nakadaira, M. Fujitsuka, O. Ito, Y. Sasaki, K. Yamamoto, and T. Erata, *Org. Lett.* **2**, 2671–2674 (2000).

33. T. Akasaka, T. Kato, K. Kobayashi, S. Nagase, K. Yamamoto, H. Funasaka, and T. Takahashi, *Nature* **374**, 600–601 (1995).

34. M. Yamada, L. Feng, T. Wakahara, T. Tsuchiya, Y. Maeda, Y. Lian, M. Kako, T. Akasaka, T. Kato, K. Kobayashi, and S. Nagase, *J. Phys. Chem. B* **109**, 6049–6051 (2005).

35. K. Kikuchi, Y. Nakao, S. Suzuki, Y. Achiba, T. Suzuki, and Y. Maruyama, *J. Am. Chem. Soc.* **116**, 9367–9368 (1994).

36. T. Akasaka, T. Kato, S. Nagase, K. Kobayashi, K. Yamamoto, H. Funasaka, and T. Takahashi, *Tetrahedron* **52**, 5015–5020 (1996).

37. T. Akasaka, S. Okubo, M. Kondo, Y. Maeda, T. Wakahara, T. Kato, T. Suzuki, K. Yamamoto, K. Kobayashi, and S. Nagase, *Chem. Phys. Lett.* **319**, 153–156 (2000).

38. T. Akasaka, S. Nagase, K. Kobayashi, T. Suzuki, T. Kato, K. Yamamoto, H. Funasaka, and T. Takahashi, *J. Chem. Soc. Chem. Commun.* 1343–1344 (1995).

39. T. Wakahara, M. Yamada, S. Takahashi, T. Nakahodo, T. Tsuchiya, Y. Maeda, T. Akasaka, M. Kako, K. Yoza, E. Horn, N. Mizorogi, and S. Nagase, *Chem. Commun.* 2680–2682 (2007).

40. M. Yamada, T. Nakahodo, T. Wakahara, T. Tsuchiya, Y. Maeda, T. Akasaka, M. Kako, K. Yoza, E. Horn, N. Mizorogi, K. Kobayashi, and S. Nagase, *J. Am. Chem. Soc.* **127**, 14570–14571 (2005).

41. T. Wakahara, A. Sakuraba, Y. Iiduka, M. Okamura, T. Tsuchiya, Y. Maeda, T. Akasaka, S. Okubo, T. Kato, K. Kobayashi, S. Nagase, and K. M. Kadish, *Chem. Phys. Lett.* **398**, 553–556 (2004).

42. T. Wakahara, Y. Iiduka, O. Ikenaga, T. Nakahodo, A. Sakuraba, T. Tsuchiya, Y. Maeda, M. Kako, T. Akasaka, K. Yoza, E. Horn, N. Mizorogi, and S. Nagase, *J. Am. Chem. Soc.* **128**, 9919–9925 (2006).

43. T. Akasaka, S. Nagase, K. Kobayashi, M. Wälchli, K. Yamamoto, H. Funasaka, M. Kako, T. Hoshino, and T. Erata, *Angew. Chem. Int. Ed. Engl.* **36**, 1643–1645 (1997).

44. K. Kobayashi, S. Nagase, Y. Maeda, T. Wakahara, and T. Akasaka, *Chem. Phys. Lett.* **374**, 562–566 (2003).

45. M. Yamada, C. Someya, T. Wakahara, T. Tsuchiya, Y. Maeda, T. Akasaka, K. Yoza, E. Horn, M. T. H. Liu, N. Mizorogi, and S. Nagase, *J. Am. Chem. Soc.* **130**, 1171–1176 (2008).

46. M. Yamada, T. Wakahara, T. Nakahodo, T. Tsuchiya, Y. Maeda, T. Akasaka, K. Yoza, E. Horn, N. Mizorogi, and S. Nagase, *J. Am. Chem. Soc.* **128**, 1402–1403 (2006).

47. T. Suzuki, Y. Maruyama, T. Kato, T. Akasaka, K. Kobayashi, S. Nagase, K. Yamamoto, H. Funasaka, and T. Takahashi, *J. Am. Chem. Soc.* **117**, 9606–9607 (1995).

48. M. Cases, M. Duran, J. Mestres, N. Martín, and M. Sol, *J. Org. Chem.* **66**, 433–442 (2001).

49. M. Prato, Q. C. Li, F. Wudl, and V. Lucchini, *J. Am. Chem. Soc.* **115**, 1148–1150 (1993).

50. T. Grösser, M. Prato, V. Lucchini, A. Hirsch, and F. Wudl, *Angew. Chem. Int. Ed. Engl.* **34**, 1343–1345 (1995).

51. B. Nuber, F. Hampel, and A. Hirsch, *Chem. Commun.* 1799–1800 (1996).

52. S. Gonzalez, N. Mártín, A. Swartz, and D. M. Guldi, *Org. Lett.* **5**, 557–560 (2003).

53. J. Averdung, H. Luftmann, I. Schlachter, and J. Mattay, *Tetrahedron* **51**, 6977–6982 (1995).

54. I. Lamparth, B. Nuber, G. Schick, A. Skiebe, T. Grösser, and A. Hirsch, *Angew. Chem. Int. Ed. Engl.* **34**, 2257–2259 (1995).

55. T. Akasaka, S. Okubo, T. Wakahara, K. Yamamoto, K. Kobayashi, S. Nagase, T. Kato, M. Kako, Y. Nakadaira, Y. Kitayama, and K. Matsuura, *Chem. Lett.* 945–946 (1999).

56. T. Zuo, L. Xu, C. M. Beavers, M. M. Olmstead, W. Fu, T. D. Crawford, A. L. Balch, and H. C. Dorn, *J. Am. Chem. Soc.* **130**, 12992–12997 (2008).

57. M. Maggini, G. Scorrano, and Prato, *J. Am. Chem. Soc.* **115**, 9798–9799 (1993).

58. M. Prato and M. Maggini, *Acc. Chem. Res.* **31**, 519–526 (1998).

59. X. Lu, X. He, L. Feng, Z. Shi, and Z. Gu, *Tetrahedron* **60**, 3713–3716 (2004).

60. B. Cao, T. Wakahara, Y. Maeda, A. Han, T. Akasaka, T. Kato, K. Kobayashi, and S. Nagase, *Chem. Eur. J.* **10**, 716–720 (2004).

61. L. Feng, X. Lu, X. He, Z. Shi, and Z. Gu, *Inorg. Chem. Commun.* **7**, 1010–1013 (2004).

62. X. Li, L. Fan, D. Liu, H. H. Y. Sung, I. D. Williams, S. Yang, K. Tan, and X. Lu, *J. Am. Chem. Soc.* **129**, 10636–10637 (2007).

63. C. M. Cardona, A. Kitaygorodskiy, A. Ortiz, M. A. Herranz, and L. Echegoyen, *J. Org. Chem.* **70**, 5092–5097 (2005).

64. T. Cai, Z. Ge, E. B. Iezzi, T. E. Glass, K. Harich, H. W. Gibson, and H. C. Dorn, *Chem. Commun.* 3594–3596 (2005).

65. L. Echegoyen, C. J. Chancellor, C. M. Cardona, B. Elliott, J. Rivera, M. M. Olmstead, and A. L. Balch, *Chem. Commun.* 2653–2655 (2006).

66. C. M. Cardona, A. Kitaygorodskiy, and L. Echegoyen, *J. Am. Chem. Soc.* **127**, 10448–10453 (2005).

67. N. Chen, E. Y. Zhang, K. Tan, C. R. Wang, and X. Lu, *Org. Lett.* **9**, 2011–2013 (2007).

68. T. Cai, C. Slebodnick, L. Xu, K. Harich, T. E. Glass, C. Chancellor, J. C. Fettinger, M. M. Olmstead, A. L. Balch, H. W. Gibson, and H. C. Dorn, *J. Am. Chem. Soc.* **128**, 6486–6492 (2006).

69. T. Cai, L. Xu, H. W. Gibson, H. C. Dorn, C. J. Chancellor, M. M. Olmstead, and A. L. Balch, *J. Am. Chem. Soc.* **129**, 10795–10800 (2007).

70. N. Martín, M. Altable, S. Filippone, A. Martín-Domenech, L. Echegoyen, and C. M. Cardona, *Angew. Chem. Int. Ed. Engl.* **45**, 110–114 (2006).

71. E. B. Iezzi, J. C. Duchamp, K. Harich, T. E. Glass, H. M. Lee, M. M. Olmstead, A. L. Balch, and H. C. Dorn, *J. Am. Chem. Soc.* **124**, 524–525 (2002).

72. H. M. Lee, M. M. Olmstead, E. Iezzi, J. C. Duchamp, H. C. Dorn, and A. L. Balch, *J. Am. Chem. Soc.* **124**, 3494–3495 (2002).
73. S. Stevenson, R. R. Stephen, T. M. Amos, V. R. Cadorette, J. E. Reid, and J. P. Phillips, *J. Am. Chem. Soc.* **127**, 12776–12777 (2005).
74. Y. Maeda, J. Miyashita, T. Hasegawa, T. Wakahara, T. Tsuchiya, T. Nakahodo, T. Akasaka, N. Mizorogi, K. Kobayashi, S. Nagase, T. Kato, N. Ban, H. Nakajima, and Y. Watanabe, *J. Am. Chem. Soc.* **127**, 12190–12191 (2005).
75. Y. Maeda, S. Sato, K. Inada, H. Nikawa, M. yamada, N. Mizorogi, T. Hasegawa, T. Tsuchiya, T. Akasaka, T. Kato, Z. Slanina and S. Nagase, *Chem. Eur. J.* **16**, 2193–2197 (2010).
76. Z. Ge, J. C. Duchamp, T. Cai, H. W. Gibson, and H. C. Dorn, *J. Am. Chem. Soc.* **127**, 16292–16298 (2005).
77. S. Stevenson, K. Harich, H. Yu, R. R. Stephen, D. Heaps, C. Coumbe, and J. P. Phillips, *J. Am. Chem. Soc.* **128**, 8829–8835 (2006).
78. S. Stevenson, M. A. Mackey, C. E. Coumbe, J. P. Phillips, B. Elliott, and L. Echegoyen, *J. Am. Chem. Soc.* **129**, 6072–6073 (2007).
79. X. Lu, J. Xu, X. He, Z. Shi, and Z. Gu, *Chem. Mater.* **16**, 953–955 (2004).
80. Y. Maeda, Y. Matsunaga, T. Wakahara, S. Takahashi, T. Tsuchiya, M. O. Ishitsuka, T. Hasegawa, T. Akasaka, M. T. H. Liu, K. Kokura, E. Horn, K. Yoza, T. Kato, S. Okubo, K. Kobayashi, S. Nagase, and K. Yamamoto, *J. Am. Chem. Soc.* **126**, 6858–6859 (2004).
81. Y. Matsunaga, Y. Maeda, T. Wakahara, T. Tsuchiya, M. O. Ishitsuka, T. Akasaka, N. Mizorogi, K. Kobayashi, S. Nagase, and K. M. Kadish, *ITE Letters on Batteries, New Technologies & Medicine* **7**, 43–49 (2006).
82. Y. Takano, M. Aoyagi, M. Yamada, H. Nikawa, Z. Slanina, N. Mizorogi, M. O. Ishitsuka, T. Tsuchiya, Y. Maeda, T. Akasaka, T. Kato, and S. Nagase, *J. Am. Chem. Soc.* **131**, 9340–9346 (2009).
83. T. Akasaka, M. T. H. Liu, Y. Niino, Y. Maeda, T. Wakahara, M. Okamura, K. Kobayashi, and S. Nagase, *J. Am. Chem. Soc.* **122**, 7134–7135 (2000).
84. T. Akasaka, T. Kono, Y. Matsunaga, T. Wakahara, T. Nakahodo, M. O. Ishitsuka, Y. Maeda, T. Tsuchiya, T. Kato, M. T. H. Liu, N. Mizorogi, Z. Slanina, and S. Nagase, *J. Phys. Chem. A* **112**, 1294–1297 (2008).
85. T. Akasaka, T. Kono, Y. Takematsu, H. Nikawa, T. Nakahodo, T. Wakahara, M. O. Ishitsuka, T. Tsuchiya, Y. Maeda, M. T. H. Liu, K. Yoza, T. Kato, K. Yamamoto, N. Mizorogi, Z. Slanina, and S. Nagase, *J. Am. Chem. Soc.* **130**, 12840–12841 (2008).

86. B. Cao, H. Nikawa, T. Nakahodo, T. Tsuchiya, Y. Maeda, T. Akasaka, H. Sawa, Z. Slanina, N. Mizorogi, and S. Nagase, *J. Am. Chem. Soc.* **130**, 983–989 (2008).

87. Y. Iiduka, T. Wakahara, K. Nakajima, T. Nakahodo, T. Tsuchiya, Y. Maeda, T. Akasaka, K. Yoza, M. T. H. Liu, N. Mizorogi, and S. Nagase, *Angew. Chem. Int. Ed. Engl.* **46**, 5562–5564 (2007).

88. Y. Iiduka, T. Wakahara, T. Nakahodo, T. Tsuchiya, A. Sakuraba, Y. Maeda, T. Akasaka, K. Yoza, E. Horn, T. Kato, M. T. H. Liu, N. Mizorogi, K. Kobayashi, and S. Nagase, *J. Am. Chem. Soc.* **127**, 12500–12501 (2005).

89. X. Lu, H. Nikawa, T. Nakahodo, T. Tsuchiya, M. O. Ishitsuka, Y. Maeda, T. Akasaka, M. Toki, H. Sawa, Z. Slanina, N. Mizorogi, and S. Nagase, *J. Am. Chem. Soc.* **130**, 9129–9136 (2008).

90. X. Lu, H. Nikawa, T. Tsuchiya, Y. Maeda, M. O. Ishitsuka, T. Akasaka, M. Toki, H. Sawa, Z. Slanina, N. Mizorogi, and S. Nagase, *Angew. Chem. Int. Ed. Engl.* **47**, 8642–8645 (2008).

91. T. Tsuchiya, R. Kumashiro, K. Tanigaki, Y. Matsunaga, M. O. Ishitsuka, T. Wakahara, Y. Maeda, Y. Takano, M. Aoyagi, T. Akasaka, M. T. H. Liu, T. Kato, K. Suenaga, J. S. Jeong, S. Iijima, F. Kimura, T. Kimura, and S. Nagase, *J. Am. Chem. Soc.* **130**, 450–451 (2008).

92. A. Hirsch, I. Lamparth, and H. R. Karfunkel, *Angew. Chem. Int. Ed. Engl.* **33**, 437–438 (1994).

93. I. Lamparth and A. Hirsch, *J. Chem. Soc. Chem. Commun.* 1727–1728 (1994).

94. J. F. Nierengarten, V. Gramlich, F. Cardullo, and F. Diederich, *Angew. Chem. Int. Ed. Engl.* **35**, 2101–2103 (1996).

95. O. Lukoyanova, C. M. Cardona, J. Rivera, L. Z. Lugo-Morales, C. J. Chancellor, M. M. Olmstead, A. Rodriguez-Fortea, J. M. Poblet, A. L. Balch, and L. Echegoyen, *J. Am. Chem. Soc.* **129**, 10423–10430 (2007).

96. P. P. Fatouros, F. D. Corwin, Z. J. Chen, W. C. Broaddus, J. L. Tatum, B. Kettenmann, Z. Ge, H. W. Gibson, J. L. Russ, A. P. Leonard, J. C. Duchamp, and H. C. Dorn, *Radiology* **240**, 756–764 (2006).

97. T. Cai, L. Xu, C. Shu, H. A. Champion, J. E. Reid, C. Anklin, M. R. Anderson, H. W. Gibson, and H. C. Dorn, *J. Am. Chem. Soc.* **130**, 2136–2137 (2008).

98. M. N. Chaur, F. Melin, A. J. Athans, B. Elliott, K. Walker, B. C. Holloway, and L. Echegoyen, *Chem. Commun.* 2665–2667 (2008).

99. L. Feng, T. Nakahodo, T. Wakahara, T. Tsuchiya, Y. Maeda, T. Akasaka, T. Kato, E. Horn, K. Yoza, N. Mizorogi, and S. Nagase, *J. Am. Chem. Soc.* **127**, 17136–17137 (2005).

100. L. Feng, T. Wakahara, T. Nakahodo, T. Tsuchiya, Q. Piao, Y. Maeda, Y. Lian, T. Akasaka, E. Horn, K. Yoza, T. Kato, N. Mizorogi, and S. Nagase, *Chem. Eur. J.* **12**, 5578–5586 (2006).
101. L. Feng, T. Tsuchiya, T. Wakahara, T. Nakahodo, Q. Piao, Y. Maeda, T. Akasaka, T. Kato, K. Yoza, E. Horn, N. Mizorogi, and S. Nagase, *J. Am. Chem. Soc.* **128**, 5990–5991 (2006).
102. N. B. Shustova, A. A. Popov, M. A. Mackey, C. E. Coumbe, J. P. Phillips, S. Stevenson, S. H. Strauss, and O. V. Boltalina, *J. Am. Chem. Soc.* **129**, 11676–11677 (2007).
103. N. Tagmatarchis, A. Taninaka, and H. Shinohara, *Chem. Phys. Lett.* **355**, 226–232 (2002).
104. C. Shu, C. Slebodnick, L. Xu, H. Champion, T. Fuhrer, T. Cai, J. E. Reid, W. Fu, K. Harich, H. C. Dorn, and H. W. Gibson, *J. Am. Chem. Soc.* **130**, 17755–17760 (2008).
105. I. E. Kareev, S. F. Lebedkin, V. P. Bubnov, E. B. Yagubskii, I. N. Ioffe, P. A. Khavrel, I. V. Kuvychko, S. H. Strauss, and O. V. Boltalina, *Angew. Chem. Int. Ed. Engl.* **44**, 1846–1849 (2005).
106. C. Shu, T. Cai, L. Xu, T. Zuo, J. Reid, K. Harich, H. C. Dorn, and H. W. Gibson, *J. Am. Chem. Soc.* **129**, 15710–15717 (2007).
107. Y. Takano, A. Yomogida, H. Nikawa, M. Yamada, T. Wakahara, T. Tsuchiya, M. O. Ishitsuka, Y. Maeda, T. Akasaka, T. Kato, Z. Slanina, N. Mizorogi, and S. Nagase, *J. Am. Chem. Soc.* **130**, 16224–16230 (2008).
108. T. Wakahara, H. Nikawa, T. Kikuchi, T. Nakahodo, G. M. A. Rahman, T. Tsuchiya, Y. Maeda, T. Akasaka, K. Yoza, E. Horn, K. Yamamoto, N. Mizorogi, Z. Slanina, and S. Nagase, *J. Am. Chem. Soc.* **128**, 14228–14229 (2006).
109. H. Nikawa, T. Kikuchi, T. Wakahara, T. Nakahodo, T. Tsuchiya, G. M. A. Rahman, T. Akasaka, Y. Maeda, K. Yoza, E. Horn, K. Yamamoto, N. Mizorogi, and S. Nagase, *J. Am. Chem. Soc.* **127**, 9684–9685 (2005).

Chapter 8

Quantum Computing with Endohedral Fullerenes

Kyriakos Porfyrakis and Simon C. Benjamin
Department of Materials, University of Oxford, Parks Road, Oxford OX1 3PH, U.K.

8.1 Introduction

This chapter begins with a short introduction to the topic of quantum information processing. We presume only a very limited familiarity with quantum mechanics; in fact, all the really essential ideas, terminology and formalism will be introduced as we go along. The introduction will discuss classical and quantum information, the qubit, entanglement, and the basic operations of quantum computer. It will include a brief look at how accelerated searching can be performed, and finally a discussion of the problem of decoherence and how to fight it. We will then introduce endohedral fullerenes, the key molecular building blocks that have been identified as promising components of a quantum technology. We will describe the synthesis of these structures, together with the key experimental demonstrations of quantum phenomena.

Advances in Carbon Nanomaterials: Science and Applications
Edited by Nikos Tagmatarchis

ISBN 978-981-426-78-78 (Hardcover), 978-981-426-78-85 (eBook)
www.panstanford.com

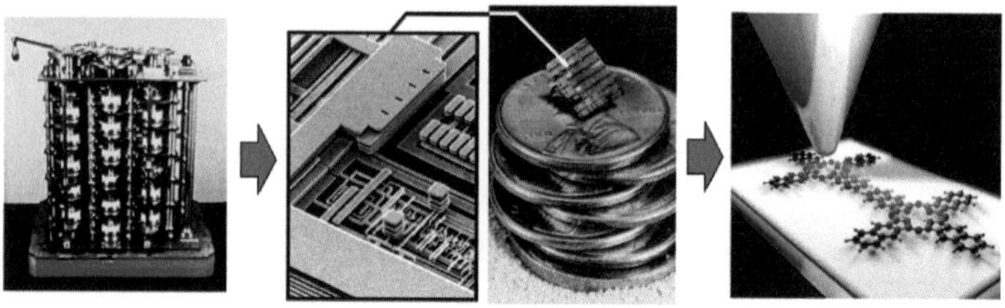

Figure 8.1. Evolution of the computer. The rightmost image shows a molecular network that might represent and manipulate information; but would it be a quantum computer? See also Color Insert.

When is a computer a *quantum* computer (see Fig. 8.1)? Our first order of business is to say what a quantum computer actually is, and what it is *not*. This basic point is something that is often misunderstood. The short definition is that *a quantum computer is a device capable of processing quantum information*. In order to understand this must think about the nature of information.

8.2 Classical Information

First let us talk about information as it was discussed by computer scientists long before the field of quantum computing arose. We will use the phrase "classical information" here, because the term *classical* should be employed whenever we want to talk about the nonquantum version of a theory — for example, classical mechanics and so on.

When computer scientists talk about information, they are dealing with how knowledge can be represented symbolically. We are using "knowledge" in a general sense that includes text and equations of course, but also images, movies and music. In practice, all these things can be represented as a stream of symbols (at least to some adequate approximation).

An important point is that any finite set of symbols can be translated into just two symbols, and this alternative representation is *efficient*. When we do this, it is conventional to use **0** and **1** as our two symbols, and to refer to any given symbol as a bit. Take the alphabet: Suppose we allow up to 64 unique symbols, to include all the upper case and lower case letters and the punctuation characters. Then we could replace each symbol with a unique string of six bits, since there are $2^6 = 64$ possible permutations. When we want to encode numbers into a series of bits, rather than translating each familiar symbol $0, 1, \ldots 8, 9$ into a series of bits, we instead use a rule system such as binary that assigns a unique number to every possible series of bits. Such a scheme has *perfect* efficiency, using no more bits than are absolutely necessary to represent a given range of numbers.

How about images? Here we can adequately encode by breaking the image up into a fine grid of points, or pixels, and assigning a pure color to each pixel. This process of approximating continuous media (images, sound, movies) as a series of symbols is called digitization, and the stored entity is said to be digital.

Since we can store any of these forms of knowledge as a stream of bits, we say that **the bit is the fundamental unit of classical information**.

Where is the physics?

Notice that we have not mentioned physics at all in our discussion of classical information. And indeed, classical information theory is a branch of applied mathematics — it does not employ the laws of physics. However, part of the motivation for viewing information as a stream of bits comes from the practical issues that we encounter when we think of building an information processing *machine*. In essence, it is easier to design a fundamental component that has two stable states than a component with ten stable states (say). The fact that we will need a larger number of these simple components is a relatively minor consideration. Let us consider how conventional computers store their bits.

8.3 Information Inside a *Classical* Computer

Modern computers store bits in different forms at different times (see Fig. 8.2). The representation depends on whether the information is being *stored* (short or long term), *transmitted* over some distance, or *processed*. Take the example of long-term storage. This

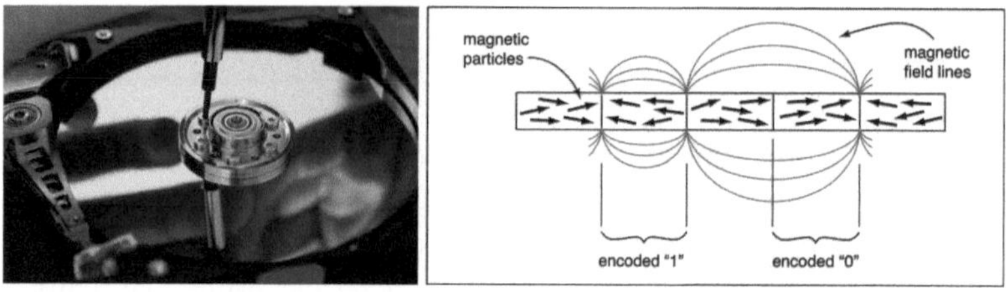

Figure 8.2. Classical information storage on a hard disk drive. See also Color Insert.

is usually done using a magnetic disk: essentially a surface that has been coated by a thin layer of magnetic particles. The surface is divided up into little areas, less than a micron in size, and each such area stores a single bit through the collective orientation of the many magnetic particles within. The value of the bit is set when a magnetic head comes and imposes an orientation on the particles by applying a strong field. Similarly, the value of the bit is read by a detecting the weak magnetic field generated by the aligned particles.

Because many particles are being used to store a single bit, there are actually a large number of states of the physical system that are all called "0" and similarly a large number that are all called "1". There are also many states that do not correspond to a clear majority of particles being aligned in either direction. These states have *no meaning* in terms of representing classical information; if the computer is operating successfully then such states will only occur transiently, as the bit value of the collective is switched from one valid state to another.

Is this wasteful? Well it gets harder to control systems as they get smaller, and read/writing the alignment of these tiny patches is the best our technology can manage at the moment! But more importantly, there is an inherent *robustness* that comes from using a large number of particles to encode a single bit. If the alignment of the group of particles suffers some degradation, perhaps because of heating or some stray magnetic field, there is still an excellent chance that we will be able to determine the original value of the bit that was stored there. Robustness to noise is, in fact, a crucial issue in any real device.

8.4 Introducing the Quantum Bit, or Qubit

Suppose we set aside technical limitations, and advantages such as robustness. What then is the fundamental requirement for a physical system to be able to store a bit *in principle*? Well we must have an entity possessing two states that can be *reliably distinguished by measurement* given a sufficiently advanced, but physically possible, technology. Notice that stability is less fundamental in the sense that, if we are fast enough we can always use an entity within the time scale of its stability. But the requirement for two distinguishable states is absolute.

But now we come to an interesting insight. The real universe is apparently governed by quantum mechanics, and when we use that theory to describe a system with two reliably distinguishable states we discover something remarkable: Any such system actually has an infinity of possible states. Anything that can store a bit is, in fact, much richer than that. The richer entity is called the quantum bit, or qubit for short, and this is the real unit of physical information.

We now begin to see what it might mean to speak of a machine that can process *quantum information*: whereas ordinary machines store and process bits, the new machine must be able to store and process these "richer" entities called qubits. Crucially, a device does not become a quantum computer merely by being composed of sufficiently small components; if a device built at the atomic scale were incapable of successfully processing qubits, then it would remain a classical computer. Indeed, processes in living systems (such as DNA replication) are sometimes seen as an information processing, but it is a classical molecular scale processing rather than quantum computing.

The theory of quantum information processing (QIP) is *the general theory* of information processing with real physical systems. Classical information processing is what happens when the system only uses a limited portion of what QIP allows. Of course, in practice this design choice has really been a necessity: we are now trying very hard to create a machine that is not limited to classical information processing, and it is very difficult!

8.5 Understanding the Qubit: The Bloch Sphere

The Schrödinger equation is the key equation in (nonrelativistic) quantum mechanics. The compact form is,

$$\hat{H}\,|\psi(t)\rangle = i\hbar\frac{d}{dt}|\psi(t)\rangle$$

Here the symbol $|\ \rangle$ is called a *ket* and is used to mean, "the state of the quantum system." The ψ inside is just a label; it is there since we often need to talk about several different states and so we need to label them. We may put t in there just to emphasize that the state of the system will, in general, depend on time, i.e., it will change or *evolve*. Obviously the $\frac{d}{dt}$ is just a time derivative, and finally the $\hat{H}$ is an operator called the *Hamiltonian* that represents the energy of the system (more about that later). So reading right-to-left the equation tells us that the way a system's state changes in time depends on the energy of that state.

Suppose that a system that has two distinct energy "eigenstates": when the system is in one of these states, a later measurement will always find the same energy. This means it could represent one classical bit.

Using our ket notation for states, let us write $|0\rangle$ and $|1\rangle$ to denote these two different states at some particular time $t = 0$. Suppose that the energies of these states are E_0 and E_1, respectively. Recall that the Hamiltonian $\hat{H}$ is precisely the operator that tells us the energy of a system; formally we write this as

$$\hat{H}\,|0\rangle = E_0|0\rangle \qquad \hat{H}\,|1\rangle = E_1|1\rangle.$$

But the Schrödinger equation says that for any state $|\psi\rangle$,

$$\hat{H}\,|\psi\rangle = i\hbar\frac{d}{dt}|\psi\rangle,$$

so in order to satisfy this equation, at later times our special states will be

$$e^{-iE_0t/\hbar}|0\rangle \quad \text{and} \quad e^{-iE_1t/\hbar}|1\rangle,$$

respectively. So they simply acquire a complex phase over time.

However, these special "solutions" are by no means the *only* solutions, the only states that are physically allowed by the

Schrödinger equation. In fact, since the Schrödinger equation is *linear*, we can make new solutions just by *adding* previous solutions. Consider a state $|\psi\rangle$ that is a *linear superposition* of our two eigenstates, weighted by any two complex numbers α and β

$$|\psi(t)\rangle = \alpha e^{-iE_0 t/\hbar}|0\rangle + \beta e^{-iE_1 t/\hbar}|1\rangle.$$

We can quickly confirm that this also satisfies the Schrödinger equation:

$$\begin{aligned}\hat{H}|\psi(t)\rangle &= \alpha e^{-iE_0 t/\hbar}\hat{H}|0\rangle + \beta e^{-iE_1 t/\hbar}\hat{H}|1\rangle \\ &= E_0\alpha e^{-iE_0 t/\hbar}|0\rangle + E_1\beta e^{-iE_1 t/\hbar}|1\rangle \\ &= i\hbar\frac{\partial}{\partial t}|\psi(t)\rangle\end{aligned}$$

This new state $|\psi\rangle$ is not an energy eigenstate; if we measure its energy it will randomly "collapse" to either of the two possible eigenstates, with probabilities proportional to $|\alpha|^2$ and $|\beta|^2$, respectively. But it is crucial to understand that the state does not contain any inherent uncertainty — until we measure it, its behavior is completely deterministic. It is simply one possible state that the system can be in.

But in that case, how many different states of a single qubit *are there*? Well, an infinite number! In fact, it takes two real numbers to specify the state of a qubit. Why not four, when α and β are both complex numbers? It turns out that there are two considerations that each remove one real number from the description of the qubit. The first is that we should normalize the state so that the probability of collapsing to $|0\rangle$ when measured, *plus* the probability of collapsing to $|1\rangle$, sums to unity. It must do one or the other! Thus $|\alpha|^2+|\beta|^2 = 1$.

The second condition has to do with the total phase. It turns out that if you multiply the state of a complete physical system by any phase factor $\exp(i\theta)$, then, in fact, the new state is identical according to every possible measurement. In other words, this external phase factor is meaningless; if $|S\rangle$ is the total state of a system then $|S\rangle$ and $\exp(i\theta)|S\rangle$ are the same thing.[1]

[1]Note that if we enlarge the system by bringing in a new particle, the total phase of each particle is now meaningful because it determines the relative phase between them. It is only the total phase of the *whole* system that is not meaningful.

Consider again our general qubit at time $t = 0$, written with two complex numbers $|\psi(t = 0)\rangle = \alpha|0\rangle + \beta|1\rangle$. In light of the above remarks it is sometimes useful to write such a state in a way that clearly has only two free parameters. For example,

$$|\psi(t = 0)\rangle = \cos\left(\frac{\theta}{2}\right)|0\rangle + \sin\left(\frac{\theta}{2}\right)\exp(i\phi)|1\rangle,$$

where we can restrict the parameters to the ranges $0 \le \theta \le \pi$ and $0 \le \phi < 2\pi$. Here we have chosen the total phase such that $|0\rangle$ has no phase prefactor, and of course $\cos^2(\theta/2) + \sin^2(\theta/2) = 1$ so the state is normalized.

Now it turns out that when a qubit is written in this form, the parameters θ and ϕ have a very beautiful geometric interpretation. They can be regarded as the angular coordinates for a point on the surface of a sphere, as in the following image.

We call this object the *Bloch sphere* a schematic of which is shown in Fig. 8.3. Every point on the surface of the sphere is a unique state of a qubit. Moreover, we will find that as qubits evolve in time, the

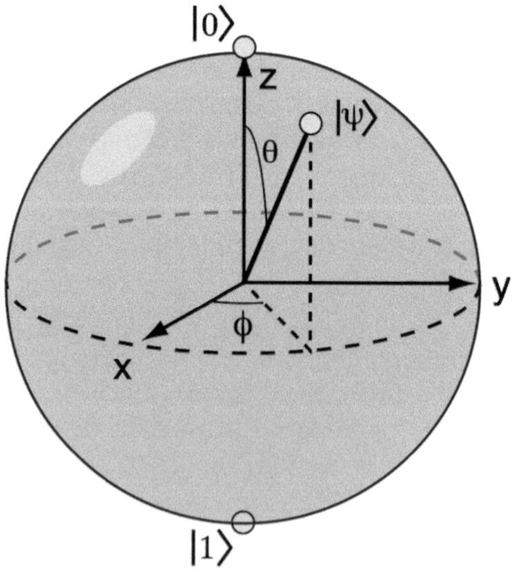

Figure 8.3. Schematic illustration of the Bloch sphere, which is commonly used to represent quantum states. See also Color Insert.

point representing the qubit's state rotates around the sphere. It is a very beautiful and helpful picture of what a qubit *is*.

8.6 More Than One Qubit: Entanglement

Suppose you prepare a physical qubit in state $|\psi_A\rangle = \alpha_A|0\rangle + \beta_A|1\rangle$, meanwhile, I prepare my own in $|\psi_B\rangle = \alpha_B|0\rangle + \beta_B|1\rangle$. How should we write the state of the combined system of two qubits? Very easily — as a product:

$$(\alpha_A|0\rangle + \beta_A|1\rangle)\,(\alpha_B|0\rangle + \beta_B|1\rangle) \tag{8.1}$$

where we understand that the ket on the left is for system A and the one on the right is for system B. For obvious reasons, Eq. (8.1) is called a *product state* of two qubits. We can multiply this out to write it as

$$\begin{aligned} &\alpha_A\alpha_B|0\rangle|0\rangle + \alpha_A\beta_B|0\rangle|1\rangle + \beta_A\alpha_B|1\rangle|0\rangle + \beta_A\beta_B|1\rangle|1\rangle \\ \text{or simply}\quad &\alpha_A\alpha_B|00\rangle + \alpha_A\beta_B|01\rangle + \beta_A\alpha_B|10\rangle + \beta_A\beta_B|11\rangle. \end{aligned} \tag{8.2}$$

Now, suppose we write a two-qubit state with some general constants, thus:

$$c_{00}|00\rangle + c_{01}|01\rangle + c_{10}|10\rangle + c_{11}|11\rangle. \tag{8.3}$$

We could pick some constants that satisfy the normalization condition $|c_{00}|^2 + |c_{01}|^2 + |c_{10}|^2 + |c_{11}|^2 = 1$ *but* that cannot be factored into the form of (8.2) and hence back to Eq. (8.1). A simple example is

$$\frac{1}{2}(|00\rangle + |01\rangle + |10\rangle - |11\rangle). \tag{8.4}$$

Because it cannot be separated into a product state, this state is said to be *entangled*. In fact, most states we could randomly write down are entangled; the product states are the exception. Product states can be understood in terms of two completely separate systems; entangled states cannot. This is a fundamentally quantum mechanical phenomenon, without a classical analog. Entanglement is crucial to QIP, since entangled states arise in all the important quantum algorithms.

8.7 Basic Components of a Processor

We will now think about the basic building blocks from which we can compose an algorithm. Both in the case of classical information processing and in QIP, we identify a set of elementary manipulations called *gates* in terms of which any algorithm can be composed. One way to approach the quantum gates is to first review the classical gates, and then generalize them. So we start with classical logic gates.

8.7.1 *Elements of a Classical Processor*

In Fig. 8.4 we depict a few of the well-known classical gates. The names of the gates make sense when we regard **0** as "no" and **1** as "yes." The NOT gate is the simplest — It simply inverts the bit that it receives, outputting a **1** when given a **0** and vice versa. The OR gate is a simple processor of information: it returns **1** if *either* of its inputs are **1**. Meanwhile, the gate called NAND (meaning NOT-AND) returns a **1** in all cases except when *both* inputs are **1**.

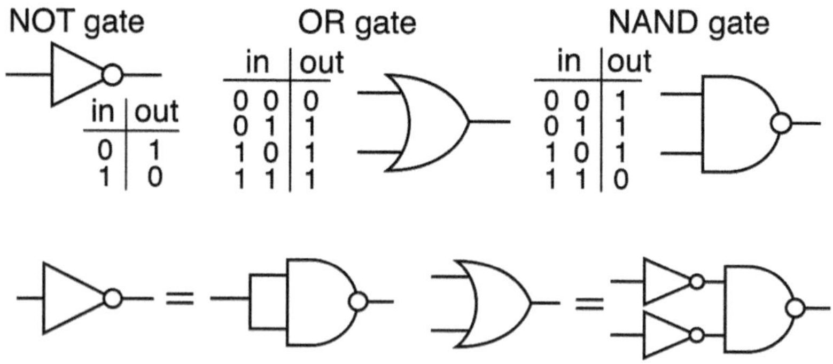

Figure 8.4. Three classical gates, and a means to construct NOT and OR purely from NAND gates.

Generally any classical computer processing task can be broken down into a series of elementary gates acting on one or two bits at a time. Now if we are to actually build a computing machine, we must work out how to physically implement each type of gate that we are going to use. Obviously, a gate involves some kind of interaction between the bits involved in the gate, and the implementation, therefore, depends on how we represent the bits physically. The usual example is high/low potentials interacting through a transistor.

But how many different kinds of gate do we need? We depict three in the figure. Is three enough for any circuit? Indeed, are three required or would one have sufficed?

A set of gates that are sufficient to efficiently implement any algorithm is called a *universal set*. It should be possible to build the other basic gates using a finite number of gates from the universal set. For classical computing, one possible universal set is the pair of gates OR and NOT. However, one can make do with just the NAND gate instead — that gate constitutes a universal set on its own.

8.7.2 *A Notation for Qubits*

Now we wish to develop an analogous picture for the essential *quantum* logic gates. Before we can discuss those gates, we need to develop a suitable notation for arrays of qubits.

Previously we introduced the *ket* notation for quantum states. Given two specific states of a qubit we can describe any state it may be in. The two states that we choose to use as our reference states form a *basis*, just as the elementary position vectors **i** and **j** form a basis for writing any vector in the x-y plane. Most often, we choose our basis states to be the energy eigenstates of an isolated qubit. We label our two eigenstates $|0\rangle$ and $|1\rangle$ and we refer to this as the *computational basis*.

It is convenient to represent the state of a qubit as a vector of length two, with the two entries corresponding to the amplitudes of $|0\rangle$ and of $|1\rangle$ within that state.

$$|0\rangle \rightarrow \begin{pmatrix} 1 \\ 0 \end{pmatrix} \qquad |1\rangle \rightarrow \begin{pmatrix} 0 \\ 1 \end{pmatrix} \qquad |\psi\rangle = \alpha|0\rangle + \beta|1\rangle \rightarrow \begin{pmatrix} \alpha \\ \beta \end{pmatrix}$$

How about an array of qubits? If we have N qubits then we need a total of 2^N basis states to describe a general state of that array. We can simply write the amplitudes as a vector, but we must always remember the basis we have chosen, including the *order* in which we are listing the basis states. Let us look at the example of a two qubit system.

$$\begin{aligned} |\psi\rangle_{two} &= c_0|0\rangle|0\rangle + c_1|0\rangle|1\rangle + c_2|1\rangle|0\rangle + c_3|1\rangle|1\rangle \\ &= c_0|00\rangle + c_1|01\rangle + c_2|10\rangle + c_3|11\rangle \\ &\rightarrow \begin{pmatrix} c_0 \\ c_1 \\ c_2 \\ c_3 \end{pmatrix} \quad \text{“ in the basis } \{|00\rangle, |01\rangle, |10\rangle, |11\rangle\} \text{ ”} \end{aligned}$$

8.7.3 *Single-Qubit Gates*

Recall that there was only one nontrivial classical gate that takes a single input bit, i.e., the NOT gate. We will we want our quantum computer to be able to do anything a classical computer could do, so let us start by writing a NOT operation in our notation. We will want $|0\rangle \Rightarrow |1\rangle$ and simultaneously $|1\rangle \Rightarrow |0\rangle$, which means

$$\begin{pmatrix} 1 \\ 0 \end{pmatrix} \Rightarrow \begin{pmatrix} 0 \\ 1 \end{pmatrix} \quad \text{and} \quad \begin{pmatrix} 0 \\ 1 \end{pmatrix} \Rightarrow \begin{pmatrix} 1 \\ 0 \end{pmatrix}$$

Operations on vectors can be written as a matrix. There is only one matrix that can do our NOT operation:

$$U_{\text{NOT}} \rightarrow \begin{pmatrix} 0 & 1 \\ 1 & 0 \end{pmatrix} \quad \text{“ in the basis } \{|0\rangle, |1\rangle\} \text{ ”}$$

Let us see what our $\hat{U}_{\text{NOT}}$ gate does (see Fig. 8.5) when we apply it to states other than $|0\rangle$ and $|1\rangle$. For a general state $|\psi\rangle = \alpha|0\rangle + \beta|1\rangle$ we find

$$\hat{U}_{\text{NOT}}|\psi\rangle = \beta|0\rangle + \alpha|1\rangle.$$

Let us take a look at this on the Bloch sphere, by writing $|\psi\rangle$ and $\hat{U}_{\text{NOT}}|\psi\rangle$ in the usual angular parameters θ and ϕ.

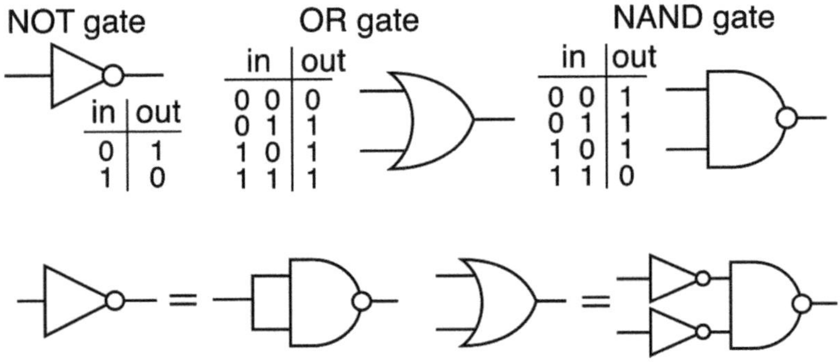

Figure 8.5. The effect of $\hat{U}_{\mathrm{NOT}}$.

$$
\begin{aligned}
|\psi\rangle &= \cos(\theta/2)|0\rangle + \sin(\theta/2)e^{i\phi}|1\rangle. \\
\hat{U}_{\mathrm{NOT}}|\psi\rangle &= \sin(\theta/2)e^{i\phi}|0\rangle + \cos(\theta/2)|1\rangle \\
&= \sin(\theta/2)|0\rangle + e^{-i\phi}\cos(\theta/2)|1\rangle \text{ discarding a global phase.}
\end{aligned}
$$

We note that the phase factor ϕ, which determines our longitude on the Bloch sphere, has gone to $-\phi$. It is helpful at this point to pick a few specific points on the Bloch sphere, and see where they end up after we apply our $\hat{U}_{\mathrm{NOT}}$. Note in particular that the states $|+\rangle \equiv (|0\rangle + |1\rangle)/\sqrt{2}$ and $|-\rangle \equiv (|0\rangle - |1\rangle)/\sqrt{2}$ do not change when operated on. After enough playing around, the picture that emerges is of a rotation by π radians (180°), around the axis that passes through $|+\rangle$ and $|-\rangle$ (which we call the x-axis).

So we have found that even the simple NOT operation is more subtle when we apply it to qubits — as we would expect since qubits are richer objects than bits. But now that we are thinking in terms of rotations, we can see that other kinds of rotation might be possible besides the simple flip caused by $\hat{U}_{\mathrm{NOT}}$. For example, can we rotate around the same axis, but by a different angle, call it γ? Let us call that operation $\hat{U}_x(\gamma)$; then what will it look like?

For small γ, it would be similar to the identity matrix $\mathbb{1}$ (since we are hardly changing the system). Meanwhile, for $\gamma \approx \pi$ it will look like $\hat{U}_{\mathrm{NOT}}$. A simple guess we could write down would be to just add together $\mathbb{1}$ and $\hat{\sigma}_x$ in a way that varies with γ, say

$$
\text{how about} \quad \hat{U}_x(\gamma) = \cos(\gamma/2)\mathbb{1} + \sin(\gamma/2)\hat{U}_{\mathrm{NOT}} \quad ?
$$

Actually this is almost correct! In fact, all we need is to insert a factor of i to make the operation satisfy a property called be *unitarity*:

$$\hat{U}_x(\gamma) = \cos(\gamma/2)\mathbb{1} + i\,\sin(\gamma/2)\hat{U}_{\mathrm{NOT}}$$

This a smoothly rotating version of our NOT operator. It contains our previous operator as a special case, up to a global phase: $\hat{U}_x(\pi) = i\hat{U}_{\mathrm{NOT}}$. Is this all we need?

No. In fact, we need to be able to create *any* rotation of the Bloch sphere, which is the same as saying any unitary matrix. This may sound like a lot of additional gates might be needed, beyond the one we have found. But in fact, all we need is to be able to rotate about *one other* axis. Readers who may have recognized $\hat{U}_{\mathrm{NOT}}$ as the Pauli matrix σ_x will not be surprised to hear that a suitable second rotation can be written down by substituting another of the Pauli matrices, for example,

$$\hat{U}_z(\gamma) = \cos(\gamma/2)\mathbb{1} + i\,\sin(\gamma/2)\hat{\sigma}_z,$$

where

$$\sigma_z = \begin{pmatrix} 1 & 0 \\ 0 & -1 \end{pmatrix}.$$

Finally, we should take note of a particular single qubit gate that is often used, the Hadamard gate. In circuits it is usually denoted by H in a box.

$$\hat{U}_{\mathrm{Hadamard}} = \frac{\sigma_x + \sigma_z}{\sqrt{2}} = \frac{1}{\sqrt{2}}\begin{pmatrix} 1 & 1 \\ 1 & -1 \end{pmatrix}$$

So we have found mathematical expressions for the kind of elementary operations that one can perform on a quantum bit. What does this mean physically? The means to perform such a gate obviously depends on the physical nature of the qubit; however, for molecular systems, very often the appropriate qubit will be a spin: either the spin of an electron or that of an atomic nucleus. Taking the case of an electron spin, we have a natural two state quantum system — by applying a static magnetic field, the spin states that are parallel and antiparallel to the field, normally called "up" and "down," have distinct energies. Then these eigenstates are the natural choice for our qubit $|0\rangle$ and $|1\rangle$, and we can rotate the spin by applying an oscillatory magnetic field. Typically in experiments of this kind, the static field might be of magnitude a few Tesla, in which case the frequency necessary to rotate the spin, i.e., to perform our gate, will be of order hundreds of GHz. By varying the phase of the oscillatory field, or by using small detunings, we can perform the complete set of rotations of our qubit, i.e., we can physically implement the gates that we have derived in this section.

For nuclear spins, the story is exactly analogous, except that the frequency required is orders of magnitude less: radio frequency rather than microwave frequency given a static field in the Tesla range.

8.7.4 *Two-Qubit Gates*

We found that the simple NOT gate is replaced by a whole range of rotations when we consider single *qubit* gates. Fortunately, if we have such a set of rotations available, then we only need a single two-qubit gate to complete the *universal set*. The typical notation for various two-qubit gates is shown in Fig. 8.6.

What properties should this gate have? Can we start from a classical gate, like AND, and generalize it? No, because the classical gates took two bits as input and gave one bit as output. If we try this with qubits, which would generally be in some superposition state, then we will be reducing the number of states in the superposition every time we apply a logical gate. Instead, we should look for a form of gate that is two-in, two-out.

How would we write such a gate, i.e., what would the operator be like? Well we have seen that the state of a two-qubit system can be written as a vector of four numbers. Our operator will transform such a vector into *another* vector of four numbers. Thus it will be a 4×4 matrix. We will need to choose a particular *basis* for this matrix when we write it down, and naturally, we will choose the basis $\{|00\rangle, |01\rangle, |10\rangle, |11\rangle\}$. Rather as we did for our quantum generalization of the NOT gate, we can now think about which output state we want, given each of the four basis states as input. We can build up our 4×4 matrix that way, providing that the complete matrix is *unitary*.

Now fortunately the property we require for our real two-qubit gate is very simple: it must be an *entangling* gate. Using this gate, we need to be able to go from an initial state that is a simple *product state* of two qubits, into a new state that cannot be written that way.

Now we will construct a suitable gate, by thinking about how a state could get entangled. Suppose that we start from state

$$|\psi\rangle = |+\rangle|0\rangle = \frac{|00\rangle + |10\rangle}{\sqrt{2}}.$$

This state has no entanglement. But we can write a similar looking state, that is entangled (*maximally* entangled in fact):

$$|\psi\rangle_{ent} = \frac{|00\rangle + |11\rangle}{\sqrt{2}}.$$

To go from one to the other, we would like a gate that maps states as follows:

$$|00\rangle \Rightarrow |00\rangle \quad |10\rangle \Rightarrow |11\rangle$$

now we do not care what happens to $|01\rangle$ or $|11\rangle$, but We will have to make a choice for them that makes the overall matrix *unitary*. The easiest choice turns out to be

$$|01\rangle \Rightarrow |01\rangle \quad |11\rangle \Rightarrow |10\rangle,$$

for which the complete matrix is

$$\hat{U}_{\text{CNOT}} = \begin{pmatrix} 1\,0\,0\,0 \\ 0\,1\,0\,0 \\ 0\,0\,0\,1 \\ 0\,0\,1\,0 \end{pmatrix} \quad \text{in the basis } \{|00\rangle, |01\rangle, |10\rangle, |11\rangle\}$$

This gate, which we have designed to be *entangling* and that satisfies the condition of being *unitary*, is a suitable two-qubit quantum gate to complete our universal set! In fact, this gate has a name, it is called the control- NOT (or c-NOT or cNOT, etc.) because it looks like a NOT operation acting on one of the qubits, but only if the other is in state $|1\rangle$. It also gets its own symbol in quantum circuit diagrams, where the $\oplus$ marks the target qubit (the one that might be NOT'ed) and the $\bullet$ marks the control qubit.

Finally, let us mention a second possible choice for our entangling two-qubit gate. This one is arguably even simpler to write:

$$\hat{U}_{\text{cphase}} = \begin{pmatrix} 1\,0\,0 & 0 \\ 0\,1\,0 & 0 \\ 0\,0\,1 & 0 \\ 0\,0\,0 & -1 \end{pmatrix} \quad \text{in the basis } \{|00\rangle, |01\rangle, |10\rangle, |11\rangle\}$$

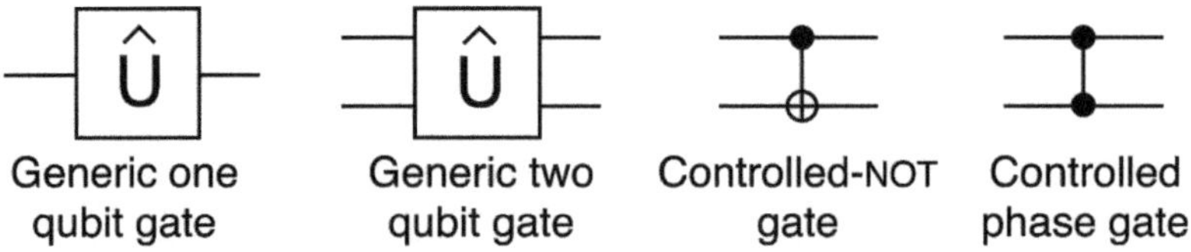

Figure 8.6. Typical notation for various two-qubit gates.

It's called the control-phase gate because you can think of it as a phase shift (or $\hat{\sigma}_z$ operation) being applied to one qubit *only if* the other is in state $|1\rangle$. But unlike the control-NOT, this time it does not matter which one you call the control and which the target — you get the same operation either way! The control-phase gate gets a special symbol, too, as shown below. It is an interesting gate to opt for when thinking how to implement operations physically, because the interactions between spins (such as the dipole–dipole interaction) naturally introduce conditional phases.

8.8 Quantum Parallelism

Now let us see what our quantum computer can do! First, We will think about how we could perform a classical algorithm on a quantum computer. This will be our starting point when generalized to a true quantum algorithm, which will hopefully run faster.

We know, of course, that a quantum system can perform classical computing — after all my laptop is ultimately governed by quantum mechanics! We could certainly describe the classical gates like AND, NAND, etc., in terms of unitary gates together with measurements. But we know that measurements can irreversibly collapse the quantum superpositions that we are expecting to use, so we might like a way of embedding a classical algorithm in a quantum device without using measurement. In other words, how can we perform operations like NAND in a reversible way?

Fortunately this was thought about quite a long time ago, before the QIP field really started. People wondered whether it is strictly necessary to erase bits to do a computation, and in answer to this, they came up with a three-in, three-out gate that can simulate any of the conventional two-in, one-out gates.

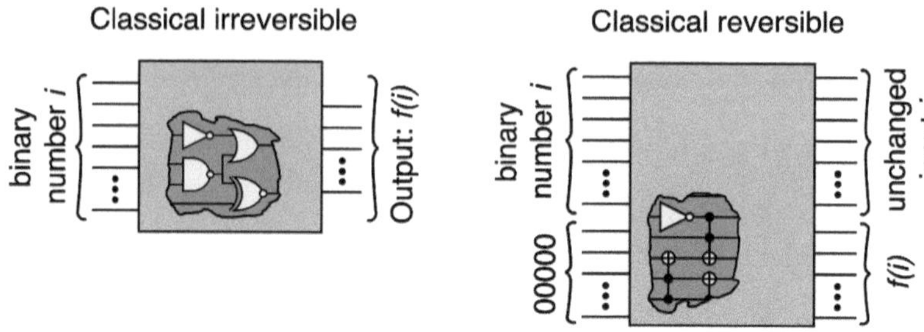

Figure 8.7. The idea of a classical *reversible* computer is a helpful bridge between conventional computers and quantum computers. See also Color Insert.

Thus, we can rewrite any classical logic circuit (and hence any classical algorithm) in terms of this gate. When we do so, it is helpful to have a separate *input register* and *output register* running right though the device. Initially, the input register obviously contains the particular input value we want to work on, and the output register is initially in an all-zero state. After the computation, the input register is unchanged, but the output register contains the output from the circuit. This is simple enough for the classical reversible computer (see Fig. 8.7), but as we will presently see, the terms *input register* and *output register* will be stretched past breaking when we cook up our quantum algorithms!

Now it proves to be straightforward to translate a reversible classical circuit into a quantum circuit: One can easily write the classical reversible gates in terms of single qubit rotations and two-qubit gates like the control-NOT. In this way, we can create a quantum logic circuit that directly performs the function of any given classical circuit. Suppose the classical circuit would take some input binary string i and produce a binary output $f(i)$; for example, the input might be two consecutive binary numbers, and the output might be their sum. Then our quantum circuit will take qubits in state $|i\rangle$ and give us output qubits in state $|f(i)\rangle$. (When writing a state in this way, we understand the numbers i and $f(i)$ to be in binary, with each digit corresponding to a successive qubit.) Formally, we can say that this algorithm will take an initial state $|In\rangle|Out\rangle = |i\rangle|000..0\rangle$ to a final state $|In\rangle|Out\rangle = |i\rangle|f(i)\rangle$

But of course that will not give us any speed up. So now let us take this circuit as a starting point and try to exploit the properties

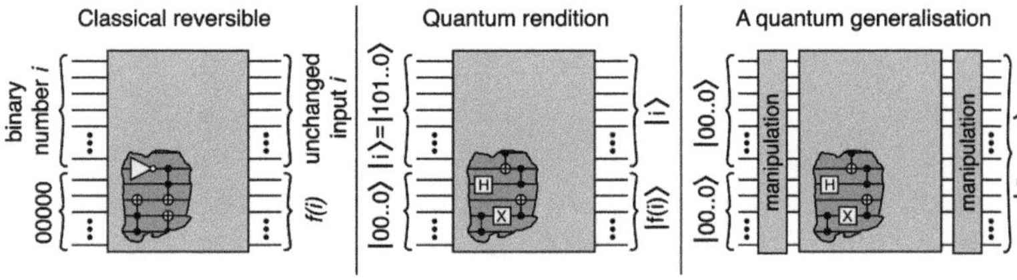

Figure 8.8. From our classical reversible computer we can obtain a design for a quantum computer that will do the same thing; then, we may be able to generalize it so that it will perform better. See also Color Insert.

of quantum mechanics more aggressively. Initially, things are going to look very promising...

Suppose that we give ourselves the ability to send in qubits in any state, not just a particular computational basis state $|i\rangle$. We can represent this in a figure by starting with qubits all in state $|0\rangle$ and performing some kind of manipulation on them *before* feeding them into the "classical" circuit. Let us take the simple case that we perform a Hadamard rotation on each qubit in the input register, taking it from state $|0\rangle$ to state $(|0\rangle + |1\rangle)/\sqrt{2}$. When we apply this single qubit rotation to each of the N qubits in the input register, we will obtain an initial state

$$C\left(\sum_{i=0}^{2^N-1} |i\rangle\right)|000...0\rangle$$

where C is just a normalization constant (actually $C = 2^{-N/2}$ in this case). So we have placed the input register into a superposition of all possible computation basis states, i.e., all the binary numbers from 0 to $2^N - 1$. What happens when we feed this state into the "classical" circuit? Well we have said that a particular state $|i\rangle|000..0\rangle$ goes to a final state $|i\rangle|f(i)\rangle$. So our superposition of initial states will go to a superposition of final states according to this rule, i.e., we will end up with

$$C\sum_{i=0}^{2^N-1} |i\rangle|f(i)\rangle$$

Look at this: This state contains *all* the possible values of input i *and* the corresponding function $f(i)$ for each one. We have evaluated $f()$ over all 2^N possibilities, with just one run of the algorithm!

This trick is called *quantum parallelism* and it certainly looks impressive. However, in order to find out anything about this state We will have to measure it — and that is the catch! Suppose we just measure this fantastically rich state in the computational basis. Well then the state will *collapse* into a particular choice of $|i\rangle$ in the input register, and $|f(i)\rangle$ in the output register. So We will learn only one value for $f()$ and what's worse, it will be a randomly chosen one! This is something we could have done on a classical computer, of course, so we have gained nothing.

This is the puzzle that designers of quantum algorithms have to deal with: to design a process from initialization through to measurement, such that the measured results tell us something we could not have found out classically (without far greater time cost).

8.8.1 *Grover's Search Algorithm*

There are several useful algorithms that have been discovered, which successfully harness the power of quantum parallelism. The most famous may be the factoring algorithm due to Peter Shor, which has application in code breaking. However, a more widely applicable example is Lev Grover's search algorithm.

This algorithm solves the following artificial problem, as well as many practical generalizations of it. Suppose that we have a function $f()$ that takes as its input a number in the range $0 \ldots 2^N$ and gives back either zero or one. However, our function *almost always* gives output zero; in fact, it gives output one for just a single special input value, call it j. Our challenge is to find that special value. Classically, we could simply search, testing each input one-by-one until we get lucky. On average, we would have to evaluate the function $2^N/2$ times to find the answer.

Grover found a way to do better. If the classical approach of systematic searching would take of order K evaluations of the function, Grover shows us how to achieve the same thing in time of order $\sqrt{K}$. Although not exponential, this is of course a massive improvement!

Now let us consider building a quantum circuit to do this job. We can begin by writing a logic circuit that would evaluate $f()$ on a conventional computer, then turn that into a reversible classical circuit, and finally into a quantum circuit (as in Fig. 8.8). This gives us a prescription for evaluating $f()$ using qubits and quantum gates. But how to generalize it, and outperform the classical machine?

There are two tricks to use. The first involves how we prepare the "input" and "output" registers at the start (although the output register is really just a single qubit here, since $f()$ just returns a single bit). Crucially, we prepare the "output" qubit in a special state, $(|0\rangle - |1\rangle)/\sqrt{2}$, rather than simply state $|0\rangle$. Now for all but one of the possible states in the input register, i.e., all but $|j\rangle$, our circuit will do nothing to the output bit. For that special state, it will perform a NOT operation. Thus, if we *had* prepared the output bit in state $|0\rangle$, it would be flipped to $|1\rangle$ — but since we prepared $(|0\rangle - |1\rangle)/\sqrt{2}$, it will change to $(|1\rangle - |0\rangle)/\sqrt{2}$. The thing to notice is that this is just the original state of the output qubit with a minus sign in front: $-(|0\rangle - |1\rangle)/\sqrt{2}$. And that is a key observation, when we consider preparing the input register in a superposition of all states (as discussed above under *quantum parallelism*) an re-running the procedure. Then, the overall effect is to put a (-1) multiplier on that one special state $|j\rangle$. That is, we obtain

$$\sum_{i=0}^{2^N-1} c_i|i\rangle \quad \text{where } c_i = C \text{ for all } i \neq j \text{ and } c_j = -C.$$

where C is just the normalization constant. Thus we face a situation where the quantum state already "knows" the answer, but we have to find a way to get at it!

This is where we do our second trick, something called inversion about the mean. Suppose we have some state

$$|A\rangle = \sum_{i=0}^{2^N-1} c_i|i\rangle,$$

where for simplicity we will take the amplitudes c_i to be real. Now it turns out there is a unitary operation $\hat{U}_{\text{inv}}$ that can transform this state as follows:

$$\hat{U}_{\text{inv}}|A\rangle = \sum_{i=0}^{2^N-1} (2\bar{c} - c_i)|i\rangle$$

where $\bar{c}$ is the average value of all the c_i in state $|A\rangle$, that is $\bar{c} = 2^{-N}\sum_{i=0}^{2^N-1} c_i$. A way to understand this is to think of each original c_i

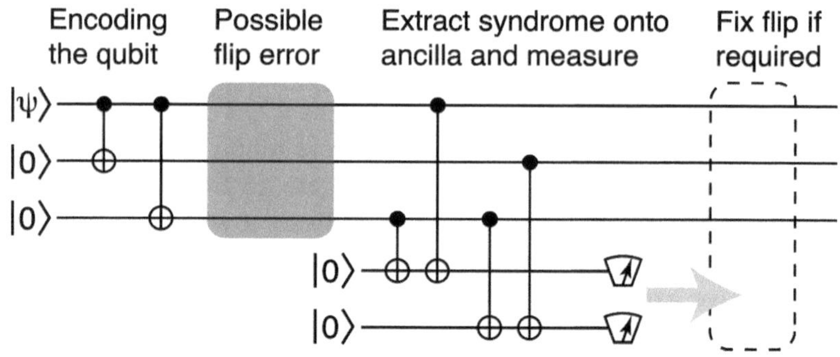

Figure 8.10. A circuit that will protect a single qubit from a flip error, by encoding it into three physical qubits.

created the state $\alpha|000\rangle + \beta|111\rangle$. How can we later check for an error, without collapsing the state? In fact, the solution, quantum error correction (QEC), uses ideas that had already been developed for correcting errors in classical data. Two of the first people to see how to apply the ideas to the quantum case were Peter Shor (the same researcher who invented the factoring algorithm) and Andrew Steane, a researcher in Oxford.

The answer is to find a circuit that extracts information (the *syndrome*) about whether there has been an error onto some additional qubits called the ancilla. Then when we measure the ancilla we learn about any error that has occurred without learning anything else about our logical qubit! In the figure, measuring the ancilla qubits will yield $|00\rangle$ if there has been no error, while $|01\rangle$, $|10\rangle$, $|11\rangle$ simply indicate that the first, second or third qubit has flipped (respectively). If we know that a specific physical qubit flipped, we can fix it by deliberately applying another flip, i.e., a $\hat{U}_{\text{NOT}}$ operation. In this way the encoded qubit is repaired without ever measuring anything about amplitudes α and β!

Of course, there are other things that can happen to a qubit besides a flip. For example, the phase relation between $|0\rangle$ and $|1\rangle$ might be disrupted. Roughly speaking the average time it takes for a qubit to suffer a flip error is often referred to as T_1, whereas the time for a phase error is denoted T_2. In most systems T_2 is far shorter. However, using the same idea outlined above it is also possible to correct phase flips — and indeed, it is possible to protect from both types of error simultaneously by using a longer encoding.

This concludes our brief introduction to quantum information processing. The ideas described here can be found in far greater detail in the book Nielsen and Chuang (2000). We now proceed to a compact description of the recent scientific achievements involving synthesizing, assembling and controlling molecular spins. While such systems constitute only one class of candidate for future quantum information technologies, they do offer some highly attractive features — not least of which is the perfect reproducibility of a given molecular building block.

8.9 Synthesis of Endohedral Fullerenes

In the previous sections, we laid down the foundations on quantum information theory and showed how one would perform universal gates with a quantum computer. In the following sections, we will focus on some remarkable molecules: endohedral fullerenes. These carbon nanomaterials have extraordinary electronic properties and have attracted considerable research on whether they could be building blocks of a solid-state quantum-information-processing device.

8.9.1 *Endohedral Metallofullerenes*

Fullerenes, due to their cage-like structure, can trap atoms inside their empty "shell." Fullerenes containing atoms or clusters in their interior are called endohedral fullerenes. Endohedral metallofullerenes are produced by the arc-discharge method. Krätschmer, Lamb, Fostiropoulos and Huffman were the first to produce macroscopic quantities of C_{60} by resistive heating of graphite rods under a He atmosphere Kratschmer *et al.* (1990). This breakthrough led to an explosion of scientific research. The first endohedral metallofullerenes were lanthanum containing fullerene cages, produced by vaporization of Lanthanum-doped graphite rods. The most stable lanthanofullerene was found to be $La@C_{82}$. Other group-3 metals (Sc, Y) and lanthanides (Ce, Gd, Pr, Nd, Ho, etc.) have since been encapsulated, mainly in C_{82} and C_{80}. In addition, group-2 metals (Ca, Sr, Ba) have been found to form endohedral metallofullerenes [Shinohara (2000)]. In all cases, there is a charge transfer from the metal to the cage, resulting in considerable modification of the electronic properties of the cage. Figure 8.11 shows a typical arc-discharge apparatus for the production of endohedral metallofullerenes.

Typically, the doped graphite rods are brought in very close proximity and direct current (100–300 A) is passed through them forming an arc between the rods. Helium pressure inside the arc chamber is maintained at 40—100 mbar. After a few hours of operation the rods are consumed. Transmission Electron Microscopy (TEM) characterization of the produced soot shows that it is mostly made of amorphous carbon and graphitic structures. It also contains typically 10–20% fullerenes. The yield of fullerenes via the arc-discharge method is very sensitive to parameters such as He pressure, current, rod size, etc. The soot produced is collected and dissolved in an organic solvent, such as toluene, in anaerobic conditions to avoid unnecessary degradation of the endohedral metallofullerenes. The fullerenes are removed from the soot by soxhlet extraction in a boiling solvent. The fullerene solution is consequently passed through a high performance liquid chromatography (HPLC) unit in order to separate the individual fullerene species.

8.9.2 *Synthesis of Endohedral Nitrogen Fullerenes*

In addition to group-2 and group-3 elements, nonmetals such as nitrogen and phosphorus as well as noble gases such as helium have all been encapsulated in fullerenes. The nonmetal elements appear to be more stable in smaller cages such as C_{60} and C_{70}. The following two production methods apply equally for entrapping a nitrogen atom in a C_{60} or C_{70} cage. $N@C_{60}$ is produced using the ion implantation method developed by Weidinger and coworkers at the Hahn-Meitner Institut in Germany [Murphy *et al.* (1996)]. Approximately 1 or 2 g of C_{60} are put into an effusion cell inside a vacuum chamber evacuated at a pressure of 10^{-6} mbar or lower. The effusion cell is heated at around 500 °C. Under these conditions the C_{60} is sublimed inside the chamber and begins to condense onto a water-cooled copper target placed above the effusion cell. At the same time the copper

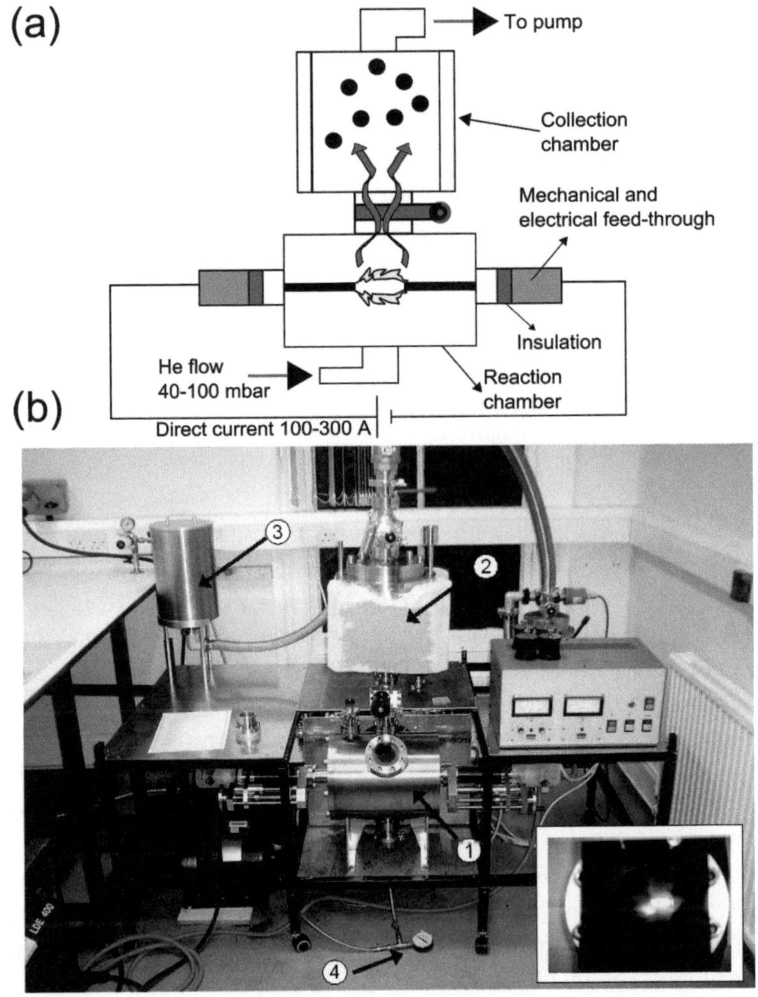

Figure 8.11. (a) Schematic illustration of an arc-reactor for the production of endohedral metallofullerenes. Two doped graphite rods are brought in close proximity and high current is passed through them. An electric arc forms and the rods begin to evaporate. The soot that is produced is carried by helium to the collection chamber where the soot condenses on the liquid nitrogen-cooled walls. (b) Arc-discharge apparatus picture. Highlighted items include 1. Main arc chamber. 2. Collection chamber. 3. Solvent reservoir. 4. He-pressure gauge. The inset in the right-hand-side corner shows an image of the arc in operation through the viewport. See also Color Insert.

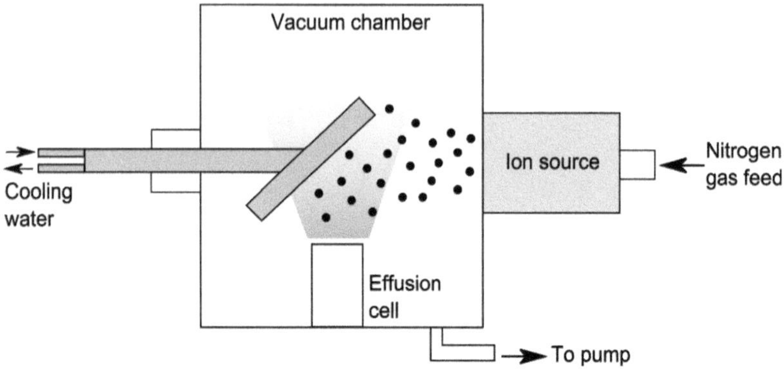

Figure 8.12. Schematic representation of an ion implantation apparatus used for the production of N@C_{60}.

target is bombarded with low energy nitrogen ions produced by an ion source. Best results are achieved using a mass-separating source, for example one producing N+ preferentially to N_2+. Typical values for the beam energy and beam current are 40 eV and 1–3 mA, respectively. The orientation of the target is such that it is located at 45° to both the effusion cell and the nitrogen ion source. Figure 8.12 shows a schematic of the ion implantation apparatus.

After a few hours of operation, the copper target is covered with a fullerene layer, several tens of micrometers thick. The copper target is subsequently immersed into an organic solvent such as CS_2 in order to extract the fullerenes. The fullerene solution is ultrasonicated for a few minutes and filtered. Between 60 and 70% of the N@C_{60}/C_{60} mixture is dissolved in CS_2, while the rest remains insoluble. The insoluble soot comprises polymerized fullerenes and destroyed fullerene cages. The filtered solution is examined by EPR (electron paramagnetic resonance) spectroscopy. The ratio of N@C_{60}/C_{60} is calculated to be 10^{-4} to 10^{-5}.

An alternative method of producing N@C_{60} is the glow discharge method. This is a rather simpler experimental setup compared with the ion implantation device. A quartz tube is equipped with two water-cooled copper electrodes at opposite ends. The chamber is filled with low-pressure (approximately 0.1 mbar) nitrogen gas. High voltage (of the order of 1 kvolt) is applied across the

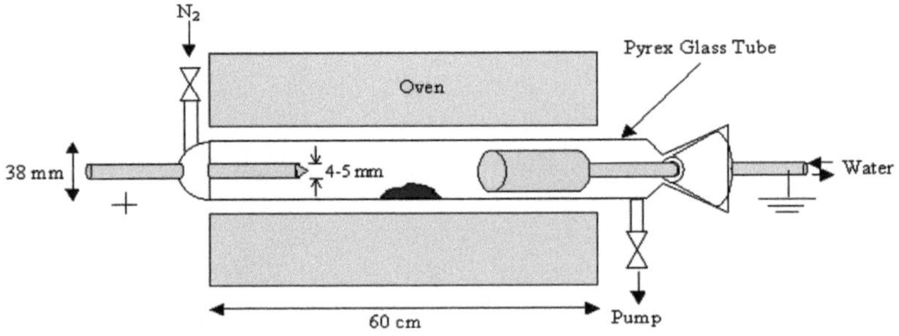

Figure 8.13. Schematic representation of the glow discharge apparatus used for the production of N@C_{60}. See also Color Insert.

electrodes resulting in the ionization of the nitrogen gas. At the same time, several tens of grams of C_{60} are put inside the quartz tube as shown in figure 8.13 The whole apparatus is then inserted in a tube oven and the system is heated up to 500 °C. C_{60} sublimes and is exposed to the ionized nitrogen gas before condensing on the copper electrodes.

At the end of the operation, the copper electrodes are immersed in organic solvents and the produced N@C_{60}/C_{60} mixture is extracted. The yield of the glow discharge method is 10^{-5} to 10^{-6} in terms of the N@C_{60}/C_{60} ratio. Hence, the advantage of simple and relatively inexpensive setup is counter-acted by an order of magnitude lower purity in the produced material.

8.10 Purification of Endohedral Fullerenes

Production of endohedral fullerenes is only the first step on the road to acquiring high-purity, individual species. As we mentioned above, multistage HPLC is the established method for fullerene isolation. This is the most crucial and laborious step in the whole process. A combination of state-of-the-art chromatography columns tailored for fullerene purification is required for the complete isolation of isomerically pure fullerenes. A typical chromatogram of the extracted and filtered fullerene solution from the arc-discharge process is shown in fig. 8.14

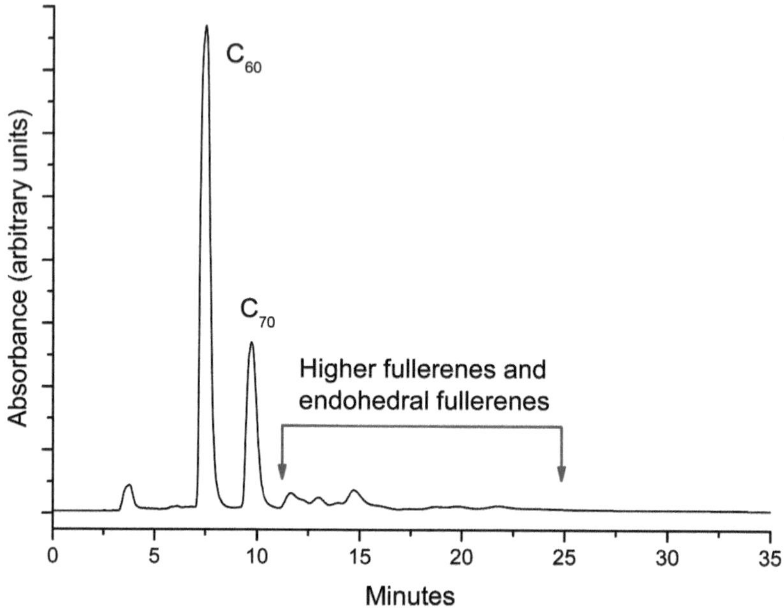

Figure 8.14. Typical HPLC chromatogram of a fullerene solution produced by the arc-discharge method. Toluene eluant, flow rate 18 ml/min, Buckyprep-M column, 20 mm × 250 mm, UV detector set at 312 nm. C_{60} is the dominant peak in the chromatogram, followed by C_{70} and smaller peaks that correspond to higher fullerenes as well as endohedral fullerenes.

It can be seen from the chromatogram that the fullerenes tend to elute with size, thus C_{60} is the first one to elute followed by C_{70} and the larger cage fullerenes, including endohedral fullerenes. The area under each peak is proportional to the mass of the fullerene species. C_{60} accounts for about 60% of the total fullerene production whereas C_{70} represents approximately 25% of the production. The remainder 15% comprises larger empty cages as well as endohedral fullerenes. Three or four stages of HPLC through a suite of reverse-phase columns is usually enough to isolate a few milligrams of high-purity endohedral species Okimoto *et al.* (2008); Akasaka *et al.* (2000); Leigh *et al.* (2005). If that process seems complicated enough, it is routine compared with the purification of N@C_{60} and related species. There are two main obstacles: first, the very low yields of the N@C_{60} production methods, and second, the fact that C_{60} and N@C_{60} are chemically almost identical. Nevertheless, via a combination of multiple injections and recycling HPLC through an appropriate column (such as the Cosmosil 5-PBB by Nacalai Tesque), it has been shown that it is possible not only to enrich but also to completely isolate $N@C_{60}$ and $N@C_{70}$ with a purity of higher than 99.5% Suetsuna *et al.* (2002); Jakes *et al.* (2003) Kanai *et al.* (2004).

8.11 Quantum Properties of Endohedral Fullerenes

The purification of endohedral fullerenes is a challenging but rewarding process. The reason is the unique electronic properties of endohedral fullerenes. More specifically, the presence of the incarcerated atom(s) alters the spin properties of the molecule. The electron or nuclear spin is a quantum property hence quantum information can be embodied in the electron/nuclear spin of the molecule. It is well known that Sc-, Y- and La-containing fullerenes have unpaired electrons Shinohara (2000). The unpaired electron spin resides mostly on the cage Morley *et al.* (2005). Nitrogen-containing fullerenes also carry quantum information embodied in the electron spin of the unpaired electrons of the nitrogen atom. In this case the spin is almost entirely inside the carbon cage (less than 5% of the spin is on the cage). The relative isolation of the electron spin from the environment makes these systems attractive for quantum computation schemes, where the lifetime of the qubits is important as we have already shown. For successful realization of quantum computing, there must be adequate immunity to decoherence: the degrading of quantum states due to interactions with the environment. Provided the coherence time is sufficiently long compared with the gate operation time, fault-tolerant error correction schemes can be implemented to overcome decoherence Steane (1996). Of the range of physical systems that have been suggested, liquid-state NMR systems have hosted the most complex quantum algorithms Vandersypen *et al.* (2001). In these systems, the qubits are embodied in the slowly decohering nuclear spins of the atoms of a molecule. However, owing to the fact that the thermal energy is always large compared with the nuclear Zeeman energy in NMR experiments, NMR-based quantum computers face a fundamental limitation in scalability and appear to be practically limited to around 10 qubits. Since scalability is one of the preconditions of effective quantum computation DiVincenzo (2000); Bennett and DiVincenzo (2000), the practical applications of NMR-based computers seem limited. EPR offers the potential to use experimentally accessible fields and temperatures to approximate pure quantum states. Endohedral fullerenes are molecular materials. Therefore, they are all identical at the most fundamental level. In addition, sophisticated chemistry can be applied to create scalable nanostructures based on these molecules. In the following section, we will focus on the spin properties of endohedral fullerenes and on their potential for quantum information processing.

8.12 N@C_{60} as a Spin Qubit

N@C_{60} has electron spin $S = 3/2$ coupled to the ^{14}N nuclear spin $I = 1$ via an isotropic hyperfine interaction. This gives rise to the rich energy level diagram shown in Fig. 8.15.

Taking into account only the first-order hyperfine interaction, the three allowed electron transitions are degenerate. For this reason, the observed continuous-wave EPR spectrum of N@C_{60} dissolved in CS_2 at room temperature (shown in Fig. 8.16) comprises three sharp resonance peaks.

The three EPR resonances are quite narrow. Their intrinsic linewidth was measured to be $\leq$ 0.3 μT. In fact, the linewidth is mainly limited by the resolution of the spectrometer and in particular the magnet stability and field homogeneity. The resolution of the spectrometer used in fig. 8.16 was about 10 μT. The ability of N@C_{60} to store quantum information effectively is demonstrated by the relaxation time T_1 and the coherence time T_2. We have studied these in different environments, and measured T_2 to be $\geq$ 0.25 ms in CS_2 solution at 160 K Morton *et al.* (2006b). Pulse sequences in a typical EPR spectrometer are of the order of 30 ns. This corresponds to more than 10^4 electron spin Rabi oscillations, before decoherence occurs. In addition to the long T_1 and T_2 times, it has been demonstrated that even in an EPR system with a 10% systematic error in single qubit operations, composite pulses can lead to fidelities

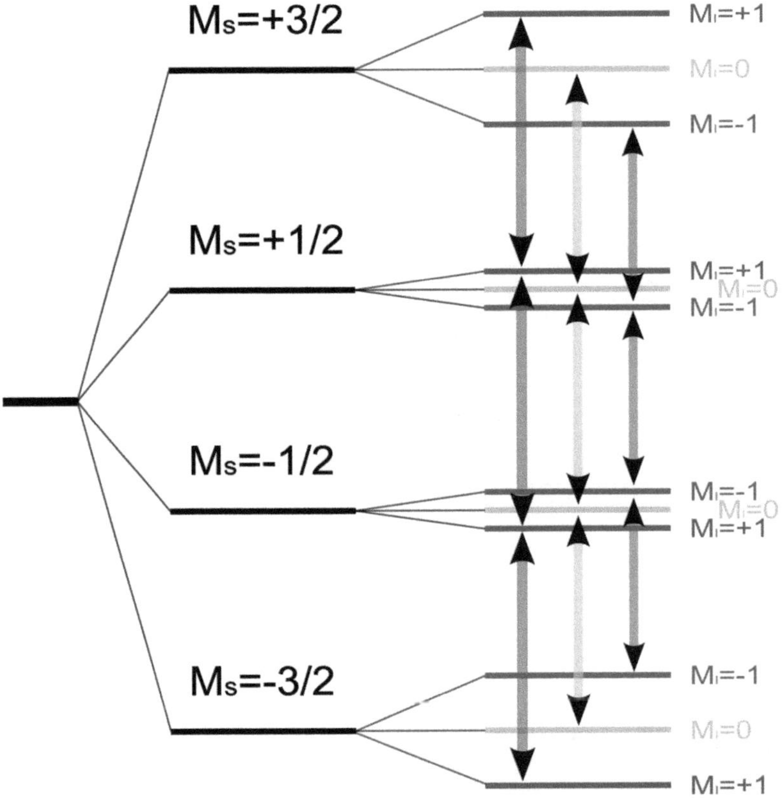

Figure 8.15. Energy level diagram of $^{14}N@C_{60}$ in a magnetic field. $^{14}N@C_{60}$ has electron spin $S = 3/2$ and nuclear spin $I = 1$. In a magnetic field, this gives rise to a 12-level structure due to the Zeeman splitting. Taking into account just the first-order hyperfine interaction, the allowed transitions (the selection rules are: $\Delta M_S = 1$ and $\Delta M_I = 0$) are triply degenerate.

between 0.999 and 0.9999 Morton *et al.* (2005). These properties of the $N@C_{60}$ system ensure that it meets all the basic criteria for fault-tolerant quantum computation. Consequently, $N@C_{60}$ has been proposed as a building block of a solid-state quantum computer Harneit (2002); Benjamin *et al.* (2006).

In addition to its excellent electron spin properties, $N@C_{60}$ is also endowed with another resource: the nuclear spin. Capable of even longer storage times of quantum information than the electron spin, the nuclear spin state can be manipulated by radio frequency pulses

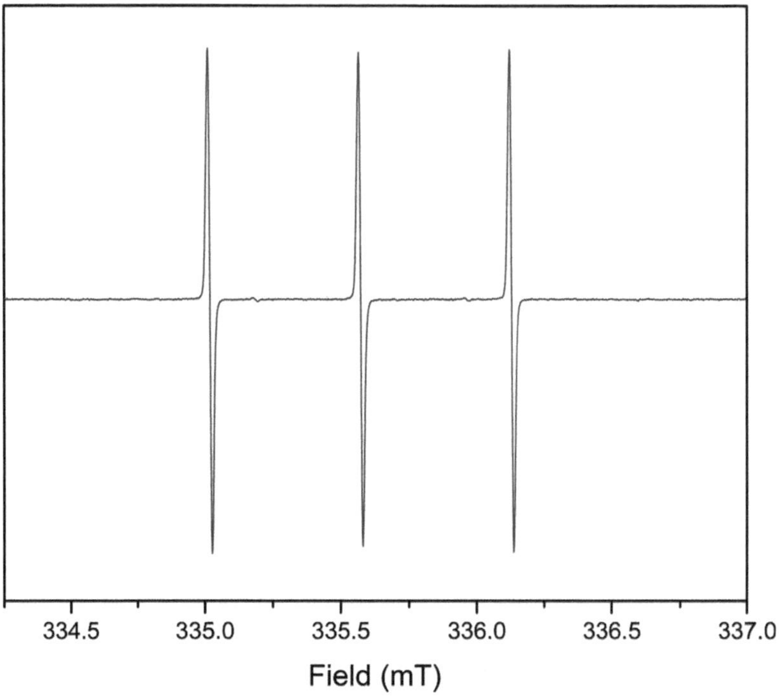

Figure 8.16. Continuous-wave EPR spectrum of $^{14}N@C_{60}$ in a CS_2 solution. The two small resonances either side of the central peak are associated with ^{15}N nuclei naturally abundant (less than 0.4%) in the sample.

as opposed to microwave pulses for the electron spin. The presence of the electron spin can be exploited to generate ultrafast phase gates, and to further protect the nuclear spin from environmental interactions by *bang-bang* decoupling Morton *et al.* (2006a). This symbiosis of electron spin and nuclear spin makes $N@C_{60}$ and its derivatives a lot more attractive for quantum information processing than many other molecular materials.

8.13 Scaling-Up of Endohedral Fullerene Nanostructures

8.13.1 *Endohedral Fullerene Dimers*

The smallest device where universal quantum gates could be applied would be a two-qubit system. For an endohedral fullerene implementation, this automatically means the linking of two endohedral molecules together via covalent or noncovalent bonds. The chemistry of fullerenes is already well established. For example, Diels–Alder cycloaddition and other reaction methods have been employed in the synthesis of fullerene adducts. Fullerene dimers, i.e., bonded pairs of fullerenes, are natural systems for the study of electronic interactions between the carbon cages. This is particularly important for endohedral fullerenes encapsulating spin-active atoms. Dipolar coupling between adjacent spins is proportional to $1/r^3$, where r is their spatial separation. Hence, in order to control the strength of the spin–spin coupling, one must control their spatial separation. In other words, chemistry can be used to control the coupling strength of the qubits. One of the simplest ways that one could use to chemically bond two fullerene molecules is to directly link the two cages. The high-speed vibration milling technique (HSVM) has been used for the synthesis of directly bonded fullerene dimers Wang *et al.* (1997); Komatsu *et al.* (2000). Using this method, C_{120}, as well as $C_{120}O$, and other similar molecules have been synthesized. Cycloaddition chemistry has been used extensively too, in order to afford a plethora of fullerene dimer molecules. Fig. 8.17 contains some examples of experimentally synthesized dimers and demonstrates how the center-to-center distance between the fullerene cages is controlled by using different bridge molecules.

The shortest distance corresponds to the directly bonded dimer and it is equal to 9.4 Å. The longest distance corresponds to the dimer with a polycyclic bridge moiety [Porfyrakis *et al.* (2007)]. That distance is calculated to be 14.8 Å. It becomes evident that the interfullerene spacing can be modulated by at least 57% using a diversity of synthetic routes and this is not the limit. Indeed other syntheses can afford even longer spacing between the fullerene cages. See, for example, [Gutierrez-Nava *et al.* (2004)].

We have seen so far that chemistry provides all the necessary tools to engineer complicated fullerene structures. So what is the progress in applying these, or similar, schemes to endohedral species?

There are two main obstacles on the road to scaling up endohedral fullerene arrays. The first one is the difficulty in producing these materials in multimilligram quantities as we highlighted

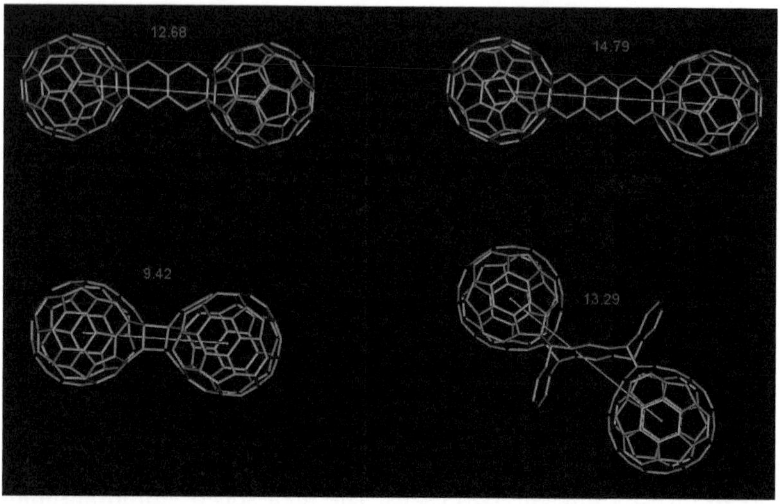

Figure 8.17. Comparison of the center-to-center distance between fullerene cages, for experimentally synthesized, covalently-bonded dimers with a variety of bridge molecules. All distances are quoted in units of Å.

earlier. The second (and an equally formidable one) is the lower thermal and photolytic stability of functionalized N@C_{60} compared with pristine N@C_{60}. It is now acknowledged that most chemical functionalizations inflict some degree of EPR signal loss on N@C_{60}. This implies that either N@C_{60} is destroyed or that the nitrogen atom escapes from the fullerene cages. This combination of synthetic difficulties might initially look like an insurmountable obstacle. However, it is possible to tune reaction conditions in such a manner that a significant "number of spins" survives the reaction. It has been recently shown that a *half-filled* endohedral fullerene dimer has been produced using a pyrrolidine functionalization scheme Zhang *et al.* (2008). The beauty of this scheme is that not only it retains about 70% of the N@C_{60} signal, but it also affords both the dimer and the monomer products by manipulation of the reagent molar ratios. This is important because the monomer can be used in a two-step reaction to yield an asymmetric fullerene dimer (for example, a ^{14}N@C_{60}-^{15}N@C_{60} dimer) in a controlled way. Also the bridge molecule can be chosen so that it acts as s photo-switch modulating the distance between the fullerene cages. A directly bonded N@C_{60}-

C_{60} dimer has also been synthesized by the HSVM method [Goedde *et al.* (2001)]. However the purity of N@C_{60} was too low to allow for in-depth spectroscopic study of the molecule.

An alternative way of approach is to use endohedral metallofullerenes as building blocks for supramolecular fullerene assembly. Although the chemistry of metallofullerenes is not so well developed (compared to C_{60}), they are beginning to be available in multimilligram quantities of high-purity materials. Some synthetic protocols for metallofullerene adducts have begun to appear in recent years Wakahara *et al.* (2006); Lukoyanova *et al.* (2007); Takano *et al.* (2008); Akasaka *et al.* (2008). Also their spin coherence properties may not be as impressive as that of N@C_{60} (indeed N@C_{60} has got the longest spin coherence out of any molecular system) but some of them, such as Y@C_{82}, are almost as impressive and would most likely be adequate for fault-tolerant quantum computing. All of the above strongly indicate that it is only a matter of time before an endohedral fullerene dimer of some form is finally synthesized.

In addition to covalent bonding, noncovalent interactions present an attractive route toward the assembly of large arrays of endohedral fullerenes for quantum information. Such interactions include hydrogen-bonding, van der Waals interactions, $\pi - \pi$ stacking interactions and coordination chemistry. Cyclodextrins, calixarenes, porphyrins, and other macrocycles can be complexed with fullerenes in order to create supramolecular arrays. Although weak in comparison with covalent bonds, it is well known that very stable structures can be achieved through the cooperative effect of such interactions. A specific advantage of these effects is thermodynamic in nature. These processes are driven with an inherent ability to "self-correct" thus decreasing the probability for incomplete or incorrect arrays Lindsey (1991).

8.13.2 *One-Dimensional and Two-Dimensional Arrays and Beyond*

The way to achieve large one-dimensional arrays of fullerenes is relatively straightforward. Fullerene molecules self-assemble into ordered arrays inside single-walled carbon nanotubes (SWNTs).

The process is spontaneous upon heating and the resulting structures are called nanotube "peapods" Smith *et al.* (1998). Metallofullerenes are thermally robust and their peapod structures are well established Hirahara *et al.* (2000); Warner *et al.* (2008). What is less understood is the peapod electronic structure and the effect of filling on the spin properties of the metallofullerenes. This has partly to do with the fact that SWNTs come with a variety of electronic properties (semi-conducting, metallic) and with many paramagnetic impurities that interfere with the magnetic properties of the encapsulated endohedral fullerenes. Thermally unstable molecules such as $N@C_{60}$ and its derivatives can be also inserted into SWNTs at low temperature and in an inert environment using supercritical fluids Khlobystov *et al.* (2004). Local spin control implies the use of electrode gates positioned in such a way that they can address a single fullerene. The technology for single spin read-out and manipulation remains elusive, however *global-addressing* schemes have been developed that require minimal assembly design control as long as the basic spin–spin interactions are characterized [Benjamin (2002)].

2-Dimensional supramolecular structures have been successfully formed on surfaces by exploiting noncovalent (mainly hydrogen bonding) interactions between the constituent molecules. Several molecular networks can form porous structures that can act as hosts for fullerenes or related molecules. The arrangement of the guest fullerene molecules is largely controlled by the size and shape of the network pores. Hexagonally packed C_{60} heptamers have been formed in a perylene tetra-carboxylic di-imide (PTCDI)-melamine network on a silver-terminated silicon surface [Theobald *et al.* (2003)]. Single C_{60} molecules have been incorporated in a trimesic acid (TMA) molecular network on graphite [Griessl *et al.* (2004)]. More recently, open-grid arrays of paired endohedral fullerenes ($Er_3N@C_{80}$) have been formed on a strontium titanate ($SrTiO_3$) "waffle" surface Deak *et al.* (2006). The molecules "fit" like eggs fitting into an egg carton. The above examples show that it is possible to arrange endohedral fullerenes in ordered two-dimensional arrays. In principle, such patterns can be extended in three-dimensional networks too, offering unparalleled possibilities for molecular quantum computing architectures.

8.14 Summary

In this chapter, we introduced the concept of a quantum computer. We learned how such a device represents a new paradigm of computation. We highlighted the differences between a classical and a quantum computer, and we demonstrated why and how a quantum computer would outperform any classical computer for some type of calculations. We learned about quantum bits, or qubits. We encountered superposition and entanglement, and we learned about universal quantum gates and how we could apply them to manipulate quantum information in a fault-tolerant manner. Quantum phenomena are inherent in atoms and molecules.

In the second part of the chapter, we became familiar with endohedral fullerenes. We learned how these molecules are synthesized, and we focused on their electronic properties. Endohedral fullerenes carry quantum information embodied in the electron and nuclear spins of their encapsulated atom(s). We explained the reasons why fullerenes are attractive as a component for a quantum information technology. We established the remarkable suitability of endohedral fullerenes for storing and manipulating quantum information. We discussed the synthetic developments in endohedral fullerene chemistry. We learned about various chemical syntheses that have been applied in order to produce arrays: both small fullerene dimers and larger one-dimensional and two-dimensional array architectures.

Clearly, there are advantages and disadvantages associated with endohedral fullerenes and their application in quantum information. We have highlighted the main questions that remain to be answered. Although there are many such questions and significant challenges that need to be overcome, research to date has demonstrated that these molecules are not just beautiful, highly symmetrical structures. There is good evidence that endohedral fullerenes will indeed find applications in future quantum technologies, including quantum computing.

References

1. Akasaka, T., Kono, T., Takematsu, Y., Nikawa, H., Nakahodo, T., Wakahara, T., Ishitsuka, M. O., Tsuchiya, T., Maeda, Y., Liu, M. T. H., Yoza, K., Kato, T., Yamamoto, K., Mizorogi, N., Slanina, Z. and Nagase, S. (2008). Does Gd@C-82 have an anomalous endohedral structure? Synthesis and single crystal X-ray structure of the carbene adduct, *Journal of the American Chemical Society* **130**, 39, pp. 12840+, doi:{10.1021/ja802156n}.
2. Akasaka, T., Okubo, S., Kondo, M., Maeda, Y., Wakahara, T., Kato, T., Suzuki, T., Yamamoto, K., Kobayashi, K. and Nagase, S. (2000). Isolation and characterization of two Pr@C-82 isomers, *Chemical Physics Letters* **319**, 1-2, pp. 153–156.
3. Benjamin, S. (2002). Quantum computing without local control of qubit–qubit interactions, *Physical Review Letters* **88**, 1, doi:{10.1103/PhysRevLett.88.017904}.
4. Benjamin, S., Ardavan, A., Andrew, G., Briggs, D., Britz, D., Gunlycke, D., Jefferson, J., Jones, M., Leigh, D., Lovett, B., Khlobystov, A., Lyon, S., Morton, J., Porfyrakis, K., Sambrook, M. and Tyryshkin, A. (2006). Towards a fullerene-based quantum computer, *Journal of Physics-Condensed Matter* **18**, 21, Sp. Iss. SI, pp. S867–S883, doi:{10.1088/0953-8984/18/21/S12}.
5. Bennett, C. and DiVincenzo, D. (2000). Quantum information and computation, *Nature* **404**, 6775, pp. 247–255.
6. Deak, D. S., Silly, F., Porfyrakis, K. and Castell, M. R. (2006). Template ordered open-grid arrays of paired endohedral fullerenes, *Journal of the American Chemical Society* **128**, 43, pp. 13976–13977, doi:{10.1021/ja0634369}.
7. DiVincenzo, D. (2000). The physical implementation of quantum computation, *Fortschritte Der Physik-Progress of Physics* **48**, 9-11, pp. 771–783.
8. Goedde, B., Waiblinger, M., Jakes, P., Weiden, N., Dinse, K. and Weidinger, A. (2001). 'Nitrogen doped' C-60 dimers (N@C-60-C-60), *Chemical Physics Letters* **334**, 1-3, pp. 12–17.
9. Griessl, S., Lackinger, M., Jamitzky, F., Markert, T., Hietschold, M. and Heckl, W. (2004). Room-temperature scanning tunneling microscopy manipulation of single C-60 molecules at the liquid–solid interface: Playing nanosoccer, *Journal of Physical Chemistry B* **108**, 31, pp. 11556–11560, doi:{10.1021/jp049521p}.

10. Gutierrez-Nava, M., Accorsi, G., Masson, P., Armaroli, N. and Nierengarten, J. (2004). Polarity effects on the photophysics of dendrimers with an oligophenylenevinylene core and peripheral fullerene units, *Chemistry—A European Journal* **10**, 20, pp. 5076–5086, doi:{10.1002/chem.200400157}.

11. Harneit, W. (2002). Fullerene-based electron-spin quantum computer, *Physical Review A* **65**, 3, Part A, doi:{10.1103/PhysRevA.65.032322}.

12. Hirahara, K., Suenaga, K., Bandow, S., Kato, H., Okazaki, T., Shinohara, H. and Iijima, S. (2000). One-dimensional metallofullerene crystal generated inside single-walled carbon nanotubes, *Physical Review Letters* **85**, 25, pp. 5384–5387.

13. Jakes, P., Dinse, K., Meyer, C., Harneit, W. and Weidinger, A. (2003). Purification and optical spectroscopy of N@C-60, *Physical Chemistry Chemical Physics* **5**, 19, pp. 4080–4083.

14. Kanai, M., Porfyrakis, K., Briggs, A. and Dennis, T. (2004). Purification by HPLC and the UV/Vis absorption spectra of the nitrogen-containing incar-fullerenes iNC(60), and iNC(70), *Chemical Communications*, 2, pp. 210–211, doi:{10.1039/b310978h}.

15. Khlobystov, A., Britz, D., Wang, J., O'Neil, S., Poliakoff, M. and Briggs, G. (2004). Low temperature assembly of fullerene arrays in single-walled carbon nanotubes using supercritical fluids, *Journal of Materials Chemistry* **14**, 19, pp. 2852–2857, doi:{10.1039/b404167d}.

16. Komatsu, K., Fujiwara, K. and Murata, Y. (2000). The fullerene cross-dimer C-130: synthesis and properties, *Chemical Communications*, 17, pp. 1583–1584.

17. Kratschmer, W., Lamb, L., Fostiropoulos, K. and Huffman, D. (1990). Solid C-60 — A new form of carbon, *Nature* **347**, 6291, pp. 354–358.

18. Leigh, D., Owen, J., Lee, S., Porfyrakis, K., Ardavan, A., Dennis, T., Pettifor, D. and Briggs, G. (2005). Distinguishing two isomers of Nd@C-82 by scanning tunneling microscopy and density functional theory, *Chemical Physics Letters* **414**, 4-6, pp. 307–310, doi:{10.1016/j.cplett.2005.08.090}.

19. Lindsey, J. (1991). self-assembly in synthetic routes to molecular devices-biological principles and chemical perspectives—A review, *New Journal of Chemistry* **15**, 2-3, pp. 153–180.

20. Lukoyanova, O., Cardona, C. M., Rivera, J., Lugo-Morales, L. Z., Chancellor, C. J., Olmstead, M. M., Rodriguez-Fortea, A., Poblet, J. M., Balch, A. L. and Echegoyen, L. (2007). "Open rather than closed" malonate methanofullerene derivatives. The formation of methanofulleroid adducts of

Y3N@C-80, *Journal of the American Chemical Society* **129**, 34, pp. 10423–10430, doi:{10.1021/ja071733n}.

21. Morley, G., Herbert, B., Lee, S., Porfyrakis, K., Dennis, T., Nguyen-Manh, D., Scipioni, R., van Tol, J., Horsfield, A., Ardavan, A., Pettifor, D., Green, J. and Briggs, G. (2005). Hyperfine structure of Sc@C-82 from ESR and DFT, *Nanotechnology* **16**, 11, pp. 2469–2473, doi:{10.1088/0957-4484/16/11/001}.
22. Morton, J., Tyryshkin, A., Ardavan, A., Benjamin, S., Porfyrakis, K., Lyon, S. and Briggs, G. (2006a). Bang-bang control of fullerene qubits using ultrafast phase gates, *Nature Physics* **2**, 1, pp. 40–43, doi:{10.1038/nphys192}.
23. Morton, J., Tyryshkin, A., Ardavan, A., Porfyrakis, K., Lyon, S. and Briggs, G. (2005). High fidelity single qubit operations using pulsed electron paramagnetic resonance, *Physical Review Letters* **95**, 20, doi:{10.1103/PhysRevLett.95.200501}.
24. Morton, J., Tyryshkin, A., Ardavan, A., Porfyrakis, K., Lyon, S. and Briggs, G. (2006b). Electron spin relaxation of N@C-60 in CS2, *Journal of Chemical Physics* **124**, 1, doi:{10.1063/1.2147262}.
25. Murphy, T., Pawlik, T., Weidinger, A., Hohne, M., Alcala, R. and Spaeth, J. (1996). Observation of atomlike nitrogen in nitrogen-implanted solid C-60, *Physical Review Letters* **77**, 6, pp. 1075–1078.
26. Nielsen, M. and Chuang, I. (2000). *Quantum Computation and Quantum Information* (Cambridge University Press, Cambridge, UK).
27. Okimoto, H., Kitaura, R., Nakamura, T., Ito, Y., Kitamura, Y., Akachi, T., Ogawa, D., Imazu, N., Kato, Y., Asada, Y., Sugai, T., Osawa, H., Matsushita, T., Muro, T. and Shinohara, H. (2008). Element-specific magnetic properties of di-erbium Er-2@C-82 and Er2C2@C-82 metallofullerenes: A synchrotron soft X-ray magnetic circular dichroism study, *Journal of Physical Chemistry C* **112**, 15, pp. 6103–6109, doi:{10.1021/jp711776j}.
28. Porfyrakis, K., Sambrook, M. R., Hingston, T. J., Zhang, J., Ardavan, A. and Briggs, G. A. D. (2007). Synthesis of fullerene dimers with controllable length, *Physica Status Solidi B-Basic Solid State Physics* **244**, 11, pp. 3849–3852.
29. Shinohara, H. (2000). Endohedral metallofullerenes, *Reports on Progress in Physics* **63**, 6, pp. 843–892.
30. Smith, B., Monthioux, M. and Luzzi, D. (1998). Encapsulated C-60 in carbon nanotubes, *Nature* **396**, 6709, pp. 323–324.
31. Steane, A. (1996). Error correcting codes in quantum theory, *Physical Review Letters* **77**, 5, pp. 793–797.

32. Suetsuna, T., Dragoe, N., Harneit, W., Weidinger, A., Shimotani, H., Ito, S., Takagi, H. and Kitazawa, K. (2002). Separation of N-2@C-60 and N@C-60, *Chemistry—A European Journal* **8**, 22, pp. 5079–5083.

33. Takano, Y., Yomogida, A., Nikawa, H., Yamada, M., Wakahara, T., Tsuchiya, T., Ishitsuka, M. O., Maeda, Y., Akasaka, T., Kato, T., Slanina, Z., Mizorogi, N. and Nagase, S. (2008). Radical Coupling Reaction of Paramagnetic Endohedral Metallofullerene La@C-82, *Journal of the American Chemical Society* **130**, 48, pp. 16224–16230, doi:{10.1021/ja802748q}.

34. Theobald, J., Oxtoby, N., Phillips, M., Champness, N. and Beton, P. (2003). Controlling molecular deposition and layer structure with supramolecular surface assemblies, *Nature* **424**, 6952, pp. 1029–1031, doi:{10.1038/nature01915}.

35. Vandersypen, L., Steffen, M., Breyta, G., Yannoni, C., Sherwood, M. and Chuang, I. (2001). Experimental realization of Shor's quantum factoring algorithm using nuclear magnetic resonance, *Nature* **414**, 6866, pp. 883–887.

36. Wakahara, T., Iiduka, Y., Ikenaga, O., Nakahodo, T., Sakuraba, A., Tsuchiya, T., Maeda, Y., Kako, M., Akasaka, T., Yoza, K., Horn, E., Mizorogi, N. and Nagase, S. (2006). Characterization of the bis-silylated endofullerene Sc3N@C-80, *Journal of the American Chemical Society* **128**, 30, pp. 9919–9925, doi:{10.1021/ja062233h}.

37. Wang, G., Komatsu, K., Murata, Y. and Shiro, M. (1997). Synthesis and x-ray structure of dumb-bell-shaped C-120, *Nature* **387**, 6633, pp. 583–586.

38. Warner, J. H., Watt, A. A. R., Ge, L., Porfyrakis, K., Akachi, T., Okimoto, H., Ito, Y., Ardavan, A., Montanari, B., Jefferson, J. H., Harrison, N. M., Shinohara, H. and Briggs, G. A. D. (2008). Dynamics of paramagnetic metallofullerenes in carbon nanotube peapods, *Nano Letters* **8**, 4, pp. 1005–1010, doi:{10.1021/nl0726104}.

39. Zhang, J., Porfyrakis, K., Morton, J. J. L., Sambrook, M. R., Harmer, J., Xiao, L., Ardavan, A. and Briggs, G. A. D. (2008). Photoisomerization of a fullerene dimer, *Journal of Physical Chemistry C* **112**, 8, pp. 2802–2804, doi:{10.1021/jp711861z}.

Chapter 9

Cell Biology of Carbon Nanotubes

Chang Guo, Khuloud Al-Jamal, Hanene Ali-Boucetta, and Kostas Kostarelos

Nanomedicine Lab, Centre for Drug Delivery Research
The School of Pharmacy, University of London, 29-39 Brunswick Square
London WC1N 1AX, United Kingdom
kostas.kostarelos@pharmacy.ac.uk

Carbon nanotubes (CNTs) were first specifically identified and described in 1991.[1] These nanoscale materials have since been widely used in a variety of fields due to their extraordinary properties, including high surface area, high mechanical strength, electronic properties, and excellent chemical and thermal stability. CNTs have also been developed and explored for a wide range of applications including in biomedicine, as biosensors, tissue engineering scaffolds, and drug delivery systems. The interaction between CNTs and mammalian cells was first observed by Pantarotto and co-workers in 2003.[2] Chemically functionalized single-walled CNTs were studied to report internalization by cells. Since then, more experimental techniques, materials, and cell types have been studied to identify the interaction between CNTs and cells *in vitro*. A variety of investigations are currently underway to study the interaction between biological systems and CNTs.

Advances in Carbon Nanomaterials: Science and Applications
Edited by Nikos Tagmatarchis

ISBN 978-981-426-78-78 (Hardcover), 978-981-426-78-85 (eBook)
www.panstanford.com

9.1 Experimental Techniques Used to Study the Interaction Between Carbon Nanotubes and Cells *In Vitro*

Carbon nanotubes (CNTs) are mainly classified as single-walled (SWNTs) and multi-walled (MWNTs) according to the number of the concentric layers of graphitic sheets rolled into cylindrical structures. Both SWNTs and MWNTs have been reported to translocate into cells using several analytical techniques, including optical microscopy, micro-Raman spectroscopy, single-particle tracking (SPT), transmission electron microscopy (TEM), flow cytometry, and fluorescence microscopy. Each technique offers its own advantages and disadvantages that will be discussed separately below.

9.1.1 *Optical Microscopy*

Optical microscopy provides imaging of CNTs in live cell cultures; however, due to low resolution, normally only large amounts of uptaken CNTs can be detected in a non-quantitative manner. Although the technique is simple (no specialized instrumentation required) and readily available in most laboratories, optical microscopy can only offer qualitative results and also suffers from the incapability to differentiate cell surface adsorption from intracellular localization of the material. Optical microscopy can be proposed as a rough, pre-screening technique to study the effect of varying CNT characteristics (e.g., surface charge, charge density, aqueous dispersibility)[3] on their interaction with cultured cells before more sophisticated and time-consuming techniques are employed.

9.1.2 *Fluorescence Microscopy Techniques*

Fluorescence microscopy is widely used to study the interaction between CNTs and cells by the following: (i) detection of the intrinsic fluorescent signals of some CNT types; (ii) imaging CNTs using X-ray fluorescence microscopy (μXRF); and (iii) detection of fluorescent probes that have been linked (covalently or non-covalently) onto the CNTs.

Since pristine SWNTs exhibit unique near-infrared intrinsic fluorescence, near-infrared fluorescence microscopy has been used to observe the cellular uptake of SWNTs in live cells first described in 2004 by the Weisman group.[4] SWNTs were seen within intracellular compartments of macrophage cells understood to be uptaken by phagocytosis. However, due to its relatively low signal intensity, near-infrared fluorescence microscopy is currently limited to qualitative detection of cellular uptake. Moreover, specialized instrumentation and expertise is also required today. μXRF was also applied to image CNT localization within macrophages by Bussy and co-workers.[5] μXRF could provide information of the CNT–cell interaction by analysis of the fluorescence signal of the catalyst metal particles bound to CNTs. This technique has been shown to have enough sensitivity to detect very low concentrations of CNTs. Thus, this technique could be employed more in the future to identify the CNT–cell interactions and the effect on cells by uptaken CNTs; however, specialized instrumentation is also needed.

Fluorescence microscopy (optical or confocal laser) can only be applied to assess the cellular uptake of CNT probed with fluorescent dyes, thus offering indirect observation of their cellular uptake. With the help of confocal laser scanning microscopy (CLSM) and organic-based fluorophores (fluorescein or members of the rhodamine, cyanine, and Alexa families) methods to label cells and subcellular compartments have also been applied to access the subcellular localization of CNTs, as shown in Table 9.1. Some of the fluorescence probes in these studies are covalently bound to nanotubes, while others are non-covalently bound to CNTs. Regarding their intracellular localization, this is generally observed with the help of added intracellular compartment markers. In only a few of these studies CNTs were reported within the nucleus of the cells with reported co-localization of nuclear stains and fluorescence from the labeled CNTs. The issue of nuclear localization of CNTs is still not conclusive and is under intense investigation by various laboratories. Most studies today report fluorescence-probed CNTs in the cytoplasm and around the perinuclear regions. Taken together, it remains difficult to conclude on the final intracellular trafficking destination of CNTs mainly due to the dramatic variation in materials used (CNT types), cells, fluorescent probes, association between CNTs, and probes and

Table 9.1. Studies of CNT cellular uptake using intracellular compartment markers

	CNT type	Dispersing agent and buffer	Cell type	Duration of CNT interaction with cells	Cell fixation solution	Markers of intracellular compartments	Conclusions	Ref.
Coated (non-covalently surface-modified) CNTs	Phospholipid PEG-coated SWNT	H_2O or physiological buffers	HL60 (human promyelocytic leukemia cells), CHO (Chinese hamster ovary cells), and 3T3 (mouse embryonic fibroblast cells)	1 h	N/A	Endosomes: FM4-64	SWNTs enter cells; uptake pathway proposed is consistent with adsorption-mediated endocytosis	6, 7
	Cy3–DNA-coated SWNT	H_2O or physiological buffers	HeLa cells (human adenocarcinoma cells)	12 h	N/A	Nuclei: DRAQ5	SWNTs transport DNA cargo	8
	Protein (SA, SpA, BSA)-coated SWNT	H_2O	HeLa cells	2–3 h	N/A	Endosomes: FM4-64	Cellular uptake via energy-dependent endocytosis pathway; endocytosed species confined inside endosomes	9
	FITC–FA–chitosan-coated SWNT	PBS solution	Hep G2 cells (human hepatocellular carcinoma cells)	Up to 5 days	4% PFA	Nuclei: DAPI	SWNTs localization in the cytoplasm (not in nucleus)	10

	Phospholipid PEG-coated SWNT	DMEM (serum-free medium)	Ntera-2 cells (human teratocarcinoma cells)	3 h	4% PFA	Nuclei: Hoechst	SWNTS readily localize within small (~2μm) vesicles in the cells	11
	Phospholipid PEG-coated SWNT	Folate-free RPMI	KB cells (human carcinoma cells)	2.5 h	Methanol at -20°C for 45 min	Nuclei: Hoechst	SWNT conjugates show high and specific binding to folate receptors	12
	AO-coated SWNT	Cell culture medium	HeLa cells	30 min up to 7 days	N/A	Lysosomes: LysoTracker	AO–SWNTs remain inside lysosomes for more than a week	13
Chemically functionalised CNTs	SWNT/MWNT: NH_3^+–CNT, $NHCOCH_3$–CNT, FITC–CNT, NH_3^+–CNT–FITC, FITC–CNT–MTX, AmB–CNT–FITC, NH_2–CNT–FITC	5% dextrose in H_2O or serum-free medium	A549 (human lung carcinoma), HeLa, Jurkat human (T lymphocyte), MOD-K (murine intestine-derived epithelial cells), *C. neoformans* (yeast), *E. Coli* (bacteria), *S. cerevisiae* (yeast)	1–4 h	4% PFA	Membranes: WGA; Nuclei: TO-PRO 3	• Cellular uptake of CNTs independent of functional group and cell type • Mechanism of CNT cellular uptake less than 50% due to energy-dependent mechanisms	14, 15

(*Contd.*)

Table 9.1. *(Continued)*

CNT type	Dispersing agent and buffer	Cell type	Duration of CNT interaction with cells	Cell fixation solution	Markers of intracellular compartments	Conclusions	Ref.
Oxidized SWNT–FB-28	H_2O	Cardiomyocytes	N/A	4% formalin	Nuclei: PI	SWNTs localize in cellular compartments	16
SWNT–PEG–FITC	N/A	HeLa cells, U2OS (human bone osteosarcoma cells), MEF (mouse embryonic fibroblasts), HT1080 (human sarcoma cells), C33A (cervical cancer cells), HEK293	Up to 7 h	4% PFA	Mitochondria: MitoTracker; Nuclei: Hoechst or DRAQ5 or DAPI	SWNTs accumulate in the nucleus, the site of ribosomal biogenesis; highly dynamic inside the cells	17
Oxidized SWNT biotinylated by streptavidin–FITC	N/A	Human smooth muscle cells (hMSCs)	6 days	N/A	Actin: phalloidin; Nuclei: DAPI	SWNTs enter cells through the cytoplasm to nuclear localization	18
Oxidized SWNT–Qdot525–EGF	PBS	HN13 cells (human head and neck squamous carcinoma cells)	1 h	3.5% PBS–formaldehyde solution	Actin: phalloidin; Nuclei: PI	Localized within cytoplasm but not in the nucleus	19

Oxidized SWNT–BSA–FITC	Cell culture medium (pH 7.2–7.4)	HEK293 (human embryonic kidney cells)	1 h	4% PFA	Membrane: WGA; Nuclei: DAPI	SWNTs translocate into cytoplasmic vesicles but not in the nucleus	20
Oxidized SWNT–BSA–fluorescein–doxorubicin	N/A	WiDr (human colon cancer cells)	4 h	N/A	Cytoplasm: BSA–fluorescein	SWNTs observed outside the nuclei, within the cytoplasm, with no co-localization with doxorubicin after internalization	21
Oxidized SWNT–HER2 IgY	Cell culture medium	SK-BR-3 (human breast carcinoma cells)	24 h	10% neutral-buffered zinc formalin	Nuclei: DAPI	HER2 IgY–SWNT complex localize on the cell membrane of SK-BR-3 cells.	22

CNT: carbon nanotubes; PEG: polyethylene glycol; BSA: bovine serum albumin; AmB: amphotericin B; MTX: methotrexate; SA: streptavidin; SpA: Staphylococcal protein A; AO: acridine orange; FA: folate acid; PFA: paraformaldehyde; PBS: phosphate buffer solution.

experimental conditions. However, what remains consistent and reproducible throughout the studies performed today is the confirmation of the original reports that CNTs exhibit the capacity to be uptaken by cells in ways and mechanisms that do not necessarily follow established binding, internalization, and trafficking patterns known for other nanoparticles.

9.1.3 *Flow Cytometry*

Flow cytometry-based assays have been proposed to assess CNT–cell associations (both cell-bound or internalized) qualitatively by measuring the increase in the sideward scattering of cells incubated with non-fluorescent CNTs.[3] Data generated using light scattering analysis established a good correlation between the increase in sideward scattering intensity and the increase in CNT intracellular accumulation, which suggested that adsorption of the CNT onto the cell membrane will eventually lead to intracellular uptake. Qualitative measurements are based on the fact that as CNTs bind to the cells, the granularity of the cells increases, which concomitantly increases the sideward scattering intensity. It is difficult to distinguish between CNTs that are bound to the cell surface or internalized by the cells because sideward scattering intensity only offers an indication of cell surface roughness.[3] However, this technique can be combined with other techniques such as optical microscopy, CLSM, or TEM to distinguish CNT cell binding from internalization.

9.1.4 *Electron Microscopy*

TEM has been widely used to study the interaction between CNTs and biological systems. Both pristine CNTs (unpurified and purified) and different types of functionalized CNTs (f-CNT) have been studied intracellularly. TEM provides the highest possible resolution, indicating the exact intracellular localization of the CNTs. However, most TEM protocols have to be performed using fixed cells, so it is difficult to follow the trafficking pathway of the CNT translocation into cells. As evidenced from Table 9.2, the intracellular localization of CNTs can be classified into three major categories: (a) CNTs, both pristine and functionalized, observed in the perinuclear region

Table 9.2. Studies using TEM to investigate the cellular uptake of CNTs

Cellular localization of CNTs	CNT type	Dispersing agent and buffer	Cell type	Duration of CNT interaction with cells	Cell fixation protocol	Conclusions	Ref.
Imaged in at the perinuclear region and inside intracellular vacuoles or vesicles	DNA-coated SWNT	dH_2O	3T3 cells	Up to 48 h	Fixed at 4°C for 24 h	SWNTs incorporate into cytoplasmic vesicles and labeled the perinuclear region of cells, but did not enter the nuclear envelope	25
	MWNT	Cell grow medium	HEK cells (human embryonic kidney cells)	Up to 48 h	Fixed in Trump's fixative at 4°C and post-fixed in 1% OsO_4 in 0.1 M sodium pgosphate buffer	MWNTs present within cytoplamic vacuoles at all time points and induced the release of the proinflammatory cytokine IL-8 in a time-dependent manner	26
	SWNT	DMEM supplemented with 5% FBS	HeLa cells	60 h	Fixed using 2.5% glutaraldehyde in 0.1 M cacodylate buffer and post fixed with 1% OsO_4	SWNT-like material in intracellular vacuoles	27

(*Contd.*)

Table 9.2. (*Continued*)

Cellular localization of CNTs	CNT type	Dispersing agent and buffer	Cell type	Duration of CNT interaction with cells	Cell fixation protocol	Conclusions	Ref.
	Oxidized SWNT–FB-28	H_2O	Cardiomyocytes	Up to 5 days	Fixed in 2.5% phosphate-buffered glutaraldehyde, pH 7.4, for 4 h at 4°C	SWNTs localized within cellular vesicles	16
	MWNT	Ultrapure sterile H_2O (pH 5.5) with Arabic gum (0.25 wt%)	A549 cells	48 h	Fixed with 2.5% glutaraldehyde, and post-fixed with OsO_4	MWNTs localized in cytoplasm, the majority of them being surrounded by a membrane	28
	MWNT, SWNT (unpurified and purified)	N/A	Macrophages	24 h	N/A	MWNTs and both purified and raw SWNTs engulfed into vacuoles that can occupy most of the cell surface cytoplasm	5
	Water-soluble MWNT by introducing an oxygen component only	IMDM (Iscove's modified Dulbecco's medium)	Fibroblast cells	2 days	Fixed in 2.5% glutaraldehyde in 0.1 M phosphate buffer (pH 7.4) for 12 h, then washed and post-fixed with 1% aqueous OsO_4 for 30 min	MWNTs enter into cells and accumulate in the cytoplasm	29

	Oxidized SWNT	Cell culture medium (pH 7.2–7.4)	HEK cells	1 h	Fixed in 2.5% glutaraldehyde in 0.1 M sodium, cacodylate buffer and rinsed, and post fixed 1 h in 2% OsO_4 with 3% potassium ferriocyanide and rinsed	SWNTs reported within endosomes	20
In the cytoplasm and within the nucleus	NH_3^+–MWNT	5% dextrose	HeLa cells	1 h	Fixed with 2% solution of uranyl acetate in water overnight at 4°C	MWNTs found to cross the plasma membrane barrier and in the nucleus	30
	SWNT	THF	HMM (human monocyte-derived macrophage)	Up to 4 days	Fixed in 4% glutaraldehyde in PIPES buffer	SWNTs enter the cytoplasm and localize within the cell nucleus	23
	Oxidized SWNT	N/A	HMSC (human mesenchymal stem cells)	Up to 6 days	Fixed in 3% gluteraldehyde, dehydrated, and sectioned at −20°C	SWNTs localize in cytoplasm and also in nucleus	18

(*Contd.*)

Table 9.2. (*Continued*)

Cellular localization of CNTs	CNT type	Dispersing agent and buffer	Cell type	Duration of CNT interaction with cells	Cell fixation protocol	Conclusions	Ref.
	Oxidized SWNT–Qdot525–EGF	PBS	HN13 cells	1 h	N/A	SWNTs localize around perinuclear region	19
	MWNT–NH_2; MWNT–COOH	dH_2O	HEK cells	Up to 48 h	Fixed in 2.5% glutaraldehyde in 0.1 M sodium cacodylate buffer (pH7.4) for 1 h at RT, and post fixed for in2% OsO_4 with 3% potassium ferrocyanide for 1 h	MWNT–COOHs and MWNT–NH_2s enter cells both through endocytosis and direct translocation	31
Non-specific intracellular regions	DNA-coated SWNT	Salt solution	Va13 (human fibroblast cells)	Overnight	Fixed in an epoxy matrix	Longer tubes on the outside of the cell membrane and shorter tubes piercing the membrane and residing in the cytosol	32
	SWNT	Serum containing (5%) medium	A549 cells	24 h	Fixed in 2.5% glutaraldehyde in 0.1 M phosphate buffer for 1 h, and post fixed in 1% OsO_4 in 0.1 M phosphate buffer for 1 h	Non-intracellular localization of SWNTs but increased number of surfactant storing lamellar bodies observed	33

FITC–FA–chitosan-coated SWNT	PBS	Hep G2 cells	1 h	Fixed in 2.5% glutaraldehyde containing 0.1 M PBS buffer for 3 h and post-fixed with 1% OsO_4 for 30 min	SWNTs located only in the cytoplasm and not in nuclei	10
SWNT and MWNT	PBS	HAEC (human aortic endothelial cells)	24 h	Fixed in Karnovsky's fixative (2.5% gluteraldehyde, 2.5% paraformaldehyde in 0.1 M sodium cacodylic buffer), post-fixed in OsO_4, mordanted in 1% tannic acid	A small number of CNTs were identified in the cytoplasm of some cells	24
80n*-MWNT	N/A	Osteoclasts	3 days	N/A	MWNTs observed inside of cells and some in the vicinity of mitochondria	34
AO-coated SWNT	Fresh culture medium containing 5% FBS	HeLa cells	30 min up to 7 days	Fixed with 2% glutaraldehyde and 1% OsO_4	AO–SWNTs remain inside lysosomes for more than a week	13

*Average diameter 80 nm; AO: acridine orange; THF: tetrahydrofuran; FA: folate acid; FBS: fetal bovine serum.

inside intracellular vacuoles or vesicles; (b) CNTs detected both in the cytoplasm and within the cell nucleus. According to these observations, CNTs were found not only to translocate across plasma membranes, but also seemed to enter the nuclear envelope. Most such studies were performed using chemically f-CNT; however, Porter *et al.*[23] imaged individual non-functionalized SWNTs within cells using low-loss energy-filtered TEM in combination with electron energy loss spectrum imaging. These techniques allowed for improved contrast between (unlabelled) SWNTs and cell organelles including the plasma membrane, vesicles, and the nucleus without staining. They showed direct evidence of the individual SWNTs crossing lipid bilayers and enter into the cytoplasm and nucleus; (c) CNTs were found into non-specific intracellular regions. An example of such study was recently reported by Simeonova *et al.*, who observed small numbers of purified (non-functionalized) SWNTs and MWNTs in the cytoplasm or along the plasma membrane of human aortic endothelial cells.[24]

9.1.5 *Micro-Raman Spectroscopy*

SWNTs show strong Raman scattering[35] evidenced by the presence of characteristic G-band peaks. CNTs uptaken within living cells could be studied by micro-Raman spectroscopy confirming their cellular uptake. This technique was first applied by Daniel *et al.* to observe the cellular uptake of CNTs in live cell cultures by comparing the Raman scattering and fluorescence spectra of SWNTs and correlating those signals to intracellular location based on area maps of the cells in comparison to optical microscopy and TEM.[25] Later, Raman spectroscopy was used to detect the cellular uptake of noncovalently surface-modified SWNTs, coated with either peptides or PEGylated lipids.[36] The advantages of micro-Raman spectroscopy are its high sensitivity and low background signal interference along with the capability for long-term detection.

9.1.6 *Intrinsic Photoluminescence (Via SPT)*

SPT is a technique used to study the diffusion of small molecules both computationally and experimentally. Although there is not high enough resolution to allow visualization of specific cellular uptake using this technique, the interaction between SWNTs and live cells can be assessed in a dynamic fashion.[37,38] Moreover, by mapping and monitoring the trajectories of SWNTs internalized into live cells as a function of time and cell topography membrane surface adsorption and desorption, diffusion, endocytosis, and exocytosis from fibroblasts (NIH-3T3 cells) has been proposed.[37,38]

9.2 Mechanisms Involved in the Cellular Uptake of CNTs

The cellular uptake of CNTs has been reported by several laboratories employing a variety of experimental techniques as discussed in Section 9.1. Table 9.3 summarizes as comprehensively as possible the different CNT types used in various studies along with the reported conclusions offered on the mechanism(s) of intracellular uptake involved in the uniformly agreed observation of CNT intracellular localization. Below, we will attempt to summarize the main such mechanisms that have been proposed (Section 9.2.1) and discuss the critical parameters that have been implicated in determining which of those mechanisms can be deemed more predominant (Section 9.2.2).

Table 9.3. Cell biology studies and the proposed mechanisms of CNT cellular uptake

Type of CNT		Cell	Experimental technique	Mechanism of uptake proposed	Ref.
Pristine CNTs	SWNT	Macrophage-like cells	Near-infrared fluorescence microscopy	Localized in small phagosomes suggesting phagocytosis pathway	4
	SWNT, MWNT	HeLa HEK cells	TEM	Localized in cytoplamic vacuoles suggesting endocytosis pathway	26, 27
	SWNT	HMM cells	Low-loss energy-filtered TEM combined with nuclei marker	Localized in the cytoplasm and also in nucleus suggesting diffusion pathway	23

(*Contd.*)

Table 9.3. (*Continued*)

Type of CNT		Cell	Experimental technique	Mechanism of uptake proposed	Ref.
	SWNT	NIH-3T3 cells	SPT	• The association between SWTN and cells is associated with several mechanisms (see Section 9.2.1.3 for details) including membrane surface adsorption and desorption, surface diffusion and endocytosis and exocytosis The cellular uptake of SWNT is reported as size-dependent	37–39
Coated (non-covalently surface-modified) CNTs	Phospholipid–PEG-coated SWNT	HL60 cells, CHO cells and 3T3 cells	CLSM combined with endosome marker	Uptake pathway is consistent with adsorption-mediated endocytosis	6
	Poly(rU)-coated SWNT	MCF7 cells	CLSM	SWNTs could penetrate the nuclear membrane suggesting a diffusion pathway	40
	DNA-coated SWNT	3T3 cells	TEM	Localized in the cytoplasmic vesicles and the perinuclear region of the cells suggesting endocytosis pathway	25
	DNA-coated SWNT; protein-coated SWNT	HeLa cells	CLSM under endocytosis-inhibiting condition	Uptake reported via an energy-dependent endocytosis pathway and the endocytosed species are confined inside endosomes	7, 8, 41
	Peptide-coated SWNT	HeLa cells	Raman scattering	Cellular uptake reported as time- and temperature-dependent suggesting endocytosis pathway	42

Table 9.3. (*Continued*)

Type of CNT		Cell	Experimental technique	Mechanism of uptake proposed	Ref.
	DNA-coated SWNT	IMR90 cells (human lung fibroblasts)	TEM and CLSM	Uptake reported as length-dependent	32
	FITC–FA–chitosan-coated SWNT	Hep G2 cells	CLSM combined with nuclei marker; TEM	Localized only in the cytoplasm and not in nuclei, suggesting endocytosis	10
Covalently modified CNT by oxidation	SWNT–PEG–FITC	HeLa , U2OS, MEF, HT1080, C33A, HEK293 cells	CLSM combined with intracellular compartment markers	Localized in the nucleus, mainly in the nucleolus suggesting diffusion	17
	Oxidized SWNT	HEK293 cells	CLSM and TEM	Localized in endosomes, suggesting uptake through an endocytosis pathway	20
	MWNT–NH_2; MWNT–COOH	HEK293 cells	TEM	Localized in endosomes and lysosomes for short term and in the nucleus at later time points, suggesting a combination of endocytosis and direct penetration	31
	Oxidized SWNT–biotin	L1210FR cells (leukemia cells)	CLSM under endocytosis-inhibiting condition	Localized inside of cells in an energy-dependent, endocytosis pathway	43
	Oxidized SWNT	BY-2 cells (Walled plant cells)	CLSM under endocytosis-inhibiting condition	• SWNTs traverse across both plant cell walls and cell membrane • SWNT/FITC is taken up by fluid-phase endocytosis	44

(*Contd.*)

Table 9.3. (*Continued*)

Type of CNT		Cell	Experimental technique	Mechanism of uptake proposed	Ref.
Covalently modified CNT by 1,3 dipolar cycloaddition	SWNT-NH-FITC	Human 3T6 and murine 3T3 fibroblasts	CLSM	Localized inside of the cells by an energy-independent, passive translocation pathway	2
	MWNT-NH_3^+	HeLa, HEK293 cells	TEM	CNTs able to cross cell membrane and accumulate in cytoplasm to reach the nucleus suggested diffusion pathway	30, 45
	SWNT/MWNT: NH_3^+-CNT, $NHCOCH_3$-CNT, FITC-CNT, NH_3^+-CNT-FITC, FITC-CNT-MTX, AmB-CNT-FITC, NH_2-CNT-FITC	A549, HeLa cells, Jurkat human, MOD-K cells, C.neoformans, *E. coli, S. cerevisiae*	CLSM combined with intracellular compartment markers and under endocytosis-inhibiting condition		

- CNTs cellular uptake reported concentration-dependent
- CNTs cellular uptake reported independent of functional group and cell type 14, 15

9.2.1 *Trafficking Pathways in the Cellular Uptake of CNT*

The exact mechanisms involved in the cellular uptake of CNT are not yet clearly elucidated and more likely are a contribution of multiple pathways. Both energy-dependent endocytosis pathways and energy-independent translocation through the plasma membrane have been reported to play a role leading to CNT cell internalization. There are several parameters that seem to play an important role in determining the intracellular localization and trafficking of CNT, among which the most critical are type of CNT surface modification and CNT dimensions (diameter and length). More than a single experimental technique and CNT type should be studied in combination to further understand those interactions.

9.2.1.1 Types of CNT endocytosis leading to internalization

The initial report of CNT cell internalization (using chemically f-CNT) was published by Pantarotto *et al.* in 2003 and observed the cellular uptake of fluorescent (fluorescein isothiocyanate [FITC]) probe-conjugated CNTs. This study reported the cellular uptake of f-CNTs even at low temperature (4°C) or in the presence of an endocytosis inhibitor (sodium azide). Based on such evidence f-CNTs were proposed to be able to translocate into cells (3T3 and 3T6 cells) under energy-independent pathways.[2] A following study by Dai and co-workers reported that PEGylated lipid-coated SWNTs were uptaken also in both adherent (HeLa) and non-adherent (HL60) cell cultures. Moreover, they observed that these lipid-coated SWNTs were co-localized intracellularly with an endosome marker (FM 4-64) at 37°C, while their uptake was blocked at low temperatures. Therefore, an energy-dependent endocytotic mechanism was proposed by these authors to account for the uptake of non-chemically f-CNT into cells.[9] The same group also studied shortened SWNTs (non-covalently) coated with ssDNA or protein (BSA) molecules to suggest that their cellular uptake follows a clathrin-dependent endocytosis pathway rather than a caveolae or lipid-rafts pathway.[8] The intracellular localization of CNTs was mainly observed by TEM and fluorescence microscopy, while other techniques such as SPT[37–39] have also been more recently employed to study the mechanism of cellular uptake of ssDNA-coated SWNTs in NIH-3T3 cells. Trajectories of non-photobleaching SWNTs were tracked during the interaction with NIH-3T3 cells in real time using optical microscopy. Thousands of individual trajectories allowed the analysis of the SWNTs trafficking pathway within these cells. Using image processing algorithms, it was proposed that within the 50.8% trajectories that identified different kinds of interactions between SWNTs and cellular compartments, around 12.7% seem to follow an endocytosis pathway.[37] It is becoming apparent that the type of molecules that are used to coat or wrap CNT to make them more dispersible in aqueous and biological media plays a critical role in the interaction with cells. Whether cellular uptake of CNTs takes place through an energy-dependent endocytosis pathway and which one of the various pathways is predominant needs further investigation using a variety of different CNT types (lipid-, polymer-, DNA-coated, etc.).

9.2.1.2 Can CNTs pierce through cell membranes as "nano-needles"?

The mechanism of CNT cellular uptake using chemically functionalized CNTs in a variety of cell types was studied by Kostarelos and co-workers.[2,14] Both SWNTs and MWNTs were functionalized using identical chemical synthesis with a wide range of molecules of increasing molecular weight (ammonium, acetamido, FITC, methotrexate, amphotericin B, and their combinations) and monitored cellular uptake in several kinds of cells (including A549, fibroblasts, HeLa, CHO, HEK293, Keratinocytes, Jurkat, *E. coli*, *C. neoformans*, and *S. cervisiae*). f-CNTs cellular uptake was observed even under endocytosis-inhibiting conditions. Based on such studies it has been suggested that f-CNTs interact with cellular membranes as "nano-needles," able to pierce the plasma membrane and translocate to the intracellular compartments in a largely energy-independent, passive diffusion mechanism. Further evidence by TEM and confocal microscopy has recently been reported in support of a "nano-needle" CNT behavior.[2,14,15] Porter *et al.* observed by TEM that SWNTs could translocate across the lipid bilayers into the neighboring cytoplasm, and also be localized inside the cell nucleus.[23] By SPT and optical microscopy, Strano *et al.* suggested around 18.4% of the trajectories following surface diffusion.[38] PEGylated SWNTs have also been reported recently in the nucleus of HeLa cells observed by fluorescence microscopy.[17] The proposition from such studies that CNTs can transport across cellular membranes and through the nuclear envelope offers further support as to their capacity to pierce through membranes; however, further investigation is needed to elucidate the exact mechanisms and possible alternative pathways involved in the intracellular trafficking of these materials.

9.2.1.3 Fate of CNTs after internalization

The Strano group applied optical microscopy and SPT to explore the fate of ssDNA-coated SWNTs following cellular uptake (NIH-3T3 cells). They reported that 49.2% trajectories following a purely convective diffusion in the flow field with no cellular interaction while the remaining 50.8% trajectories followed different trafficking pathways, including 6.2% membrane surface adsorption, 18.4% surface diffusion, 12.7% endocytosis, 5.9% exocytosis, and 7.4% desorption from the membrane. That was the first published evidence indicating CNT exocytosis after cellular internalization.[37] In an alternative paradigm, recent studies by Kagan *et al.* have reported the possibility for enzymatic degradation CNTs[46]; however, this work has been carried out only chemically, in the absence of interaction with cells. Further data on the degradation mechanisms of CNTs *in vitro* and *in vivo* are very much needed. Nevertheless, information about the fate of CNTs following cellular internalization is still scarce and at very early stages, with further investigation in this area clearly needed.

9.2.2 *Parameters Involved in the Cellular Uptake of CNTs*

9.2.2.1 Surface modification of CNT: non-covalent coating versus chemical conjugation

Different approaches to modify the CNT surface result in different degrees of aqueous dispersibility, stability in cell media, and type of interaction with cellular membranes and other intracellular components. Kam *et al.* reported an energy-dependent endocytosis pathway for the cellular uptake of SWNTs coated with large molecular weight biopolymers,[7,8] while others found that energy-independent cell internalization was taking place extensively by f-CNT chemically conjugated with small molecular weight functional groups.[14] It seems that the interaction between cells and large biopolymers linked to CNT by non-covalent coating or chemical conjugation is a critical factor that favors energy-dependent endocytotic mechanisms. On the other hand, f-CNTs functionalized with small molecules are able to translocate inside the cytoplasm by energy-independent endocytotic mechanisms that favor piercing of the plasma membrane via lipid exchange. It would be interesting and useful to determine the characteristics (e.g., molecular weight, charge, hydrophobicity) of the molecules used to surface-modify nanotubes in correlation with the cell internalization mechanisms that these will dictate.

9.2.2.2 CNT diameter and length

It is still not clear whether the diameter of CNTs (determined by the number of the concentric carbon layers) is involved in the mechanisms leading to cellular uptake, since both SWNTs and MWNTs have been reported to be able to internalize into cells. The effect of CNT length on cellular uptake has been also been studied using SWNTs. One publication has suggested length-dependent cellular uptake based on evidence that as different lengths of SWNTs (average lengths of 660 $\pm$ 40 nm, 430 $\pm$ 35 nm, 320 $\pm$ 30 nm, and 130 $\pm$ 18 nm studied) were compared, CNT of 320 $\pm$ 30 nm provided the highest cellular uptake.[39] More studies on the effect of CNT dimensions on cellular uptake are needed, even though they can be challenging since other parameters (such as aggregation in biological media, wide length, and diameter distributions among CNT samples) will exert significant impact on the studied effects; therefore, great caution is advised.

9.2.2.3 Concentration of CNT

The cellular uptake of CNTs has been reported to be dependent on their concentration interacting with cells, the higher the concentration of dispersible CNTs the higher the cellular uptake.[15,17] However, great care should be taken to make sure that cell internalization occurs at concentrations below the toxicity threshold. For example, actin cytoskeleton disruption accompanied with altered VE-cadherin localization and a concomitant diminished viability of human aortic endothelial cells has been found to be related to high concentrations of CNTs.[24] Cheng and co-workers[17] reported an interesting phenomenon of reversible accumulation of FITC-labeled PEGylated SWNTs (FITC–PEG–SWNTs) within the nucleus of several mammalian cell lines (Table 9.1), by studying their intracellular trafficking and fate. By comparing the fluorescence intensity of intracellular CNTs and extracellular CNTs, they observed that the intranuclear distribution of SWNTs depended on the extracellular concentration of SWNTs and the translocation of CNTs in and out of cells at similar rates. Even though such results are intriguing, the underlying mechanisms of cellular internalization and exocytosis need to be verified.

9.2.2.4 Cell type

Some of the cell types that have been reported to internalize CNTs are shown in Tables 9.1 and 9.2. Our group and others have reported the cellular internalization of different f-CNTs in a wide variety of cell types, including mammalian cells including fibroblasts exhibiting deficient phagocytosis, fungi, yeast, and bacterial cells.[2,4,14] More recently, other cell types have also been reported to uptake CNTs including plant cells.[44] It seems that CNT exhibit a capacity to internalize in cells irrespective of cell type; however, more work needs to be performed to correlate the internalization of different CNTs with cell types.

9.2.2.5 Duration of CNT interaction with cells

The cellular uptake of CNTs has been reported to be dependent on incubation time; the longer the incubation with CNTs the higher the degree of cellular uptake of CNTs.[31] The incubation times between CNTs and cells in different studies are shown in Tables 9.1 and 9.2. In one report, cellular uptake in the cytoplasm and nucleus was reported after 6 days of incubation[18]; however, this parameter will greatly depend on other CNT characteristics such as the stability of the dispersion in biological media, the surface modification of the CNTs (e.g., surface charge) that may accelerate interaction with cell cultures. More systematic studies using adherent and non-adherent cell cultures with different types of CNTs are needed to elucidate the importance of this parameter.

9.3 Conclusion

The cell biology of CNTs has become an increasingly interesting area of research both at the basic biological level and also due to the variety of potential biomedical applications using CNTs. The field has experienced an exponential increase in the number of studies and laboratories using CNTs in contact with various cell types that will surely increase in the next few years. We have already learnt that the type and nature of molecules used to modify the surface of CNT play a determinant role in their initial interaction with cells and their subsequent intracellular trafficking and translocation. From the basic cell biology point of view, the now numerous reports on the capacity of CNT structures to pierce cellular membranes and translocate directly through to the cytoplasm offer a new insight into the way fabulous nanostructures interact with lipid membranes and at the same time a novel tool to transport small molecules intracellularly. This is only the beginning in a research area that promises to exploit CNTs both as a tool for basic cell biology and a useful nanodevice for the delivery of therapeutic or diagnostic agents.

References

1. M. Monthioux and V. L. Kuznetsov, *Carbon* **44**(9), 1621–1623 (2006).
2. D. Pantarotto *et al., Chem. Commun.* (1), 16–17 (2004).
3. D. Cai *et al., Nanotechnology* **19**(34), 345102–345111 (2008).
4. P. Cherukuri *et al., J. Am. Chem. Soc.* **126**(48), 15638–15639 (2004).
5. C. Bussy *et al., Nano Lett.* **8**(9), 2659–2663 (2008).
6. N. W. S. Kam, W. Kim, and H. J. Dai, *Abs. Pap. Am. Chem. Soc.,* **227**, U508–U508 (2004).
7. N. W. Kam, Z. Liu, and H. Dai, *J. Am. Chem. Soc.* **127**(36), 12492–12493 (2005).
8. N. W. Kam *et al., Proc. Natl. Acad. Sci. U. S. A.* **102**(33), 11600–11605 (2005).
9. N. W. S. Kam and H. J. Dai, *J. Am. Chem. Soc.* **127**(16), 6021–6026 (2005).
10. B. Kang *et al., Nanotechnology* **19**(37), 375103–375110 (2008).
11. R. P. Feazell *et al., J. Am. Chem. Soc.* **129**(27), 8438–8439 (2007).

12. S. Dhar *et al.*, *J. Am. Chem. Soc.* **130**(34), 11467–11476 (2008).
13. X. Zhang *et al.*, *Chemistry* **16**(2), 556–561 (2009).
14. K. Kostarelos *et al.*, *Nat. Nanotechnol.* **2**(2), 108–113 (2007).
15. L. Lacerda *et al.*, *Adv. Mater.* **19**(14), 1789–1789 (2007).
16. R. Krajcik *et al.*, *Biochem. Biophys. Res. Commun.* **369**(2), 595–602 (2008).
17. J. Cheng *et al.*, *ACS Nano* **2**(10), 2085–2094 (2008).
18. E. Mooney *et al.*, *Nano Lett.* **8**(8), 2137–2143 (2008).
19. A. A. Bhirde *et al.*, *ACS Nano* **3**(2), 307–316 (2009).
20. Q. X. Mu *et al.*, *ACS Nano* **3**(5), 1139–1144 (2009).
21. E. Heister *et al.*, *Carbon* **47**(9), 2152–2160 (2009).
22. Y. Xiao *et al.*, *Bmc Cancer* **9**, 1–11 (2009).
23. A. E. Porter *et al.*, *Nat. Nanotechnol.* **2**(11), 713–717 (2009).
24. V. G. Walker *et al.*, *Toxicol. Appl. Pharmacol.* **3**, 319–328 (2009).
25. D. A. Heller *et al.*, *Adv. Mater.* **17**(23), 2793–2799 (2005).
26. N. A. Monteiro-Riviere *et al.*, *Toxicol. Lett.* **155**(3), 377–384 (2005).
27. H. N. Yehia *et al.*, *J. Nanobiotechnology* **5**, 8 (2007).
28. A. Simon-Deckers *et al.*, *Toxicology* **253**(1–3), 137–146 (2008).
29. J. Meng *et al.*, *Colloids Surf. B Biointerfaces* **71**(1), 148–153 (2009).
30. D. Pantarotto *et al.*, *Angew. Chem. Int. Ed. Engl.* **43**(39), 5242–5246 (2004).
31. Q. Mu, D. Broughton, and B. Yan, *Nano Lett.* **9**(12), 4370–4375 (2009).
32. M. L. Becker *et al.*, *Adv. Mater.* **19**(7), 939–945 (2007).
33. M. Davoren *et al.*, *Toxicol In Vitro* **21**(3), 438–448 (2007).
34. N. Narita *et al.*, *Nano Lett.* **9**(4), 1406–1413 (2009).
35. R. Saito, G. Dresselhaus, M. S. Dresselhaus, *Physical Properties of Carbon Nanotubes*, Imperial College Press, London, 1998.
36. Z. Liu *et al.*, *J. Am. Chem. Soc.* **130**(41), 13540–13541 (2008).
37. H. Jin, D. A. Heller, and M. S. Strano, *Nano Lett.* **8**(6), 1577–1585 (2008).
38. M. S. Strano and H. Jin, *ACS Nano* **2**(9), 1749–1752 (2008).
39. H. Jin *et al.*, *ACS Nano* **3**(1), 149–158 (2009).
40. Q. Lu *et al.*, *Nano Lett.* **4**(12), 2473–2477 (2004).
41. N. W. S. Kam, Z. A. Liu, and H. J. Dai, *Angew. Chem. Int.l Ed. Engl.* **45**(4), 577–581 (2006).
42. S. F. Chin *et al.*, *Exp. Biol. Med.* **232**(9), 1236–1244 (2007).

43. J. Chen *et al., J. Am. Chem. Soc.* **130**(49), 16778–16785 (2008).
44. Q. L. Liu *et al., Nano Lett.* **9**(3), 1007–1010 (2009).
45. K. Kostarelos *et al., J. Drug Deliv. Sci. Technol.* **15**(1), 41–47 (2005).
46. B. L. Allen *et al., Nano Lett.* **8**(11), 3899–3903 (2008).

Index

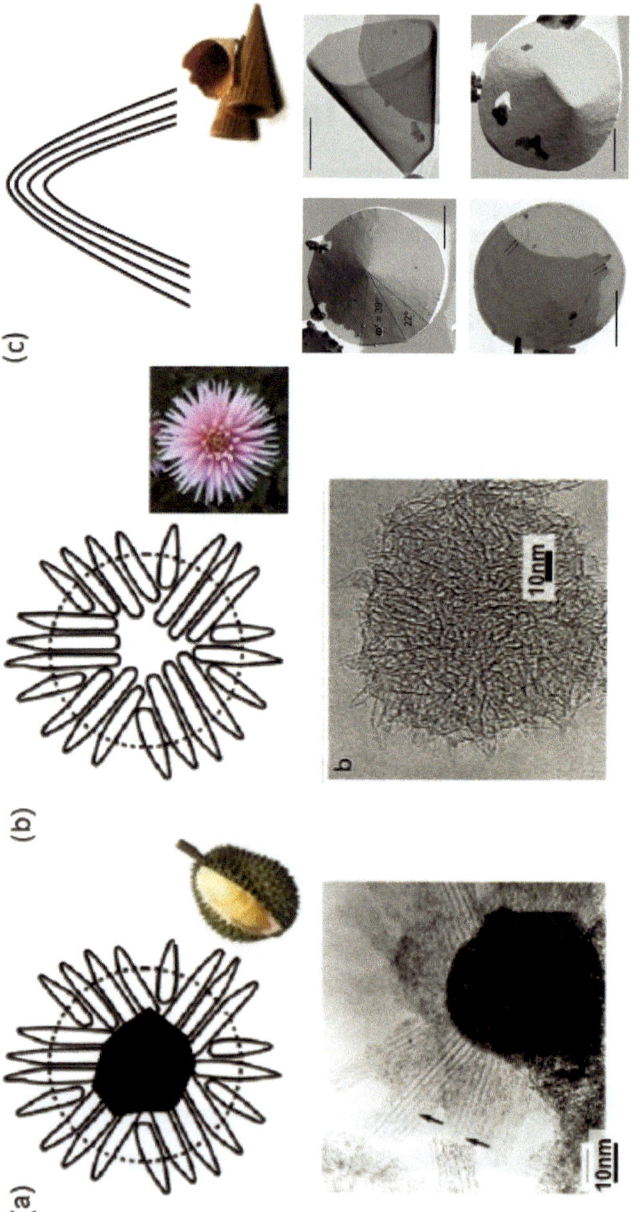

Figure 1.13

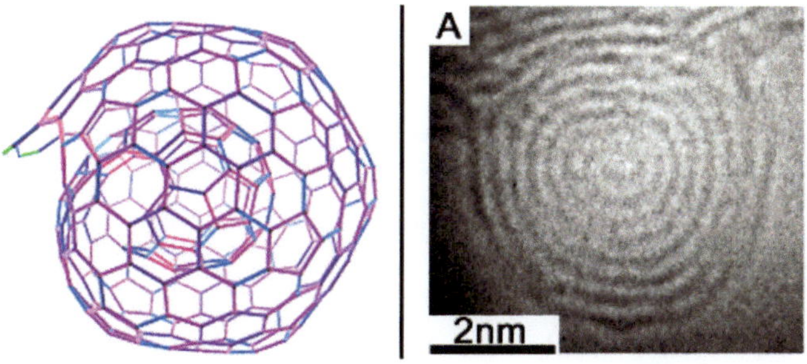

Figure 1.23

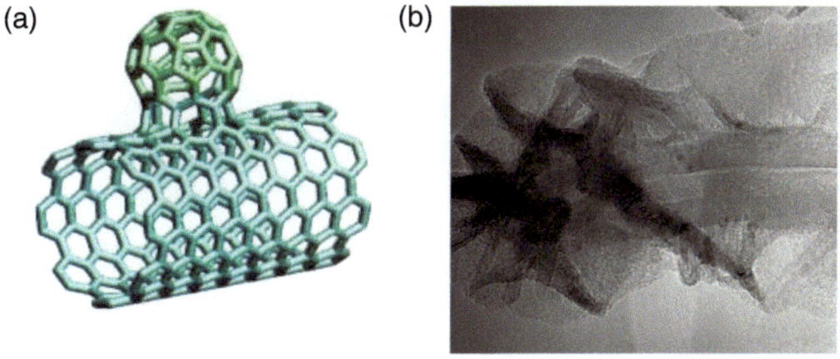

Figure 1.26

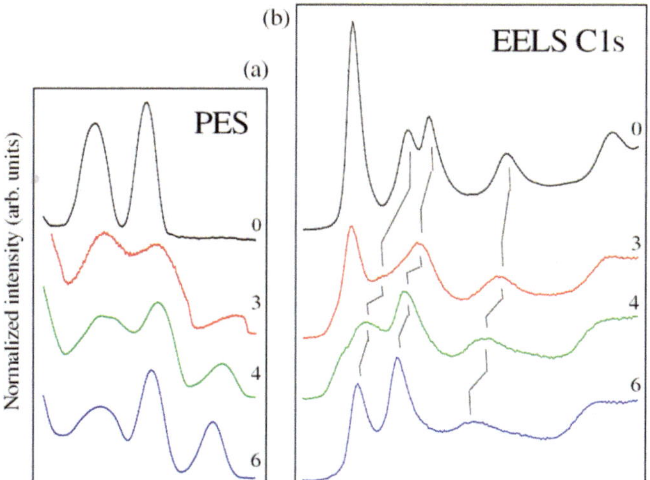

Figure 2.8

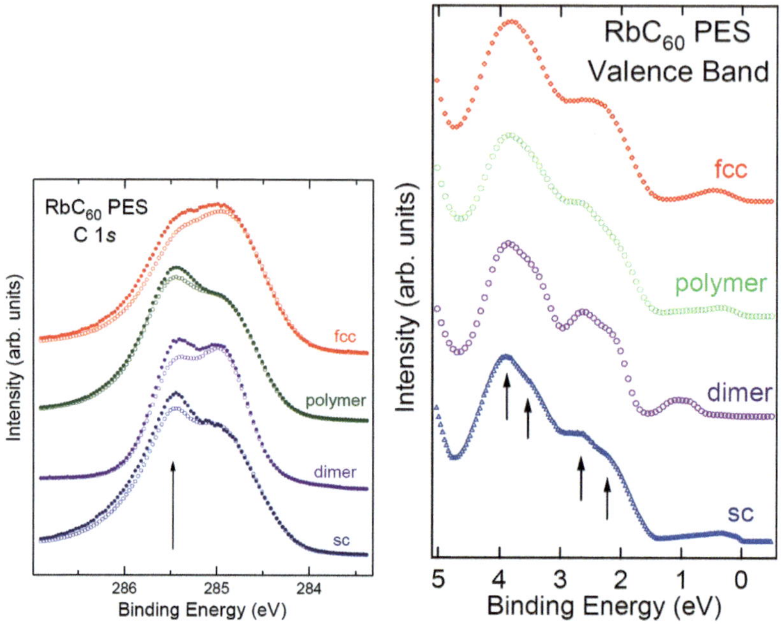

Figure 2.9

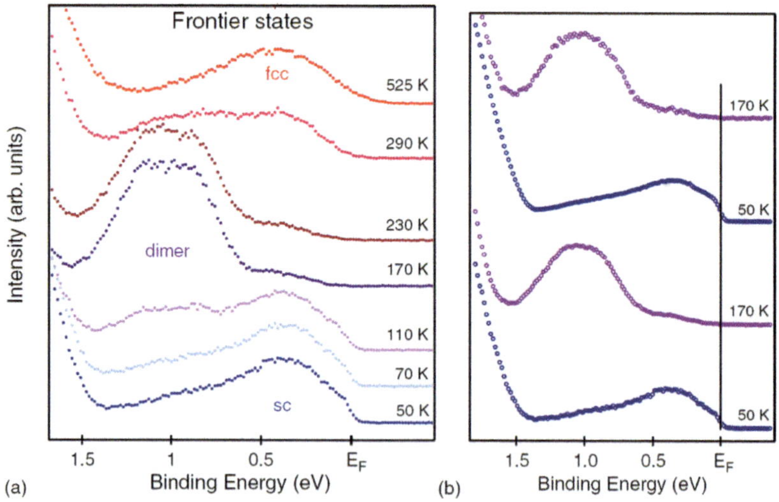

Figure 2.10

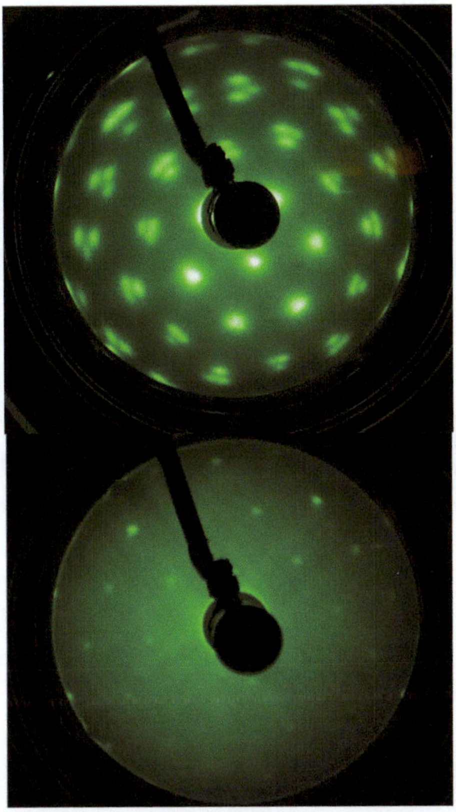

Figure 2.11

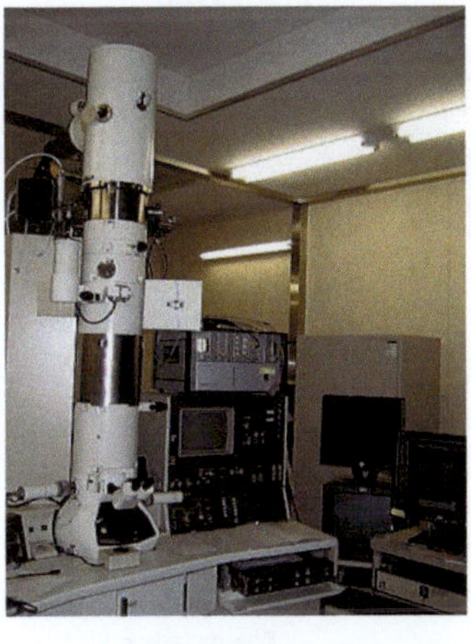

Figure 3.1

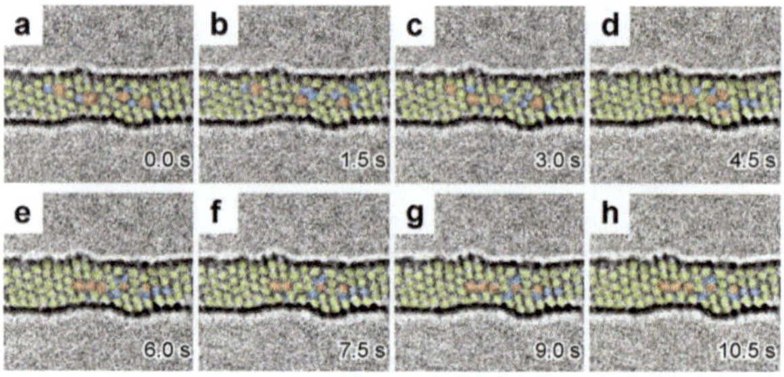

Figure 3.4

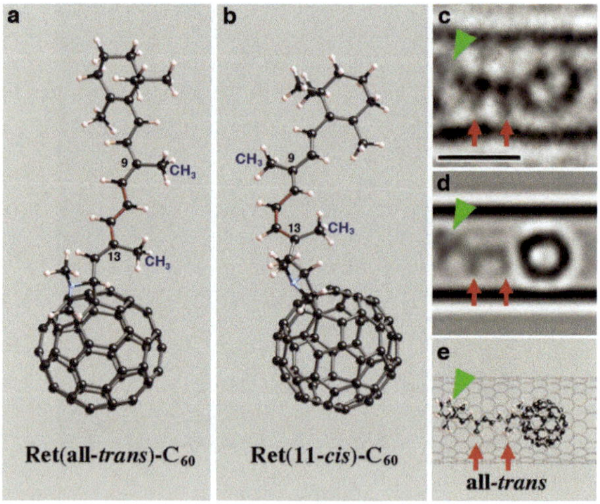

Figure 3.5

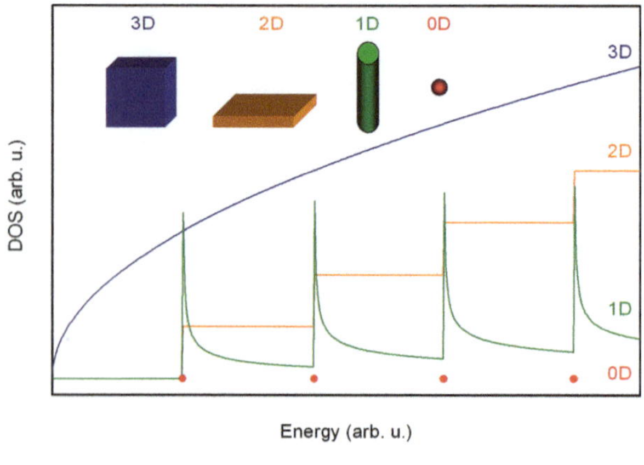

Figure 4.1

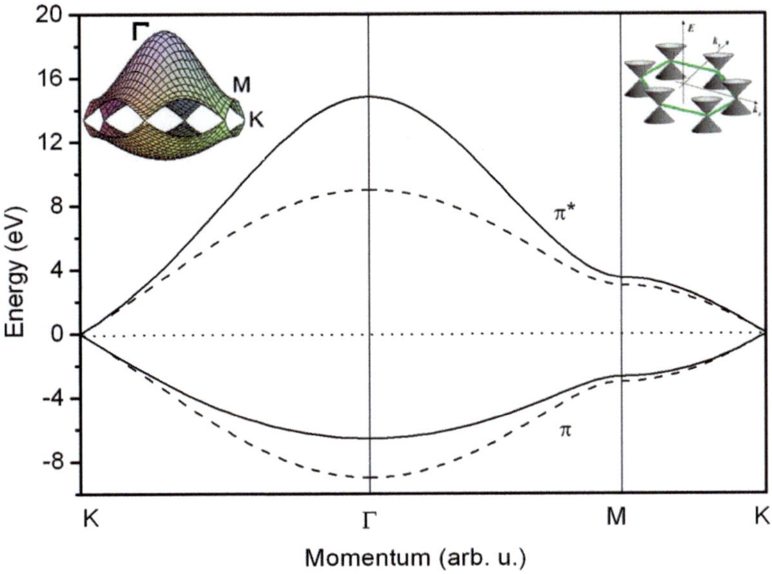

Figure 4.2

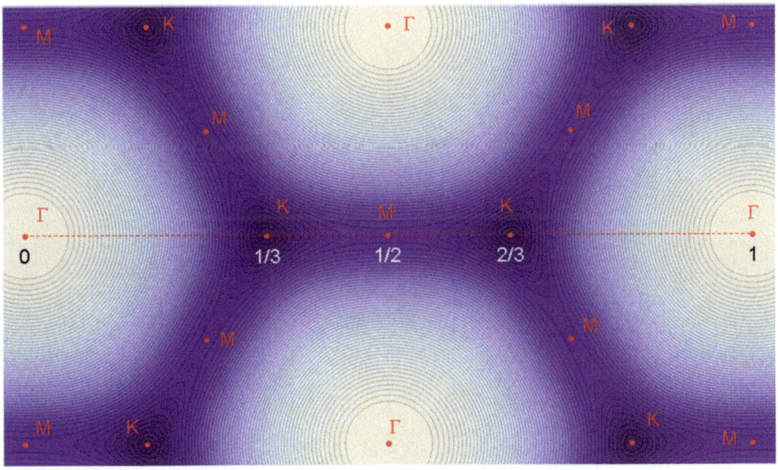

Figure 4.5

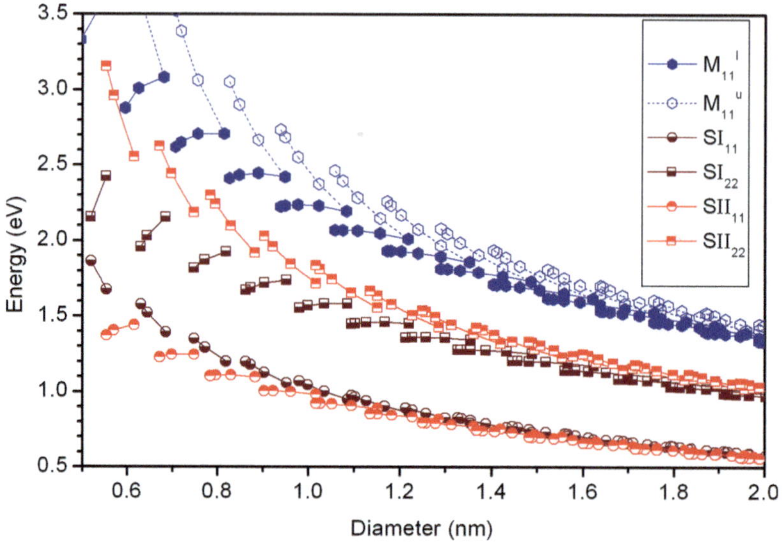

Figure 4.8

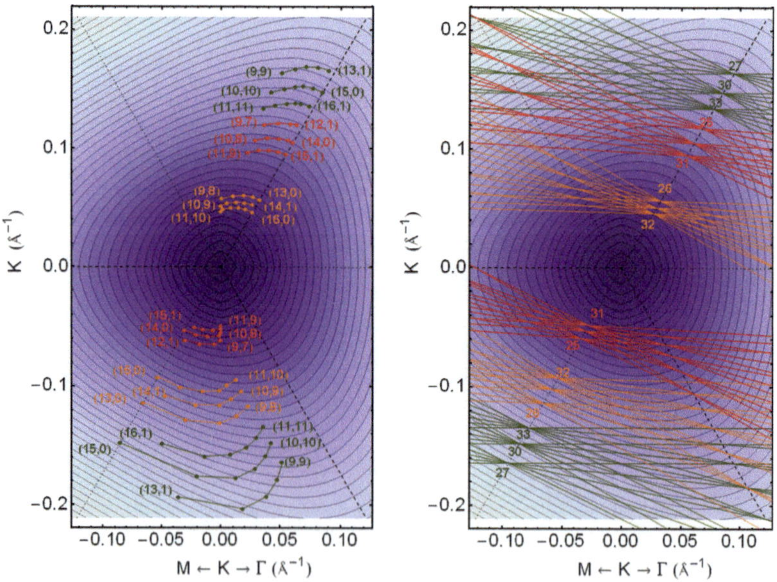

Figure 4.9

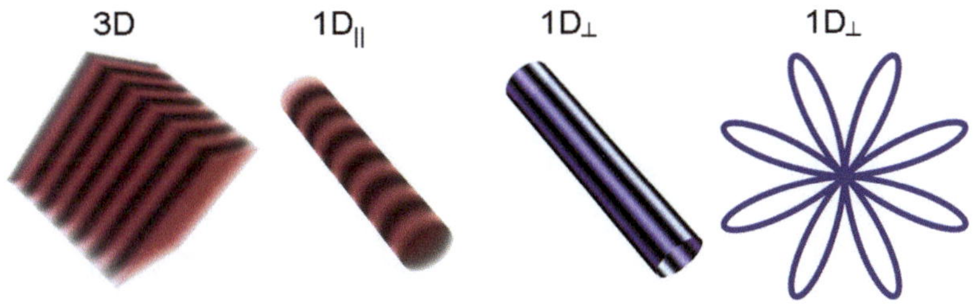

Figure 4.13

Figure 5.12

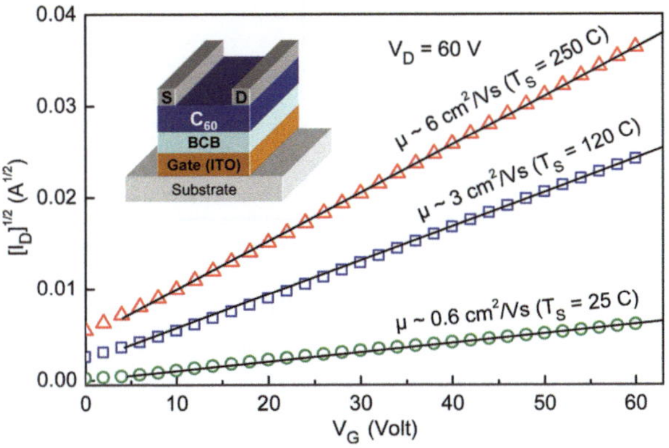

Figure 5.13

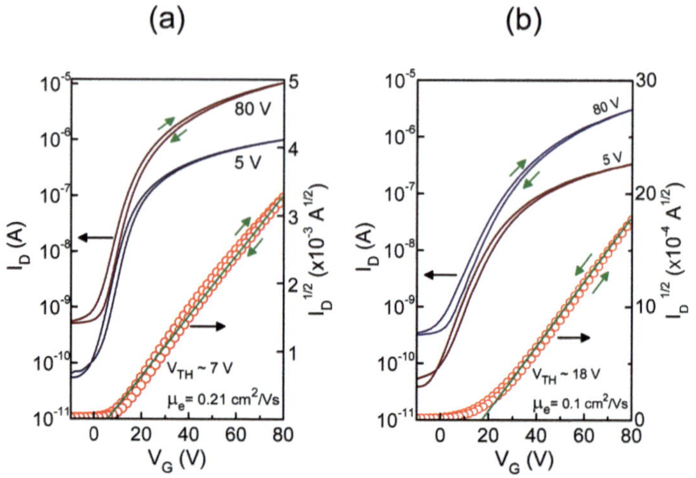

Figure 5.14

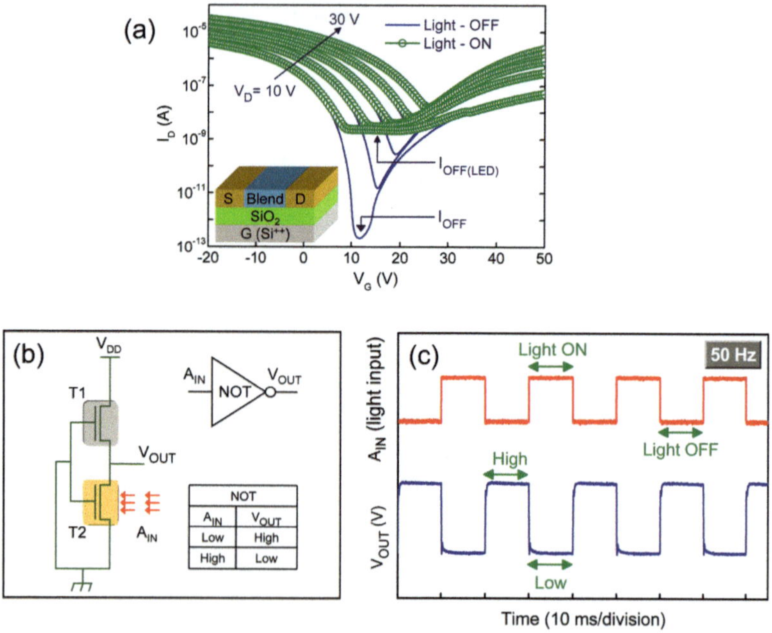

NOT	
A_{IN}	V_{OUT}
Low	High
High	Low

Figure 5.20

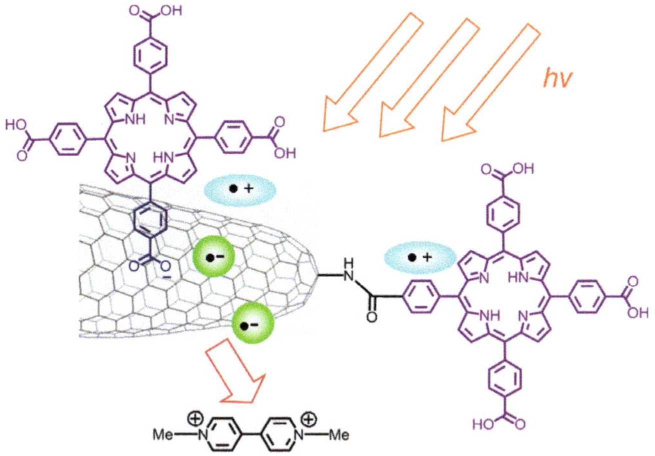

Figure 6.10

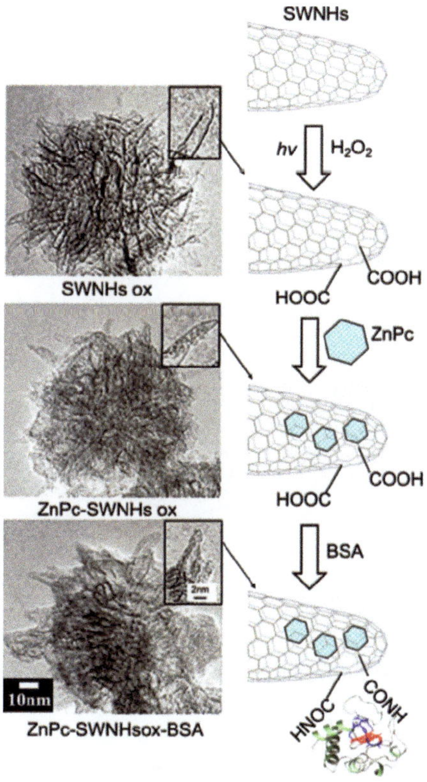

Figure 6.14

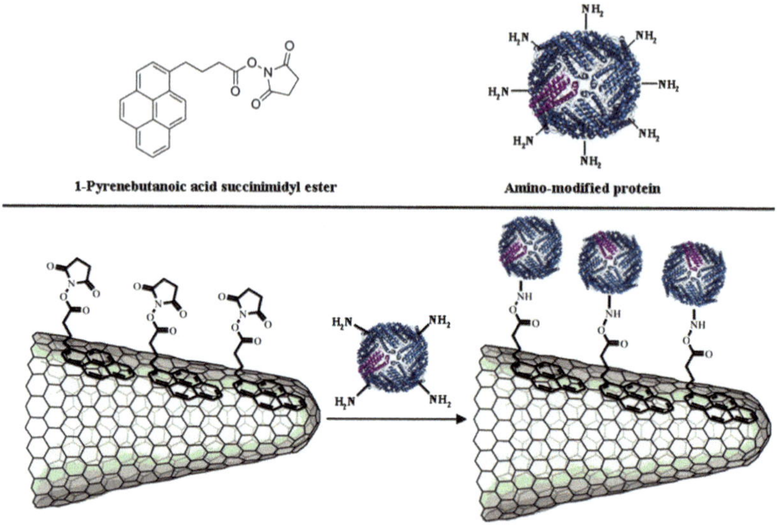

Figure 6.15

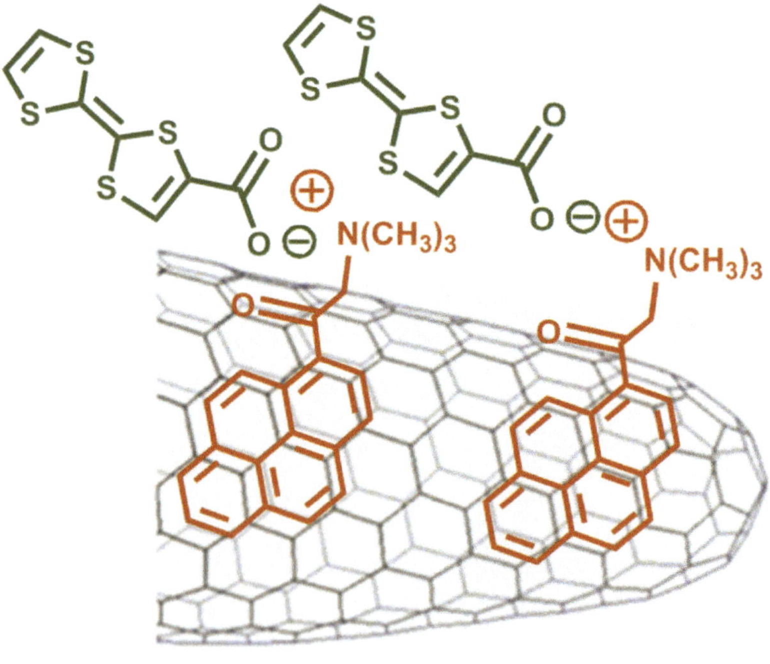

Figure 6.17

Figure 8.1

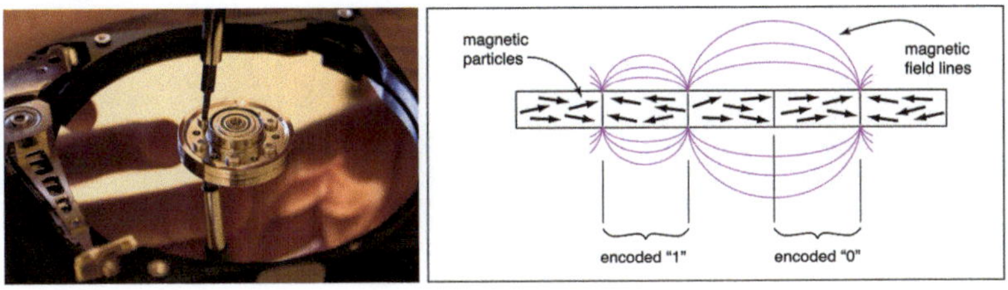

Figure 8.2

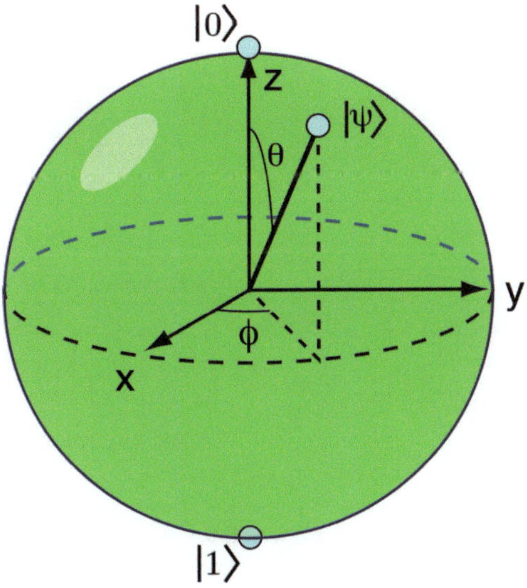

Figure 8.3

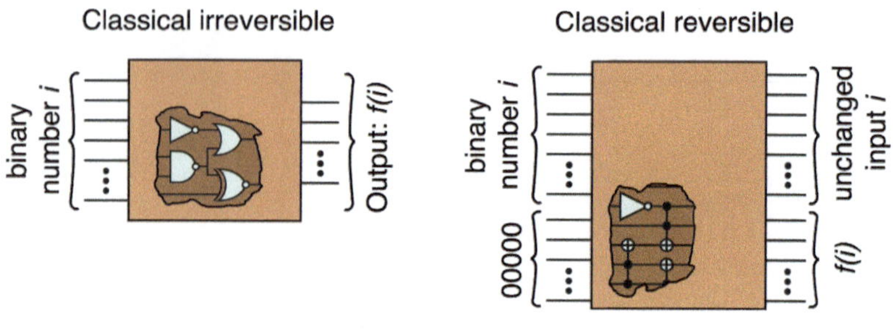

Figure 8.7

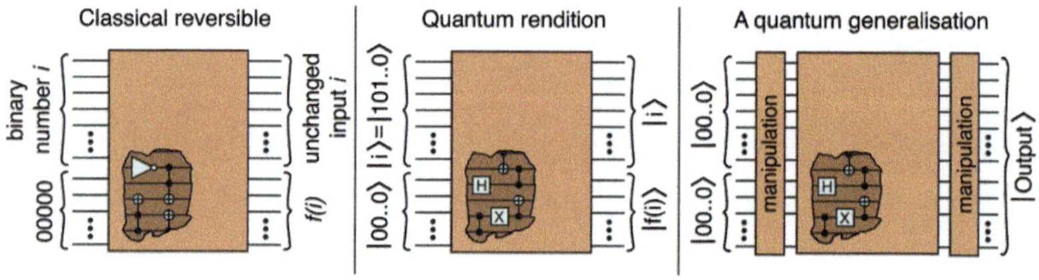

Figure 8.8

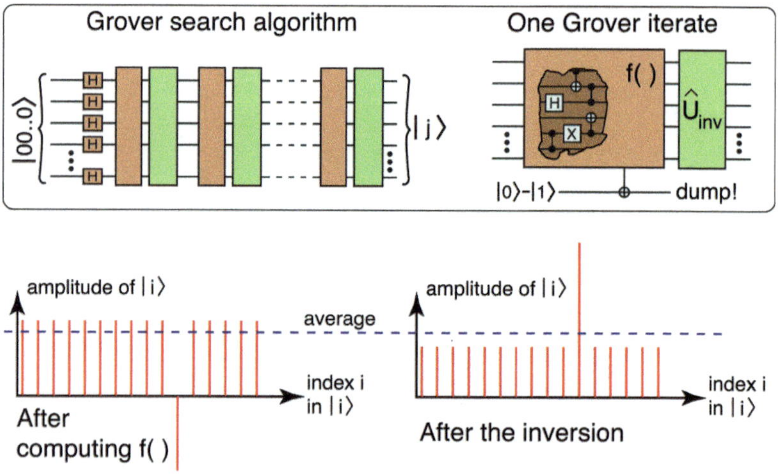

Figure 8.9

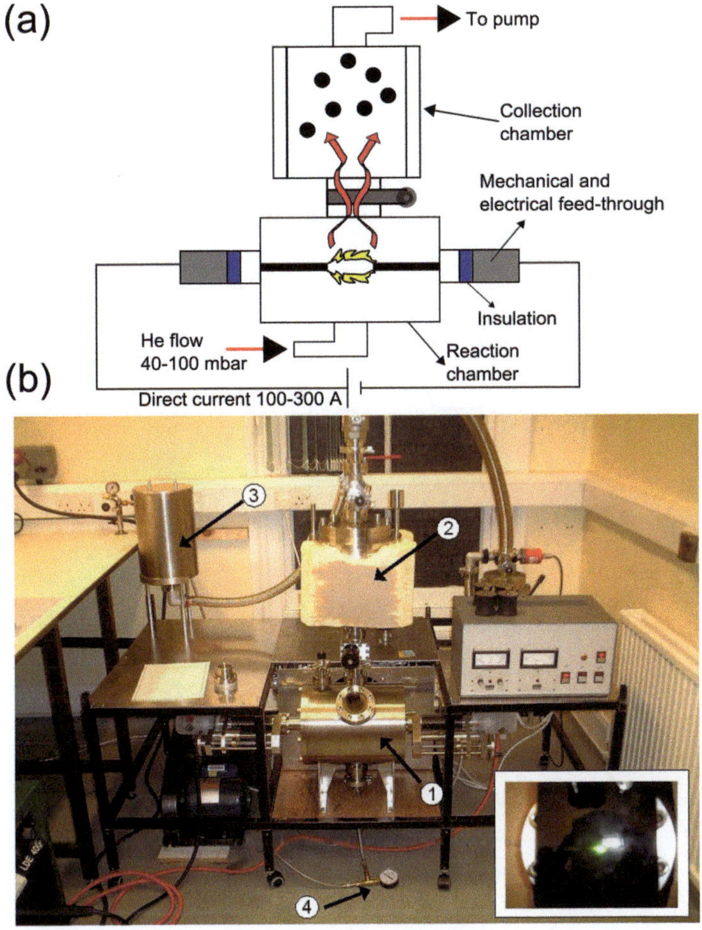

Figure 8.11

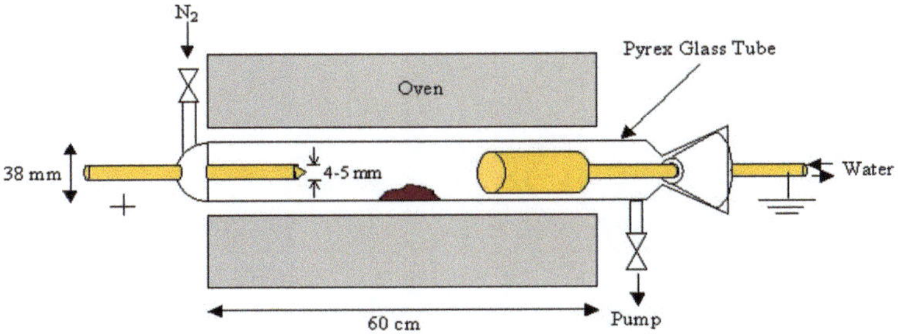

Figure 8.13

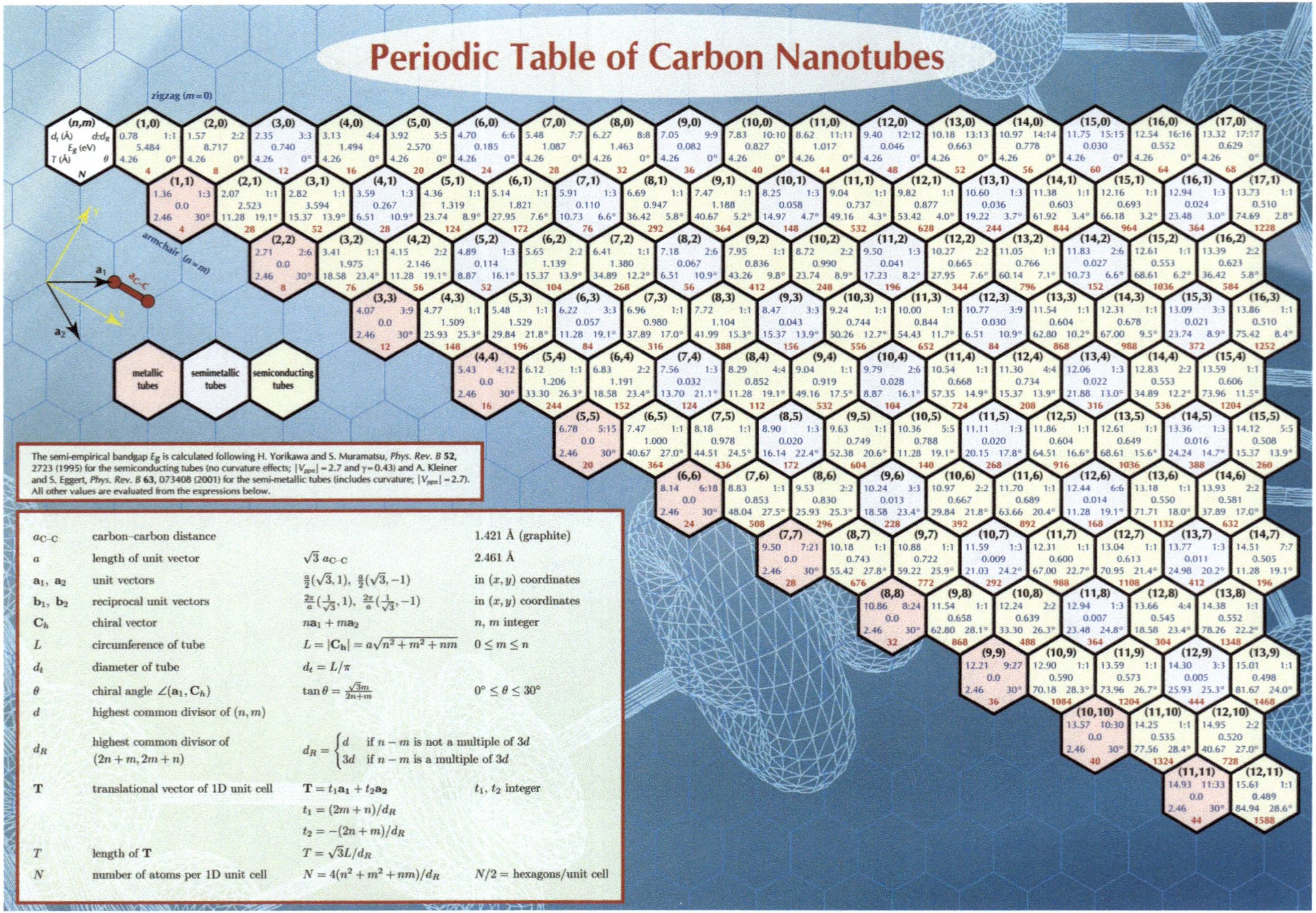

Figure 1.5 A larger version of the table is freely downloadable from www.panstanford.com/books/9789814267878.